AF074615

PostgreSQL 16 Administration Cookbook

Solve real-world Database Administration challenges with 180+ practical recipes and best practices

Gianni Ciolli
Boriss Mejías
Jimmy Angelakos
Vibhor Kumar
Simon Riggs

BIRMINGHAM—MUMBAI

PostgreSQL 16 Administration Cookbook

Copyright © 2023 Packt Publishing

All rights reserved. No part of this book may be reproduced, stored in a retrieval system, or transmitted in any form or by any means, without the prior written permission of the publisher, except in the case of brief quotations embedded in critical articles or reviews.

Every effort has been made in the preparation of this book to ensure the accuracy of the information presented. However, the information contained in this book is sold without warranty, either express or implied. Neither the authors, nor Packt Publishing or its dealers and distributors, will be held liable for any damages caused or alleged to have been caused directly or indirectly by this book.

Packt Publishing has endeavored to provide trademark information about all of the companies and products mentioned in this book by the appropriate use of capitals. However, Packt Publishing cannot guarantee the accuracy of this information.

Senior Publishing Product Manager: Gebin George

Acquisition Editor – Peer Reviews: Swaroop Singh

Project Editor: Parvathy Nair

Content Development Editors: Davide Oliveri, Elliot Dallow, Soham Amburle

Copy Editor: Safis Editing

Technical Editor: Aniket Shetty

Proofreader: Safis Editing

Indexer: Manju Arasan

Presentation Designer: Ganesh Bhadwalkar

Developer Relations Marketing Executive: Vignesh Raju

First published: December 2023

Production reference: 1301123

Published by Packt Publishing Ltd.
Grosvenor House
11 St Paul's Square
Birmingham
B3 1RB, UK.

ISBN 978-1-83546-058-0

www.packt.com

Boriss, Gianni, Jimmy, and Vibhor are grateful to Simon Riggs, for having been the main author of all the past editions of this book. They hope that, by joining forces, they were able to continue that high standard.

Contributors

About the authors

Gianni Ciolli is Vice President and Field CTO at EDB; he was Global Head of Professional Services at 2ndQuadrant until it was acquired by EDB. Gianni has been a PostgreSQL consultant, trainer, and speaker at many PostgreSQL conferences in Europe and abroad for more than 10 years. He has a PhD in Mathematics from the University of Florence. He has worked with free and Open-Source software since the 1990s and is active in the community. He lives between Frankfurt and London and plays the piano in his spare time.

Gianni has learned a lot from his colleagues and customers over the years and would like to thank them.

Boriss Mejías is a Senior Solutions Architect at EDB, building on his experience as PostgreSQL consultant and trainer at 2ndQuadrant. He has been working with Open-Source software since the beginning of the century, contributing to several projects both with code and community work. He has a PhD in Computer Science from the Université catholique de Louvain, and an Engineering degree from Universidad de Chile. Complementary to his role as Solutions Architect, he gives PostgreSQL training and is a regular speaker at PostgreSQL conferences. He loves spending time with his family and playing air guitar.

I would like to thank my co-authors for the great collaboration in writing this book. It has been a great experience. I would also like to thank the PostgreSQL community for everything I have learned from them during all these years. Special thanks to my family for all their support, laughs, and daily fun.

Jimmy Angelakos is a Systems and Database Architect and recognized PostgreSQL expert, with a wealth of experience gained from his career in Software Architecture and his key roles at 2nd-Quadrant and EDB. He studied Computer Science at the University of Aberdeen and has worked with, and contributed to, Open Source tools for 25+ years. He is passionate about participating in the community, and is an active member of PostgreSQL Europe and an occasional contributor to the PostgreSQL project. He is a regular speaker at conferences and events focused on databases and Open Source software, sharing his insights with the community.

No one is an island, and none of this would have been possible without the mentoring, knowledge sharing, and guidance that the PostgreSQL community has so generously provided to me over the years.

Vibhor Kumar, Global VP at EDB, is a pioneering data tech leader. He manages a global team of engineers, optimizing clients' Postgres databases for peak performance and scalability. He advises Fortune 500 clients, including many financial institutes, on innovating and transforming their data platforms. His past experience spans IBM, BMC Software, and CMC Ltd. He holds a BSc in Computer Science from the University of Lucknow and a Master's from the Army Institute of Management. As a certified expert in numerous technologies, he often shares his insights on DevOps the cloud, and database optimization through blogging and speaking at events.

I'm thankful to everyone who supported this project. Special thanks to my wife, Nandini Karkare, for her constant support and love. I'm also grateful to my colleagues and co-authors for their insights and contributions and to Marc Linster for his mentorship. This book is a result of our collective efforts. Thank you all for being part of this journey.

Simon Riggs is a Major Developer of PostgreSQL since 2004. Formerly, Simon was Founder and CEO of 2ndQuadrant, acquired by EDB in 2020.

Simon has contributed widely to PostgreSQL, initiating new projects, contributing ideas, and committing many important features, as well as working directly with database architects and users on advanced solutions.

About the reviewers

Marcelo Diaz is a Software Engineer with more than 15 years of experience, with a special focus on PostgreSQL. He is passionate about Open-Source software and has promoted its application in critical and high-demand environments, working as a software developer and consultant for both private and public companies. He currently works very happily at Cybertec and as a technical reviewer for Packt Publishing. He enjoys spending his leisure time with his daughter, Malvina, his wife, Romina, and their pets. He also likes to play "fulbo", but currently he enjoys it more watching Messi on TV.

Martín Marqués began his career as a DBA and Software Developer at a local university in Argentina over 20 years ago. He dedicated 13 years to these roles, during which he provided training using custom materials to various agencies. Later, he transitioned to a technical support role, specializing in remote DBA services and consulting for clients at 2ndQuadrant.

In recent years, Martín shifted into a management role within technical support at EnterpriseDB. In the past year, he has taken on the position of Engineering Manager for five EnterpriseDB products.

Afroditi Loukidou is a PostgreSQL and Open-Source enthusiast, currently working as a Technical Lead at EnterpriseDB. She has studied Industrial Informatics and holds an MSc in Computer Networks. Her journey with PostgreSQL started at 2ndQuadrant and went on with EDB, where she has gained a wealth of experience working as a PostgreSQL engineer assisting smaller and bigger customers with PostgreSQL operational aspects, maintenance, tuning, upgrades and more. In her role as a Technical Lead, she also gets exposure to more architectural aspects and larger-scale projects of varied complexity and has always found this book to be a great resource to turn to. She lives in London and loves music, mountaineering, and generally spending time in nature.

Learn more on Discord

To join the Discord community for this book – where you can share feedback, ask questions to the author, and learn about new releases – follow the QR code below:

https://discord.gg/pQkghgmgdG

Table of Contents

Preface   xxxix

Chapter 1: First Steps   1

Introducing PostgreSQL 16 2

    What makes PostgreSQL different? • 2

        Robustness • 4

        Security • 4

        Ease of use • 5

        Extensibility • 5

        Performance and concurrency • 6

        Scalability • 6

        SQL and NoSQL data models • 6

        Popularity • 7

        Commercial support • 7

        Research and development funding • 8

How to get PostgreSQL 8

    How to do it... • 8

    How it works... • 10

    There's more... • 11

Connecting to the PostgreSQL server 11

    Getting ready • 11

    How to do it... • 13

How it works... • 13

There's more... • 14

See also • 15

Enabling access for network/remote users 15

How to do it... • 15

How it works... • 16

There's more... • 17

See also • 18

Using the pgAdmin 4 GUI tool 18

How to do it... • 19

How it works... • 22

See also • 23

Using the psql query and scripting tool 24

Getting ready • 24

How to do it... • 24

How it works... • 27

There's more... • 28

See also • 29

Changing your password securely 29

How to do it... • 29

How it works... • 30

Avoiding hardcoding your password 30

Getting ready • 30

How to do it... • 30

How it works... • 31

There's more... • 32

Using a connection service file 32

How to do it... • 32

How it works... • 33

There's more... • 33

Troubleshooting a failed connection 33

    How to do it... • 33

    There's more... • 35

PostgreSQL in the cloud 35

    Getting ready • 35

    How to do it... • 35

    How it works... • 40

    There's more... • 40

PostgreSQL with Kubernetes 41

    Getting ready • 41

    How to do it... • 42

    How it works... • 43

    There's more... • 44

PostgreSQL with TPA 45

    Getting ready • 45

    How to do it... • 46

    There's more • 49

Chapter 2: Exploring the Database   51

What type of server is this? 52

    How to do it... • 52

    There's more... • 53

What version is the server? 53

    How to do it... • 53

    How it works... • 54

    There's more... • 55

What is the server uptime? 56

    How to do it... • 56

    How it works... • 56

    See also • 57

Locating the database server files 57
  Getting ready • 57
  How to do it... • 57
  How it works... • 58
  There's more... • 59

Locating the database server's message log 60
  Getting ready • 61
  How to do it... • 61
  How it works... • 62
  There's more... • 62

Locating the database's system identifier 63
  Getting ready • 63
  How to do it... • 63
  How it works... • 64

Listing databases on the database server 64
  How to do it... • 65
  How it works... • 66
  There's more... • 67

How many tables are there in a database? 68
  How to do it... • 68
  How it works... • 69
  There's more... • 71

How much disk space does a database use? 71
  How to do it... • 72
  How it works... • 72

How much memory does a database currently use? 72
  How to do it... • 73
  How it works... • 74

How much disk space does a table use? 74
  How to do it... • 74
  How it works... • 75
  There's more... • 76

Which are my biggest tables? 76
   How to do it... • 76
   How it works... • 77

How many rows are there in a table? 77
   How to do it... • 77
   How it works... • 78

Quickly estimating the number of rows in a table 79
   How to do it... • 79
   How it works... • 80
   There's more... • 81

Listing extensions in this database 82
   How to do it... • 82
   How it works... • 83
   There's more... • 83
   See also • 83

Understanding object dependencies 84
   Getting ready • 84
   How to do it... • 85
   How it works... • 86
   There's more... • 86

Chapter 3: Server Configuration   89

Read the fine manual (RTFM) 90
   How to do it... • 90
   How it works... • 90
   There's more... • 91

Planning a new database 91
   Getting ready • 91
   How to do it... • 91
   How it works... • 92
   There's more... • 93

Setting the configuration parameters for the database server 93

Getting ready • 93

How to do it... • 94

How it works... • 97

There's more... • 97

Setting the configuration parameters in your programs 99

How to do it... • 100

How it works... • 101

There's more... • 101

Finding the configuration settings for your session 102

How to do it... • 102

How it works... • 104

Finding parameters with non-default settings 104

How to do it... • 105

How it works... • 105

There's more... • 105

Setting parameters for particular groups of users 106

How to do it... • 106

How it works... • 106

A basic server configuration checklist 107

Getting ready • 107

How to do it... • 107

There's more... • 108

Adding an external module to PostgreSQL 109

Getting ready • 110

How to do it... • 110

Installing modules using a software installer • 110

Installing modules from PGXN • 111

Installing modules from source code • 112

How it works... • 112

Using an installed module/extension 113
    Getting ready • 113
    How to do it... • 113
    How it works... • 114

Managing installed extensions 114
    How to do it... • 114
    How it works... • 116
    There's more... • 117

Chapter 4: Server Control 119

An overview of controlling the database server 120

Starting the database server manually 121
    Getting ready • 121
    How to do it... • 121
    How it works... • 123

Stopping the server safely and quickly 124
    How to do it... • 124
    How it works... • 125
    See also • 125

Stopping the server in an emergency 125
    How to do it... • 126
    How it works... • 126

Reloading server configuration files 126
    How to do it... • 126
    How it works... • 128
    There's more... • 129

Restarting the server quickly 129
    How to do it... • 129
    There's more... • 130

Preventing new connections 131
    How to do it... • 131
    How it works... • 132

Restricting users to only one session each 132

    How to do it... • 132

    How it works... • 133

Pushing users off the system 134

    How to do it... • 134

    How it works... • 135

Deciding on a design for multitenancy 136

    How to do it... • 136

    How it works... • 137

Using multiple schemas 137

    Getting ready • 137

    How to do it... • 138

    How it works... • 139

Giving users their own private databases 140

    Getting ready • 140

    How to do it... • 140

    How it works... • 141

    There's more... • 141

    See also • 142

Running multiple servers on one system 142

    Getting ready • 142

    How to do it... • 142

    How it works... • 143

Setting up a connection pool 144

    Getting ready • 144

    How to do it... • 144

    How it works... • 146

    There's more... • 146

Accessing multiple servers using the same host and port 148

    Getting ready • 148

    How to do it... • 148

There's more... • 150

Running multiple PgBouncer on the same port to leverage multiple cores 150

Getting ready • 150

How to do it... • 151

How it works... • 152

Chapter 5: Tables and Data   153

Choosing good names for database objects 154

Getting ready • 154

How to do it... • 154

There's more... • 155

Handling objects with quoted names 156

Getting ready • 157

How to do it... • 157

How it works... • 158

There's more... • 158

Identifying and removing duplicates 159

Getting ready • 159

How to do it... • 160

How it works... • 162

There's more... • 163

Preventing duplicate rows 164

Getting ready • 164

How to do it... • 164

How it works... • 167

There's more... • 167

Duplicate indexes • 167

Uniqueness without indexes • 167

A real-world example – IP address range allocation • 168

A real-world example – a range of time • 169

Finding a unique key for a set of data 169

Getting ready • 170

How to do it... • 170

How it works... • 172

Generating test data 172

How to do it... • 172

How it works... • 175

There's more... • 175

See also • 176

Randomly sampling data 176

How to do it... • 177

How it works... • 178

Loading data from a spreadsheet 180

Getting ready • 180

How to do it... • 181

How it works... • 182

There's more... • 183

Loading data from flat files 183

Getting ready • 183

How to do it... • 183

How it works... • 185

There's more... • 186

Making bulk data changes using server-side procedures with transactions 187

Getting ready • 187

How to do it... • 188

There's more... • 188

Dealing with large tables with table partitioning 191

How to do it... • 192

How it works... • 193

There's more... • 193

Finding good candidates for partition keys 194
Getting ready • 194
How to do it... • 194
There's more... • 195

Consolidating data with MERGE 195
Getting ready • 195
How to do it... • 196
There's more... • 197

Deciding when to use JSON data types 197
Getting ready • 197
How to do it... • 198
Example: moving sparse columns to JSON • 199
Example: expose JSON data using a view • 201
There's more... • 202

Chapter 6: Security   205

An overview of PostgreSQL security 206
Typical user roles • 207

The PostgreSQL superuser 207
How to do it... • 207
How it works... • 208
There's more... • 208
Other superuser-like attributes • 208
See also • 208

Revoking user access to tables 208
Getting ready • 208
How to do it... • 208
How it works... • 210
There's more... • 210
Database creation scripts • 210
Default search path • 211
Securing views • 211

Granting user access to a table 212
Getting ready • 212
How to do it… • 213
How it works… • 213
There's more… • 214

Granting user access to specific columns 214
Getting ready • 214
How to do it… • 214
How it works… • 215
There's more… • 215

Granting user access to specific rows 216
Getting ready • 216
How to do it… • 216
How it works… • 218
There's more… • 218

Creating a new user 218
Getting ready • 218
How to do it… • 219
How it works… • 219
There's more… • 219

Temporarily preventing a user from connecting 220
Getting ready • 220
How to do it… • 220
How it works… • 220
There's more… • 220
    Limiting the number of concurrent connections by a user • 220
Revoking a user's database access • 221
How it works… • 221
    Forcing NOLOGIN users to disconnect • 222

Removing a user without dropping their data 222
Getting ready • 222
How to do it… • 222

Table of Contents

    How it works... • 223

Checking whether all users have a secure password 223

    How to do it... • 223

    How it works... • 224

Giving limited superuser powers to specific users 224

    Getting ready • 225

    How to do it... • 225

        Assigning backup privileges to a user • 226

    How it works... • 226

    There's more... • 226

Auditing database access 227

    Getting ready • 227

    Auditing access • 227

    Auditing SQL statements • 228

    Auditing table access • 230

    Managing the audit log • 230

    Auditing data changes • 231

Always knowing which user is logged in 233

    Getting ready • 234

    How to do it... • 234

    How it works... • 234

    There's more... • 235

        Not inheriting user attributes • 235

Integrating with LDAP 235

    Getting ready • 235

    How to do it... • 235

    How it works... • 236

    There's more... • 236

        Setting up the client to use LDAP • 236

        Replacement for the User Name Map feature • 236

    See also • 236

Connecting using encryption (SSL / GSSAPI) 236

    Getting ready • 237

    How to do it... • 237

    How it works... • 237

    There's more... • 237

        Getting the SSL key and certificate • 238

        Setting up a client to use SSL • 238

        Checking server authenticity • 239

Using SSL certificates to authenticate 240

    Getting ready • 240

    How to do it... • 240

    How it works... • 241

    There's more... • 241

        Avoiding duplicate SSL connection attempts • 241

        Using multiple client certificates • 242

        Using the client certificate to select a database user • 242

    See also • 242

Mapping external usernames to database roles 243

    Getting ready • 243

    How to do it... • 243

    How it works... • 244

    There's more... • 244

Using column-level encryption 244

    Getting ready • 244

    How to do it... • 245

    How it works... • 248

    There's more... • 248

        For really sensitive data • 248

        For really, really, really sensitive data • 249

    See also • 249

Setting up cloud security using predefined roles 249
    Getting ready • 250
    How to do it... • 250
    How it works... • 251
    There's more... • 252

Chapter 7: Database Administration   255

Writing a script that either succeeds entirely or fails entirely 257
    How to do it... • 257
    How it works... • 258
    There's more... • 260

Writing a psql script that exits on the first error 261
    Getting ready • 261
    How to do it... • 262
    How it works... • 262
    There's more... • 263

Using psql variables 263
    Getting ready • 263
    How to do it... • 263
    How it works... • 264
    There's more... • 264

Placing query output into psql variables 264
    Getting ready • 264
    How to do it... • 265
    How it works... • 265
    There's more... • 266

Writing a conditional psql script 266
    Getting ready • 266
    How to do it... • 266
    How it works... • 267
    There's more... • 267

Investigating a psql error 267

    Getting ready • 269

    How to do it... • 269

    There's more... • 269

Setting the psql prompt with useful information 269

    Getting ready • 269

    How to do it... • 270

    How it works... • 271

Using pgAdmin for DBA tasks 271

    Getting ready • 271

    How to do it... • 272

    How it works... • 276

    There's more... • 276

Scheduling jobs for regular background execution 276

    Getting ready • 277

    How to do it... • 277

    How it works... • 278

    There's more... • 278

Performing actions on many tables 279

    Getting ready • 280

    How to do it... • 280

    How it works... • 281

    There's more... • 282

Adding/removing columns on a table 284

    How to do it... • 284

    How it works... • 285

    There's more... • 286

Changing the data type of a column 290

    Getting ready • 290

    How to do it... • 290

    How it works... • 291

    There's more... • 292

Changing the definition of an enum data type 294

Getting ready • 294

How to do it... • 294

How it works... • 295

There's more... • 297

Adding a constraint concurrently 297

Getting ready • 298

How to do it... • 298

How it works... • 299

There's more... • 300

Adding/removing schemas 300

How to do it... • 300

There's more... • 301

Using schema-level privileges • 302

Moving objects between schemas 302

How to do it... • 302

How it works... • 303

There's more... • 303

Adding/removing tablespaces 303

Getting ready • 303

How to do it... • 304

How it works... • 306

There's more... • 307

Putting pg_wal on a separate device • 307

Tablespace-level tuning • 308

Moving objects between tablespaces 308

Getting ready • 308

How to do it... • 308

How it works... • 309

There's more... • 310

Accessing objects in other PostgreSQL databases 311

    Getting ready • 312

    How to do it... • 312

    How it works... • 314

    There's more... • 315

Accessing objects in other foreign databases 316

    Getting ready • 317

    How to do it... • 317

    How it works... • 318

    There's more... • 318

Making views updatable 319

    Getting ready • 319

    How to do it... • 321

    How it works... • 325

    There's more... • 327

Using materialized views 329

    Getting ready • 329

    How to do it... • 329

    How it works... • 330

    There's more... • 331

Using GENERATED data columns 331

    How to do it... • 332

    How it works... • 332

    There's more... • 332

Using data compression 333

    Getting ready • 333

    How to do it... • 333

    How it works... • 334

    There's more... • 335

Chapter 8: Monitoring and Diagnosis — 337

Cloud-native monitoring — 339

Providing PostgreSQL information to monitoring tools — 341
- Finding more information about generic monitoring tools • 343

Real-time viewing using pgAdmin — 343
- Getting ready • 343
- How to do it... • 343

Monitoring the PostgreSQL message log — 345
- Getting ready • 345
- How to do it... • 346
- How it works... • 346
- There's more... • 347

Checking whether a user is connected — 347
- Getting ready • 347
- How to do it... • 347
- How it works... • 347
- There's more... • 347

Checking whether a computer is connected — 348
- How to do it... • 348
- There's more... • 348

Repeatedly executing a query in psql — 348
- How to do it... • 349
- There's more... • 349

Checking which queries are running — 349
- Getting ready • 349
- How to do it... • 350
- How it works... • 350
- There's more... • 350
  - *Catching queries that only run for a few milliseconds* • 350
  - *Watching the longest queries* • 351
  - *Watching queries from ps* • 352

See also • 352

Monitoring the progress of commands 352

Getting ready • 352

How to do it... • 352

How it works... • 354

There's more... • 354

Checking which queries are active or blocked 354

Getting ready • 354

How to do it... • 354

How it works... • 355

There's more... • 355

Knowing who is blocking a query 356

Getting ready • 356

How to do it... • 356

How it works... • 357

Killing a specific session 357

How to do it... • 357

How it works... • 357

There's more... • 358

Using statement_timeout to clean up queries that take too long to run • 358

Killing idle in-transaction sessions • 358

Knowing whether anybody is using a specific table 359

Getting ready • 359

How to do it... • 359

How it works... • 359

There's more... • 359

Knowing when a table was last used 360

Getting ready • 360

How to do it... • 360

How it works... • 362

Monitoring I/O statistics 362

Getting ready • 363

How to do it... • 363

How it works... • 364

There's more... • 365

Usage of disk space by temporary data 365

Getting ready • 365

How to do it... • 365

How it works... • 367

There's more... • 367

Finding out whether a temporary file is in use anymore • 368

Logging temporary file usage • 368

Understanding why queries slow down 368

Getting ready • 368

How to do it... • 369

How it works... • 369

There's more... • 370

Do queries return significantly more data than they did earlier? • 370

Do queries also run slowly when they run alone? • 370

Is the second run of the same query also slow? • 371

Table and index bloat • 371

See also • 372

Analyzing the real-time performance of your queries 372

Getting ready • 372

How to do it... • 373

How it works... • 373

There's more... • 373

Tracking important metrics over time 374

Getting ready • 374

How to do it... • 374

How it works... • 375

There's more... • 377

Chapter 9: Regular Maintenance 379

Controlling automatic database maintenance 380

Getting ready • 380

How to do it... • 380

How it works... • 383

There's more... • 386

See also • 387

Avoiding auto-freezing 387

How to do it... • 387

Removing issues that cause bloat 389

Getting ready • 389

How to do it... • 389

How it works... • 390

There's more... • 390

Actions for heavy users of temporary tables 390

How to do it... • 390

How it works... • 391

Identifying and fixing bloated tables and indexes 392

Getting ready • 392

How to do it... • 393

How it works... • 395

There's more... • 398

Monitoring and tuning a vacuum 398

Getting ready • 398

How to do it... • 398

How it works... • 399

There's more... • 402

Maintaining indexes 403

Getting ready • 403

How to do it... • 404

How it works... • 405

There's more... • 405

Finding unused indexes 406

How to do it... • 406

How it works... • 406

Carefully removing unwanted indexes 407

Getting ready • 408

How to do it... • 408

How it works... • 409

Planning maintenance 409

How to do it... • 409

How it works... • 410

There's more... • 411

Chapter 10: Performance and Concurrency 413

Finding slow SQL statements 414

Getting ready • 414

How to do it... • 415

How it works... • 417

There's more... • 417

Finding out what makes SQL slow 418

Getting ready • 418

How to do it... • 418

There's more... • 421

Locking problems • 422

EXPLAIN options • 422

Not enough CPU power or disk I/O capacity for the current load • 422

See also • 423

Reducing the number of rows returned 423

How to do it... • 423

There's more... • 424

Simplifying complex SQL queries 426

    Getting ready • 426

    How to do it... • 427

    There's more... • 431

        Using materialized views • 433

        Using set-returning functions for some parts of queries • 434

Speeding up queries without rewriting them 434

    How to do it... • 434

        Increasing work_mem • 434

        Setting recursive_worktable_factor • 435

        More ideas with indexes • 438

    There's more... • 440

        Time-series partitioning • 440

        Using a view that contains TABLESAMPLE • 440

        In case of many updates, set fillfactor on the table • 441

        Rewriting the schema – a more radical approach • 441

Discovering why a query is not using an index 441

    Getting ready • 441

    How to do it... • 442

    How it works... • 443

    There's more... • 443

Forcing a query to use an index 443

    Getting ready • 444

    How to do it... • 444

    There's more... • 446

Using parallel query 446

    How to do it... • 447

    How it works... • 447

Using Just-In-Time (JIT) compilation 449

    Getting ready • 449

    How it works... • 449

Creating time-series tables using partitioning 452
How to do it... • 452

How it works... • 453

There's more... • 454

Using optimistic locking to avoid long lock waits 455
How to do it... • 455

How it works... • 456

There's more... • 456

Reporting performance problems 457
How to do it... • 458

There's more... • 458

Chapter 11: Backup and Recovery   459

Understanding and controlling crash recovery 460
How to do it... • 460

How it works... • 462

There's more... • 462

Planning your backups 463
How to do it... • 464

There's more... • 465

Hot logical backup of one database 465
How to do it... • 466

How it works... • 466

There's more... • 468

See also • 468

Hot logical backup of all databases 468
How to do it... • 469

How it works... • 469

See also • 469

Backup of database object definitions 469
How to do it... • 470

There's more... • 470

A standalone hot physical backup 471

    Getting ready • 471

    How to do it... • 471

    How it works... • 471

    There's more... • 473

Hot physical backups with Barman 474

    Getting ready • 475

    How to do it... • 476

    How it works... • 480

    There's more... • 481

Recovery of all databases 483

    Getting ready • 483

    How to do it... • 484

        Logical – from the custom dump taken with pg_dump -F c • 484

        Logical – from the script dump created by pg_dump -F p • 484

        Logical – from the script dump created by pg_dumpall • 485

        Physical – from a standalone backup • 485

        Physical – with Barman • 486

    How it works... • 488

    There's more... • 488

Recovery to a point in time 490

    Getting ready • 490

    How to do it... • 490

    How it works... • 491

    There's more... • 492

    See also • 493

Recovery of a dropped/damaged table 493

    How to do it... • 494

        Logical – from the custom dump taken with pg_dump -F c • 494

        Logical – from the script dump • 495

        Physical • 496

    How it works... • 496

    See also • 497

Recovery of a dropped/damaged database 497
How to do it... • 497
Logical – from the custom dump -F c • 497
Logical – from the script dump created by pg_dump • 497
Logical – from the script dump created by pg_dumpall • 498
Physical • 498

Extracting a logical backup from a physical one 498
Getting ready • 498
How to do it... • 499
There's more... • 499

Improving the performance of logical backup/recovery 499
Getting ready • 499
How to do it... • 499
How it works... • 500
There's more... • 501

Improving the performance of physical backup/recovery 501
Getting ready • 501
How to do it... • 502
How it works... • 502
There's more... • 503
See also • 503

Validating backups 503
Getting ready • 504
How to do it... • 504
How it works... • 505
There's more... • 506

Chapter 12: Replication and Upgrades   509

Replication concepts 511
Topics • 511
Basic concepts • 512
History and scope • 512

Practical aspects • 514

  Data loss • 515

  Single-master replication • 515

  Multinode architectures • 516

  Multi-master replication • 516

  Other approaches to replication • 517

Replication best practices 517

  Getting ready • 517

  How to do it... • 517

  There's more... • 519

Setting up streaming replication 519

  Getting ready • 520

  How to do it... • 520

  How it works... • 521

  There's more... • 522

Setting up streaming replication security 523

  Getting ready • 524

  How to do it... • 524

  How it works... • 524

  There's more... • 525

Hot Standby and read scalability 525

  Getting ready • 526

  How to do it... • 526

  How it works... • 528

Managing streaming replication 529

  Getting ready • 529

  How to do it... • 529

  There's more... • 530

  See also • 531

Using repmgr 531

  Getting ready • 531

  How to do it... • 532

How it works... • 534

There's more... • 534

Using replication slots 534

Getting ready • 534

How to do it... • 535

There's more... • 536

See also • 536

Setting up replication with TPA 536

Getting ready • 536

How to do it... • 536

How it works... • 537

There's more... • 538

Setting up replication with CloudNativePG 538

Getting ready • 538

How to do it... • 538

How it works... • 539

There's more... • 539

Monitoring replication 539

Getting ready • 539

How to do it... • 541

There's more... • 543

Performance and synchronous replication (sync rep) 544

Getting ready • 544

How to do it... • 545

How it works... • 547

There's more... • 548

Delaying, pausing, and synchronizing replication 548

Getting ready • 549

How to do it... • 549

There's more... • 549

See also • 551

Logical replication 551

    Getting ready • 552

    How to do it… • 552

    How it works… • 554

    There's more… • 555

EDB Postgres Distributed 556

    Getting ready • 556

    How to do it… • 557

    How it works… • 559

    There's more… • 560

Archiving transaction log data 560

    Getting ready • 560

    How to do it… • 561

    There's more… • 562

    See also • 563

Upgrading minor releases 563

    Getting ready • 563

    How to do it… • 563

    How it works… • 564

    There's more… • 564

Major upgrades in-place 565

    Getting ready • 565

    How to do it… • 566

    How it works… • 566

    There's more… • 567

Major upgrades online 567

    How to do it… • 567

    How it works… • 568

Other Books You May Enjoy   571

Index   575

Preface

PostgreSQL is a powerful, open source database management system with an enviable reputation for high performance and stability. With many new features in its arsenal, PostgreSQL 16 allows you to scale up your PostgreSQL infrastructure. With this book, you'll take a step-by-step, recipe-based approach to effective PostgreSQL administration.

This book will get you up and running with all the latest features of PostgreSQL 16 while helping you explore the entire database ecosystem. You will acquire skills to address a range of challenges encountered in database administration, including tasks like table creation, view management, performance enhancement, and database security. As you make progress, the book will draw attention to important topics such as monitoring roles, validating backups, regular maintenance, and recovery of your PostgreSQL 16 database. This will help you understand roles, ensuring high availability, concurrency, and replication. Along with updated recipes, this book touches upon important areas such as using generated columns, the MERGE statement, deploy automation, PostgreSQL on the private or public cloud, and much more.

By the end of this PostgreSQL book, you'll have gained the knowledge you need to manage your PostgreSQL 16 database efficiently, both in the cloud and on-premises.

Who this book is for

This PostgreSQL 16 book is for database administrators, database architects, database developers, and anyone with an interest in planning and running live production databases using PostgreSQL 16. Those looking for hands-on solutions to any problem associated with PostgreSQL 16 administration will also find this book useful. Some experience with handling PostgreSQL databases will help you to make the most out of this book; however, it is a useful resource even if you are just beginning your Postgres journey.

What this book covers

Chapter 1, *First Steps*, introduces you to PostgreSQL 16; it explains how to download and install PostgreSQL 16, connect to a PostgreSQL server, enable server access to the network or remote users, use graphical administration tools or the `psql` query and scripting tool, change your password securely, avoid hardcoding your password, use a connection service file, and troubleshoot a failed connection. This chapter also covers how to use PostgreSQL in the cloud, including Kubernetes, and with TPA (Trusted Postgres Architect).

Chapter 2, *Exploring the Database*, shows how to identify the version of the database server you are using, and the server uptime. It helps you locate the database server files, the database server message log, and the database's system identifier. It explains how to list the databases on the database server, and it contains recipes to find out the number of tables in your database, how much memory and disk space is used by each database and table, what the biggest tables are, how many rows a table has, both exactly and in a quicker estimate, and how to understand object dependencies.

Chapter 3, *Server Configuration*, starts by showing where the documentation is; then it discusses how to plan a new database and how to view and change the settings of the database server in various ways: at the global level, at the database level, from within the application, in a session, and depending on what user is logged in. The chapter ends with three recipes on how to add, use, and manage PostgreSQL extensions.

Chapter 4, *Server Control*, includes recipes on how to start and stop the database server manually, and how to reload the server configuration. Regarding connections, we show how to prevent new ones, limit them per user, and terminate them to push a given user off the system. The chapter ends with some recipes on topics that are pertinent to scalability and resource allocation, such as the various multi-tenancy options based on separate instances, databases, or schemas, and setting up a connection pool using one or more PgBouncer instances.

Chapter 5, *Tables and Data*, begins with two recipes on database object names, followed by twelve recipes with solutions of practical problems such as duplicate rows, finding good candidates for unique constraints, generating test data and sample data, loading data from spreadsheets or flat files, applying large data changes, using partitioning to handle large data, consolidating data with the MERGE statement, and properly using JSON data types.

Chapter 6, *Security*, provides a security overview, followed by recipes on the PostgreSQL superuser, revoking user access to a table and granting user access to a table and to specific columns or rows.

We then discuss creating a new user, temporarily preventing a user from connecting, removing a user without dropping their data, checking whether all users have a secure password, giving limited superuser powers to specific users, auditing changes, knowing who is currently connected, integrating with LDAP, connecting using SSL, encrypting sensitive data, and setting up cloud security.

Chapter 7, Database Administration, starts with recipes on writing scripts where all commands either succeed or fail, or exit on the first error, use `psql` variables to store data or to conditionally alter the flow of the script, and set up a more useful prompt. We then see how to schedule maintenance jobs, perform actions on many tables, add and remove columns in tables, change the data type of a column, or the definition of an enumerative type, add constraints concurrently, add and remove schemas or tablespaces, and move objects between them, access objects in other databases, not only running PostgreSQL, enabling data updates on views, use materialized views, generated data columns, and data compression.

Chapter 8, Monitoring and Diagnosis, begins with an overview of monitoring, including on the cloud, and how to fetch relevant monitoring data from PostgreSQL, including from the logs. It provides recipes that answer questions such as whether a user is connected, what they are running, whether they are active or blocked, who they are being blocked by, whether anybody is using a specific table, when the table was last used, how much disk space is being used by temporary data and whether it is still in use or not, and why your queries could be slowing down. It also demonstrates how to monitor the progress of commands, monitor the PostgreSQL log, and produce a daily summary report, monitor PostgreSQL I/O statistics, kill a specific session, kill idle in-transaction sessions, and analyze the performance of your queries and track important metrics over time.

Chapter 9, Regular Maintenance, provides useful recipes on how to control automatic database maintenance, avoid auto-freezing, avoid transaction wraparound, offer solutions for heavy users of temporary tables, identify and fix bloated tables and indexes, maintain indexes, find unused indexes, carefully remove unwanted indexes, and plan maintenance.

Chapter 10, Performance and Concurrency, covers topics such as how to find slow SQL statements, collect regular statistics from `pg_stat*` views, discover what makes SQL slow, reduce the number of rows returned, simplify complex SQL, and speed up queries without rewriting them. It also delves into understanding why some queries do not use an index, how to force a query to use an index, and how to reap the benefits of parallel queries, understand JIT, use optimistic locking, and report performance problems. Additionally, you will learn about the new parallel query features, `TABLESAMPLE` and time-series partitioning.

Chapter 11, Backup and Recovery, provides useful information about the backup and recovery of your PostgreSQL database through recipes on how to understand and control crash recovery and how to plan backups. Additionally, you will learn about the hot logical backup of one database, the hot logical backup of all databases, the hot logical backup of all tables in a tablespace, the backup of database object definitions, the standalone hot physical database backup, the hot physical backup, and continuous archiving. It also includes topics such as the recovery of all databases, recovery to a point in time, the recovery of a dropped or damaged table, the recovery of a dropped or damaged database, the recovery of a dropped or damaged tablespace, how to improve the performance of backup/recovery, and incremental/differential backup and restore.

Chapter 12, Replication and Upgrades, explains that replication isn't magic, although it can be pretty cool. It's even cooler when it works, and that's what this chapter is all about. This chapter covers replication concepts such as replication best practices, how to set up streaming log replication, both physical and logical, how to manage hot standby, synchronous replication, how to upgrade to a new minor release, in-place major upgrades, major upgrades online, setting up replication using the CloudNativePG Kubernetes operator and Trusted Postgres Architect (TPA), and Postgres Distributed with multiple writable nodes.

To get the most out of this book

In order for this book to be useful, you need access to a PostgreSQL client that is allowed to execute queries on a server. Ideally, you'll also be the server administrator. Full client and server packages for PostgreSQL are available for most popular operating systems at https://www.postgresql.org/download/. All the examples here are executed at the Command Prompt, usually running the psql program. This makes them applicable to most platforms. It's straightforward to do most of these operations by using a GUI tool for PostgreSQL, such as pgAdmin:

- pgAdmin: https://www.pgadmin.org/download/

Download the color images

We also provide a PDF file that has color images of the screenshots/diagrams used in this book. You can download it here: https://packt.link/gbp/9781835460580.

Conventions used

There are a number of text conventions used throughout this book.

Code in text: Indicates code words in text, database table names, folder names, filenames, file extensions, pathnames, dummy URLs, user input, and Twitter handles. Here is an example: "Many experienced PostgreSQL DBAs will prefer to execute their own VACUUM commands."

A block of code is set as follows:

```
autovacuum = on
track_counts = on
```

Any command-line input or output is written as follows:

```
VACUUM (DISABLE_PAGE_SKIPPING);
```

 Warnings or important notes appear like this.

 Tips and tricks appear like this.

Sections

In this book, you will find several headings that appear frequently (Getting ready, How to do it..., How it works..., There's more..., and See also).

To give clear instructions on how to complete a recipe, use these sections as follows:

Getting ready

This section tells you what to expect in the recipe and describes how to set up any software or any preliminary settings required for the recipe.

How to do it...

This section contains the steps required to follow the recipe.

How it works...

This section usually consists of a detailed explanation of what happened in the previous section.

There's more...

This section consists of additional information about the recipe in order to make you more knowledgeable about the recipe.

See also

This section provides helpful links to other useful information for the recipe.

Get in touch

Feedback from our readers is always welcome.

General feedback: If you have questions about any aspect of this book, mention the book title in the subject of your message and email us at customercare@packtpub.com.

Errata: Although we have taken every care to ensure the accuracy of our content, mistakes do happen. If you have found a mistake in this book, we would be grateful if you would report this to us. Please visit www.packtpub.com/support/errata, selecting your book, clicking on the Errata Submission Form link, and entering the details.

Piracy: If you come across any illegal copies of our works in any form on the Internet, we would be grateful if you would provide us with the location address or website name. Please contact us at copyright@packt.com with a link to the material.

If you are interested in becoming an author: If there is a topic that you have expertise in and you are interested in either writing or contributing to a book, please visit authors.packtpub.com.

Share your thoughts

Once you've read *PostgreSQL 16 Administration Cookbook*, we'd love to hear your thoughts! Scan the QR code below to go straight to the Amazon review page for this book and share your feedback.

https://packt.link/r/1835460585

Your review is important to us and the tech community and will help us make sure we're delivering excellent quality content.

Download a free PDF copy of this book

Thanks for purchasing this book!

Do you like to read on the go but are unable to carry your print books everywhere?

Is your eBook purchase not compatible with the device of your choice?

Don't worry, now with every Packt book you get a DRM-free PDF version of that book at no cost.

Read anywhere, any place, on any device. Search, copy, and paste code from your favorite technical books directly into your application.

The perks don't stop there, you can get exclusive access to discounts, newsletters, and great free content in your inbox daily

Follow these simple steps to get the benefits:

1. Scan the QR code or visit the link below

https://packt.link/free-ebook/9781835460580

2. Submit your proof of purchase
3. That's it! We'll send your free PDF and other benefits to your email directly

1
First Steps

PostgreSQL is a feature-rich, general-purpose database-management system. It's a complex piece of software, but every journey begins with the first step.

We'll start with your first connection. Many people fall at the first hurdle, so we'll try not to skip past that too swiftly. We'll quickly move on to enabling remote users, and from there, we will move on to getting access through GUI administration tools.

We will also introduce the `psql` query tool, which is the tool used to load our sample database, as well as many other examples in the book.

For additional help, we've included a few useful recipes that you may need for reference.

In this chapter, we will cover the following recipes:

- Introducing PostgreSQL
- How to get PostgreSQL
- Connecting to the PostgreSQL server
- Enabling access for network/remote users
- Using the `pgAdmin` GUI tool
- Using the `psql` query and scripting tool
- Changing your password securely
- Avoiding hardcoding your password
- Using a connection service file
- Troubleshooting a failed connection
- PostgreSQL in the cloud

- PostgreSQL with Kubernetes
- PostgreSQL with TPA

Introducing PostgreSQL 16

PostgreSQL is an advanced SQL database server, available on a wide range of platforms. One of the clearest benefits of PostgreSQL is that it is open source, meaning that you have a very permissive license to install, use, and distribute it without paying anyone any fees or royalties. On top of that, PostgreSQL is known as a database that stays up for long periods and requires little or no maintenance, in most cases. Overall, PostgreSQL provides a very low total cost of ownership.

PostgreSQL is also known for its huge range of advanced features, developed over the course of more than 30 years of continuous development and enhancement. Originally developed by the Database Research Group at the University of California, Berkeley, PostgreSQL is now developed and maintained by a huge army of developers and contributors. Many of these contributors have full-time jobs related to PostgreSQL, working as designers, developers, database administrators, and trainers. Some, but not many, of these contributors work for companies that specialize in support for PostgreSQL. No single company owns PostgreSQL, nor are you required (or even encouraged) to register your usage.

PostgreSQL has the following main features:

- Excellent SQL standards compliance, up to SQL:2023
- Client-server architecture
- A highly concurrent design, where readers and writers don't block each other
- Highly configurable and extensible for many types of applications
- Excellent scalability and performance, with extensive tuning features
- Support for many kinds of data models, such as relational, post-relational (arrays and nested relations via record types), document (JSON and XML), and key/value

What makes PostgreSQL different?

The PostgreSQL project focuses on the following objectives:

- Robust, high-quality software with maintainable, well-commented code
- Low-maintenance administration for both embedded and enterprise use
- Standards-compliant SQL, interoperability, and compatibility
- Performance, security, and high availability

What surprises many people is that PostgreSQL's feature set is more similar to Oracle or SQL Server than it is to MySQL. The only connection between MySQL and PostgreSQL is that these two projects are open source; apart from that, the features and philosophies are almost totally different.

One of the key features of Oracle, since Oracle 7, has been snapshot isolation, where readers don't block writers and writers don't block readers. You may be surprised to learn that PostgreSQL was the first database to be designed with this feature, and it offers a complete implementation. In PostgreSQL, this feature is called **Multiversion Concurrency Control** (**MVCC**), and we will discuss this in more detail later in the book.

PostgreSQL is a general-purpose database management system. You define the database that you want to manage with it. PostgreSQL offers you many ways in which to work. You can either use a normalized database model, augmented with features such as arrays and record subtypes, or use a fully dynamic schema with the help of JSONB and an extension named hstore. PostgreSQL also allows you to create your own server-side functions in any of a dozen different languages, including a formal notion of **transform** to ensure data is properly converted.

PostgreSQL is highly extensible, so you can add your own data types, operators, index types, and functional languages. You can even override different parts of the system, using plugins to alter the execution of commands, or add a new query optimizer.

All of these features offer a huge range of implementation options to software architects. There are many ways out of trouble when building applications and maintaining them over long periods of time. Regrettably, we simply don't have space in this book for all the cool features for developers; this book is about administration, maintenance, and backup.

In the early days, when PostgreSQL was still a research database, the focus was solely on the cool new features. Over the last 20 years, enormous amounts of code have been rewritten and improved, giving us one of the largest and most stable software servers available for operational use.

Who is using PostgreSQL? Prominent users include Apple, BASF, Genentech, Heroku, IMDB, Skype, McAfee, NTT, the UK Met Office, and the US National Weather Service. Early in 2010, PostgreSQL received well in excess of 1,000,000 downloads per year, according to data submitted to the European Commission, which concluded that *"PostgreSQL is considered by many database users to be a credible alternative."* PostgreSQL has gone on from there to be even more popular.

We need to mention one last thing: when PostgreSQL was first developed, it was named **Postgres**, and therefore, many aspects of the project still refer to the word *Postgres* – for example, the default database is named postgres, and the software is frequently installed using the postgres user ID. As a result, people shorten the name PostgreSQL to simply Postgres and, in many cases, use the two names interchangeably.

PostgreSQL is pronounced *post-grez-q-l*. Postgres is pronounced *post-grez*.

Some people get confused and refer to it as *Postgre or Postgre SQL*, which are hard to say and likely to confuse people. Two names are enough, so don't use a third one!

The following sections explain the key areas in more detail.

Robustness

PostgreSQL is robust, high-quality software, supported by testing for both features and concurrency. By default, the database provides strong disk-write guarantees, and developers take the risk of data loss very seriously in everything they do. Options to trade robustness for performance exist, although they are not enabled by default.

All actions on the database are performed within transactions, protected by a transaction log that will perform automatic crash recovery in case of software failure.

Databases may optionally be created with data block checksums to help diagnose hardware faults. Multiple backup mechanisms exist, with full and detailed **Point-in-Time Recovery (PITR)** if you need a detailed recovery. A variety of diagnostic tools are available as well.

Database replication is supported natively. Synchronous replication can provide greater than *5 nines* (99.999%) of availability and data protection, if properly configured and managed, or even higher with appropriate redundancy.

Security

Access to PostgreSQL is controllable via host-based access rules. Authentication is flexible and pluggable, allowing for easy integration with any external security architecture. The latest **Salted Challenge Response Authentication Mechanism (SCRAM)** provides full 256-bit protection.

Full SSL-encrypted access is supported natively for both user access and replication. A full-featured cryptographic function library is available for database users.

PostgreSQL provides role-based access privileges to access data, by command type. PostgreSQL also provides **Row-Level Security (RLS)** for privacy, medical, and military-grade security.

Functions can execute with the permissions of the definer, while views may be defined with security barriers to ensure that security is enforced ahead of other processing.

All aspects of PostgreSQL are assessed by an active security team, while known exploits are categorized and reported at http://www.postgresql.org/support/security/.

Ease of use

Clear, full, and accurate documentation exists as a result of a development process where documentation changes are required.

The documentation can easily be found on the PostgreSQL website at https://www.postgresql.org/docs/. In this book, we will refer many times to the URLs of specific sections of that documentation.

Another option is to install a copy of the exact same documentation on your laptop, in the PDF or HTML format, for offline use. You can do it easily on most operating systems by installing the appropriate package, as in this Ubuntu/Debian example:

```
$ sudo apt-get install postgresql-doc-16
```

Hundreds of small changes occur with each release, which smooth off any rough edges of usage, supplied directly by knowledgeable users.

PostgreSQL works on small and large systems in the same way and across operating systems.

Client access and drivers exist for every language and environment, so there is no restriction on what type of development environment is chosen now or in the future.

The SQL standard is followed very closely; there is no weird behavior, such as silent truncation of data.

Text data is supported via a single data type that allows the storage of anything from 1 byte to 1 gigabyte. This storage is optimized in multiple ways, so 1 byte is stored efficiently, and much larger values are automatically managed and compressed.

PostgreSQL has a clear policy of minimizing the number of configuration parameters, and with each release, we work out ways to auto-tune the settings.

Extensibility

PostgreSQL is designed to be highly extensible. Database extensions can be easily loaded by using CREATE EXTENSION, which automates version checks, dependencies, and other aspects of configuration.

PostgreSQL supports user-defined data types, operators, indexes, functions, and languages.

Many extensions are available for PostgreSQL, including the **PostGIS** extension, which provides world-class **Geographical Information System (GIS)** features.

Performance and concurrency

PostgreSQL 16 can achieve significantly more than 1,000,000 reads per second on a 4-socket server, and it benchmarks at more than 50,000 write transactions per second with full durability, depending upon your hardware. With advanced hardware, even higher levels of performance are possible.

PostgreSQL has an advanced optimizer that considers a variety of join types, utilizing user data statistics to guide its choices. PostgreSQL provides the widest range of index types of any commonly available database server, fully supporting all data types.

PostgreSQL provides **MVCC**, which enables readers and writers to avoid blocking each other.

Taken together, the performance features of PostgreSQL allow a mixed workload of transactional systems and complex search and analytical tasks. This is important because it means we don't always need to unload our data from production systems and reload it into analytical data stores just to execute a few ad hoc queries. PostgreSQL's capabilities make it the database of choice for new systems, as well as the correct long term choice in almost every case.

Scalability

PostgreSQL 16 scales well on a single node, with multiple CPU sockets. PostgreSQL efficiently runs up to hundreds of active sessions and thousands of connected sessions when using a session pool. Further scalability is achieved in each annual release.

PostgreSQL provides multi-node read scalability using the **Hot Standby** feature. Transparent multi-node write scalability is under active development. The starting point for this is EDB Postgres Distributed (formerly *Bi-directional replication*, which will be discussed in *Chapter 12, Replication and Upgrades*), as it allows transparent and efficient synchronization of reference data across multiple servers. Other forms of write scalability have existed for more than a decade, starting from the PL/Proxy language, Greenplum and Citus.

SQL and NoSQL data models

PostgreSQL follows the SQL standard very closely. SQL itself does not force any particular type of model to be used, so PostgreSQL can easily be used for many types of models at the same time, in the same database.

PostgreSQL can be used as a relational database, in which case we can utilize any level of denormalization, from the full **Third Normal Form** (**3NF**) to the more normalized star schema models. PostgreSQL extends the relational model to provide arrays, row types, and range types.

A document-centric database is also possible using PostgreSQL's text, XML, and binary JSON (JSONB) data types, supported by indexes optimized for documents and by full-text search capabilities.

Key/value stores are supported using the `hstore` extension.

Popularity

When MySQL was taken over by a commercial database vendor some years back, it was agreed in the EU monopoly investigation that followed that PostgreSQL was a viable competitor. That's certainly been true, with the PostgreSQL user base expanding consistently for more than a decade.

Various polls have indicated that PostgreSQL is the favorite database for building new, enterprise-class applications. The PostgreSQL feature set attracts serious users who have serious applications. Financial services companies may be PostgreSQL's largest user group, although governments, telecommunication companies, and many other segments are strong users as well. This popularity extends across the world; Japan, Ecuador, Argentina, and Russia have very large user groups, as do the US, Europe, and Australasia.

Amazon Web Services' chief technology officer, Dr. Werner Vogels, described PostgreSQL as *"an amazing database,"* going on to say that *"PostgreSQL has become the preferred open source relational database for many enterprise developers and start-ups, powering leading geospatial and mobile applications."* More recently, AWS has revealed that PostgreSQL is their fastest-growing service.

Commercial support

Many people have commented that strong commercial support is what enterprises need before they can invest in open source technology. Strong support is available worldwide from a number of companies.

The authors of this book work for **EnterpriseDB** (**EDB**), the largest company providing commercial support for open source PostgreSQL, offering 24/7 support in English with bug-fix resolution times.

Many other companies provide strong and knowledgeable support to specific geographic regions, vertical markets, and specialized technology stacks.

PostgreSQL is also available as a hosted or cloud solution from a variety of companies, since it runs very well in cloud environments.

A full list of companies is kept up to date at the following URL: http://www.postgresql.org/support/professional_support/.

Research and development funding

PostgreSQL was originally developed as a research project at the University of California, Berkeley in the late 1980s and early 1990s. Further work was carried out by volunteers until the late 1990s. Then, the first professional developer became involved. Over time, more and more companies and research groups became involved, supporting many professional contributors. Further funding for research and development was provided by the National Science Foundation.

The project also received funding from the *EU FP7 Programme* in the form of the *4CaaST* project for cloud computing and the *AXLE* project for scalable data analytics. *AXLE* deserves a special mention because it was a three-year project aimed at enhancing PostgreSQL's business intelligence capabilities, specifically for very large databases. The project covered security, privacy, integration with data mining, and visualization tools and interfaces for new hardware.

Other funding for PostgreSQL development comes from users who directly sponsor features and companies that sell products and services based around PostgreSQL.

Many features are contributed regularly by larger commercial companies, such as EDB.

How to get PostgreSQL

PostgreSQL is 100% open source software and is freely available to use, alter, or redistribute in any way you choose. Its license is an approved open source license, very similar to the **Berkeley Software Distribution** (BSD) license, although only just different enough that it is now known as **The PostgreSQL License** (TPL). You can see the license here: https://opensource.org/licenses/PostgreSQL.

How to do it...

PostgreSQL is already being used by many different application packages, so you may find it already installed on your servers. Many Linux distributions include PostgreSQL as part of the basic installation or include it with the installation disk.

One thing to be wary of is that the included version of PostgreSQL may not be the latest release. It will typically be the latest major release that was available when that operating system release was published. There is usually no good reason to stick to that version – there is no increased stability implied there—and later production versions are just as well supported by the various Linux distributions as the earlier versions.

If you don't have a copy yet or the latest version, you can download the source code or binary packages for a variety of operating systems from http://www.postgresql.org/download/.

Installation details vary significantly from platform to platform, and there aren't any special tricks or recipes to mention. Just follow the installation guide, and away you go! We've consciously avoided describing the installation processes here to make sure we don't garble or override the information published to assist you.

EDB has provided the main macOS/Windows installer for PostgreSQL for many years, which can be accessed here: https://www.enterprisedb.com/downloads/postgres-postgresql-downloads. This gives you the option of installing both client and server software so that you can try it out on your laptop:

Figure 1.1: The PostgreSQL Setup Wizard

The installer shown in *Figure 1.2* also allows you to install just the client software, allowing you to work with remote database servers, such as *PostgreSQL in the cloud*:

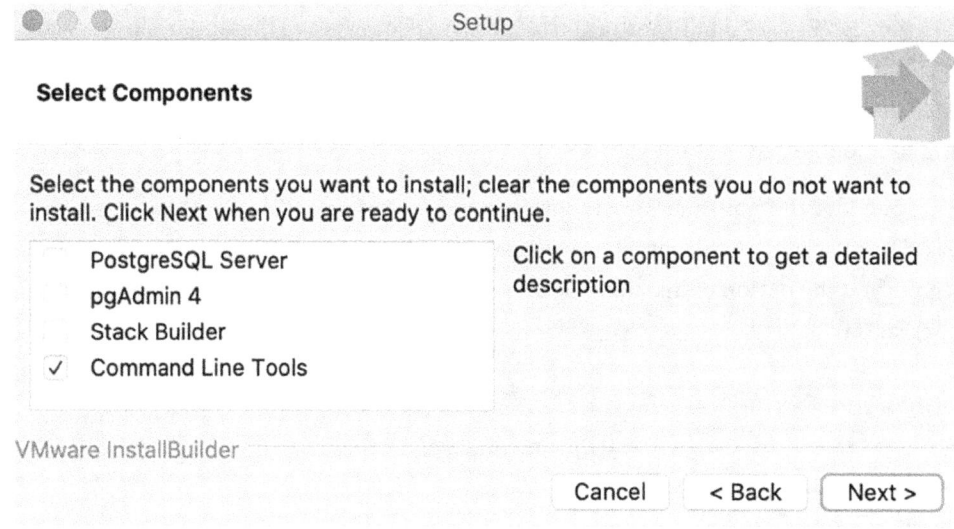

Figure 1.2: Selecting components to install

If you would like to receive email updates of the latest news, you can subscribe to the PostgreSQL announce mailing list, which contains updates from all the vendors that support PostgreSQL. You'll get a few emails each month about new releases of core PostgreSQL, related software, conferences, and user group information. It's worth keeping in touch with these developments.

> **Note**
>
> For more information about the PostgreSQL announcement mailing list, visit http://archives.postgresql.org/pgsql-announce/.

How it works...

Many people ask questions such as, *How can this be free? Are you sure I don't have to pay someone? Who gives this stuff away for nothing?*

Open source applications such as PostgreSQL work on a community basis, where many contributors perform tasks that make the whole process work. For many of these people, their involvement is professional, rather than merely a hobby, and they can do this because there is generally great value for both the contributors and their employers alike.

You might not believe it. You don't have to, because it just works!

There's more...

Remember that PostgreSQL is more than just the core software. There is a huge range of websites that offer add-ons, extensions, and tools for PostgreSQL. You'll also find an army of bloggers who provide useful tricks and discoveries that will help you in your work.

Besides these, a range of professional companies can offer you help when you need it.

Connecting to the PostgreSQL server

How do we access PostgreSQL?

Connecting to the database is the first experience of PostgreSQL for most people, so we want to make it a good one. Let's do it now and fix any problems we have along the way. Remember that a connection needs to be made secure, so there may be some hoops for us to jump through to ensure that the data we wish to access is secure.

Before we can execute commands against the database, we need to connect to the database server to give us a session.

Sessions are designed to be long-lived, so you connect once, perform many requests, and eventually disconnect. There is a small overhead during the connection. It may become noticeable if you connect and disconnect repeatedly, so you may wish to investigate the use of **connection pools**. Connection pools allow pre-connected sessions to be quickly served to you when you wish to reconnect. We will discuss them in *Chapter 4, Server Control*.

Getting ready

First, cache your database. If you don't know where it is, you'll probably have difficulty accessing it. There may be more than one database, and you'll need to know the right one to access and have the authority to connect to it.

You need to specify the following parameters to connect to PostgreSQL:

- A host or host address
- A port
- A database name
- A user
- A password (or other means of authentication; but only if requested)

To connect, there must be a PostgreSQL server running on that host and listening to the port with that number. On that server, a database and a user with the specified names must also exist. Furthermore, the host must explicitly allow connections from your client (as explained in the *Enabling access for network/remote users* recipe), and you must also pass the authentication step using the method the server specifies – for example, specifying a password won't work if the server has requested a different form of authentication. Note that you might not need to provide a password at all if PostgreSQL can recognize that your user is already authenticated by the OS; this is called **peer authentication**. After showing an example in this recipe, we will discuss it fully in the next recipe: *Enabling access for network/remote users* (despite not being a network/remote connection method).

Almost all PostgreSQL interfaces use the `libpq` interface library. When using `libpq`, most of the connection parameter handling is identical, so we can discuss that just once.

If you don't specify the preceding parameters, PostgreSQL looks for values set through environment variables, which are as follows:

- PGHOST or PGHOSTADDR
- PGPORT (set this to 5432 if it is not set already)
- PGDATABASE
- PGUSER
- PGPASSWORD (this is definitely not recommended by us, nor by the PostgreSQL documentation, even if it still exists)

If you somehow specify the first four parameters but not the password, PostgreSQL looks for a password file, as discussed in the *Avoiding hardcoding your password* recipe.

Some PostgreSQL interfaces use the client-server protocol directly, so the ways in which the defaults are handled may differ. The information we need to supply won't vary significantly, so check the exact syntax for that interface.

Connection details can also be specified using a connection string, as in this example:

```
psql "user=myuser host=myhost port=5432 dbname=mydb password=mypasswd"
```

or alternatively using a **Uniform Resource Identifier** (**URI**) format, as follows:

```
psql postgresql://myuser:mypasswd@myhost:5432/mydb
```

Both examples specify that we will connect the `psql` client application to the PostgreSQL server at the `myhost` host, on port 5432, with the database name `mydb`, user `myuser` and password `mypasswd`.

> **Note**
>
> If you do not specify mypasswd in the preceding URI, you may be prompted to enter the password.

How to do it...

In this example, Afroditi is a database administrator who needs to connect to PostgreSQL to perform some maintenance activities. She can SSH to the database server using her own username afroditi, and DBAs are given sudo privileges to become the postgres user, so she can simply launch psql as the postgres user:

```
afroditi@dbserver1:~$ sudo -u postgres psql
psql (16.0 (Debian 16.0-1.pgdg120+1))
Type "help" for help.

postgres=#
```

Note that psql was launched as the postgres user, so it used the postgres user for the database connection, and that psql on Linux attempts a Unix socket connection by default. Hence, this matches peer authentication.

How it works...

PostgreSQL is a client-server database. The system it runs on is known as the **host**. We can access the PostgreSQL server remotely, through the network. However, we must specify host, which is a hostname, or hostaddr, which is an IP address. We can specify a host as localhost if we wish to make a TCP/IP connection to the same system. Rather than using TCP/IP to localhost, it is usually better to use a Unix socket connection, which is attempted if the host begins with a slash (/) and the name is presumed to be a directory name (the default is /tmp).

On any system, there can be more than one database server. Each database server listens to exactly one well-known network port, which cannot be shared between servers on the same system. The default port number for PostgreSQL is 5432, which has been registered with the **Internet Assigned Numbers Authority** (**IANA**) and is uniquely assigned to PostgreSQL (you can see it used in the /etc/services file on most *nix servers). The port number can be used to uniquely identify a specific database server, if any exist. IANA (http://www.iana.org) is the organization that coordinates the allocation of available numbers for various internet protocols.

A database server is also sometimes known as a **database cluster** because the PostgreSQL server allows you to define one or more databases on each server. Each connection request must identify exactly one database, identified by its dbname. When you connect, you will only be able to see the database objects created within that database.

A database user is used to identify the connection. By default, there is no limit on the number of connections for a particular user. In the *Enabling access for network/remote users* recipe, we will cover how to restrict that. In more recent versions of PostgreSQL, users are referred to as login roles, although many clues remind us of the earlier nomenclature, and that still makes sense in many ways. A login role is a role that has been assigned the CONNECT privilege.

Each connection will typically be authenticated in some way. This is defined at the server level: client authentication will not be optional at connection time if the administrator has configured the server to require it.

Once you've connected, each connection can have one active transaction at a time and one fully active statement at any time.

The server will have a defined limit on the number of connections it can serve, so a connection request can be refused if the server is oversubscribed.

There's more...

If you are already connected to a database server with psql and you want to confirm that you've connected to the right place and in the right way, you can execute some, or all, of the following commands. Here is the command that shows the current_database:

```
SELECT current_database();
```

The following command shows the current_user ID:

```
SELECT current_user;
```

The next command shows the IP address and port of the current connection, unless you are using Unix sockets, in which case both values are NULL:

```
SELECT inet_server_addr(), inet_server_port();
```

A user's password is not accessible using general SQL, for obvious reasons.

You may also need the following:

```
SELECT version();
```

This is just one of several ways to check the database software version; please refer to the *What version is the server?* recipe in *Chapter 2, Exploring the Database*. You can also use the new `psql` meta-command, `\conninfo`. This displays most of the preceding information in a single line:

```
postgres=# \conninfo
You are connected to database postgres, as user postgres, via socket in /var/run/postgresql, at port 5432.
```

See also

There are many other snippets of information required to understand connections. Some of them are mentioned in this chapter, and others are discussed in *Chapter 6, Security*. For further details, refer to the PostgreSQL server documentation, which we provided a link to earlier in this chapter.

Enabling access for network/remote users

PostgreSQL comes in a variety of distributions. In many of these, you will note that remote access is initially disabled as a security measure. You can do this quickly, as described here, but you really should read the chapter on security soon.

How to do it...

By default, on Linux PostgreSQL gives access to clients who connect using Unix sockets, provided that the database user is the same as the system's username, as in the example from the previous recipe.

A socket is effectively a filesystem path that processes running on the same host can use for two-way communication. The PostgreSQL server process can see the OS username under which the client is running, and authenticate the client based on that. This is great, but unfortunately only applies to the special case when the client and the server are running on the same host. For all the remaining cases, we need to show you how to enable all the other connection methods.

Note

In this recipe, we mention configuration files, which can be located as shown in the *Finding the configuration settings for your session* recipe in *Chapter 3, Server Configuration*.

The steps are as follows:

1. Add or edit this line in your postgresql.conf file:

   ```
   listen_addresses = '*'
   ```

2. Add the following line as the first line of pg_hba.conf to allow access to *all* databases for *all* users with an encrypted password:

   ```
   # TYPE    DATABASE    USER    CIDR-ADDRESS    METHOD
   host      all         all     0.0.0.0/0       scram-sha-256
   ```

3. After changing listen_addresses, we restart the PostgreSQL server, as explained in the *Updating the parameter file* recipe in *Chapter 3, Server Configuration*.

> **Note**
>
> This recipe assumes that postgresql.conf does not include any other configuration files, which is the case in a default installation. If changing listen_addresses in postgresql.conf does not seem to work, perhaps that setting is overridden by another configuration file. Check out the recipe we just mentioned for more details.

How it works...

The listen_addresses parameter specifies which IP addresses to listen to. This allows you to flexibly enable and disable listening on interfaces of multiple **Network Interface Cards** (**NICs**) or virtual networks on the same system. In most cases, we want to accept connections on all NICs, so we use *, meaning all IP addresses. But the user could also specify the IP address of a given interface. For instance, you might decide to be listening only on connections coming through a specific VPN.

The pg_hba.conf file contains a set of host-based authentication rules. Each rule is considered in sequence until one rule matches the incoming connection and is applied for authentication, or the attempt is specifically rejected with a reject method, which is also implemented as a rule.

The rule that we added to the pg_hba.conf file means that a remote connection that specifies any user or database on any IP address will be asked to authenticate using a SCRAM-SHA-256-encrypted password. The following are the parameters required for SCRAM-SHA-256-encrypted passwords:

- Type: For this, `host` means a remote connection.
- Database: For this, `all` means for all databases. Other names match exactly, except when prefixed with a plus (+) symbol, in which case we mean a group role rather than a single user. You can also specify a comma-separated list of users or use the @ symbol to include a file with a list of users. You can even specify `sameuser` so that the rule matches when you specify the same name for the user and database.
- User: For this, `all` means for all users. Other names match exactly, except when prefixed with a plus (+) symbol, in which case we mean a group role rather than a single user. You can also specify a comma-separated list of users, or use the @ symbol to include a file with a list of users.
- CIDR-ADDRESS: This consists of two parts: an IP address and a subnet mask. The subnet mask is specified as the number of leading bits of the IP address that make up the mask. Thus, /0 means 0 bits of the IP address so that all IP addresses will be matched. For example, 192.168.0.0/24 would mean matching the first 24 bits, so any IP address of the 192.168.0.x form would match. You can also use `samenet` or `samehost`.
- Method: For this, `scram-sha-256` means that PostgreSQL will ask the client to provide a password encrypted with SCRAM-SHA-256. A common choice is `peer`, which is enabled by default and described in the *There's more...* section of this recipe. Another common (and discouraged!) setting is `trust`, which effectively means no authentication. Other authentication methods include GSSAPI, SSPI, LDAP, RADIUS, and PAM. PostgreSQL connections can also be made using SSL, in which case client SSL certificates provide authentication. See the *Using SSL certificates to authenticate the client* recipe in *Chapter 6, Security*, for more details.

Don't use the `password` authentication method in `pg_hba.conf` as this sends the password in plain text (it has been deprecated for years). This is not a real security issue if your connection is encrypted with SSL, but there are normally no downsides with SCRAM-SHA-256 anyway, and you have extra security for non-SSL connections.

There's more...

We have mentioned `peer` authentication as a method that allows password-less connections via Unix sockets. It is enabled by default, but only on systems that have Unix sockets, meaning that it does not exist on Windows: this is why the Windows installer asks you to insert the administrator password during installation.

When using a Unix socket connection, the client is another process running on the same host; therefore, Postgres can reliably get the OS username under which the client is running. The logic of peer authentication is to allow a connection attempt if the client's OS username is identical to the database username being used for the connection. Hence, if there is a database user with exactly the same name as an OS user, then that user can benefit from password-less authentication.

It is a safe technique, which is why it is enabled by default. In the special case of the postgres user, you can connect as a database superuser in a password-less way. This is not a security breach because PostgreSQL actually runs as the postgres OS user, so if you log in to the server with that user, then you are allowed to access all the data files.

When installing PostgreSQL on your Linux laptop, an easy way to enable your own access is to create a database user identical to your OS username, and also a database with the same name. This way, you can use psql with your normal OS user for password-less connections. Of course, to create the user and the database you need first to connect as the predefined postgres superuser, which you can do by running psql as the postgres OS user.

In earlier versions of PostgreSQL, access through the network was enabled by adding the -i command-line switch when you started the server. This is still a valid option, but now it is just equivalent to specifying the following:

```
listen_addresses = '*'
```

So, if you're reading some notes about how to set things up and this is mentioned, be warned that those notes are probably long out of date. They are not necessarily wrong, but it's worth looking further to see whether anything else has changed.

See also

Look at the installer and/or OS-specific documentation to find the standard location of the files.

Using the pgAdmin 4 GUI tool

Graphical administration tools are often requested by system administrators. PostgreSQL has a range of tool options. In this book, we'll cover pgAdmin 4.

pgAdmin 4 is a client application that sends and receives SQL to and from PostgreSQL, displaying the results for you. The admin client can access many database servers, allowing you to manage a fleet of servers. The tool works in both standalone app mode and within web browsers.

How to do it...

pgAdmin 4 is usually named just pgAdmin. The 4 at the end has a long history but isn't that important. It is more of an "epoch" than a release level; pgAdmin 4 replaces the earlier pgAdmin 3. Instructions to download and install it can be found at https://www.pgadmin.org/.

When you start pgAdmin, you will be prompted to register a new server.

Give your server a name on the **General** tab, and then click the **Connection** tab, as shown in the screenshot, and fill in the five basic connection parameters, as well as the other information. You should uncheck the **Save password?** option:

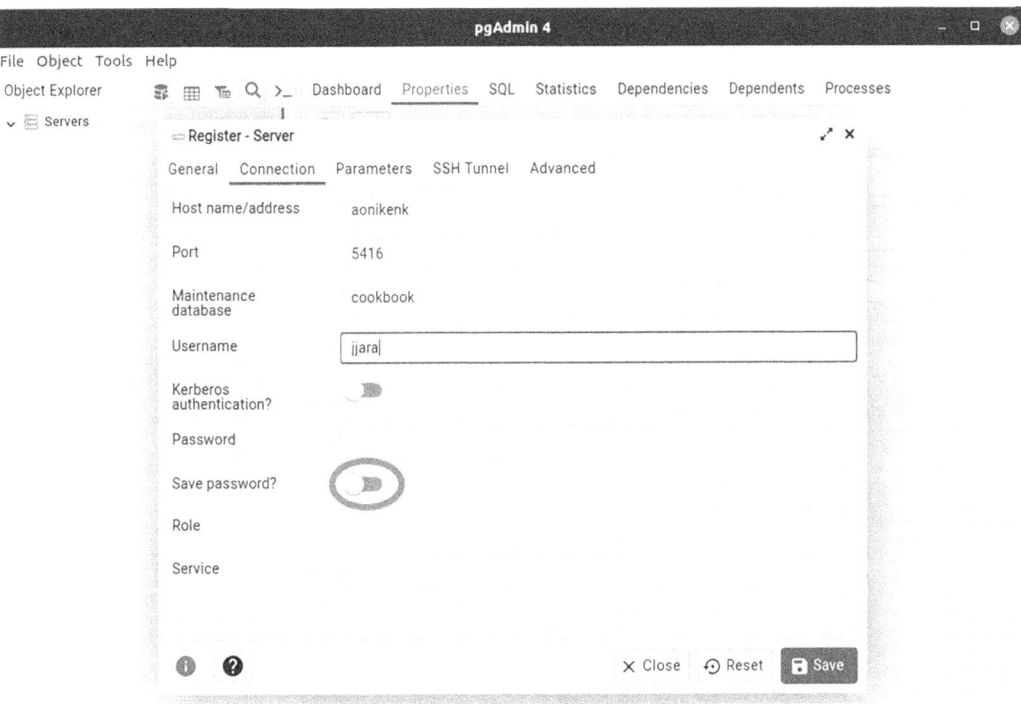

Figure 1.3: The server connection properties

If you have many database servers, you can group them together. I suggest keeping any replicated servers together in the same server group. Give each server a sensible name.

Once you've added a server, pgAdmin will connect to it and display information about it, using the information that you have added.

The default screen is **Dashboard**, which presents a few interesting graphs based on the data it polls from the server. That's not very useful, so click on the **Statistics** tab.

You will then get access to the main browser screen, with the object tree view on the left and statistics on the right, as shown in the following screenshot:

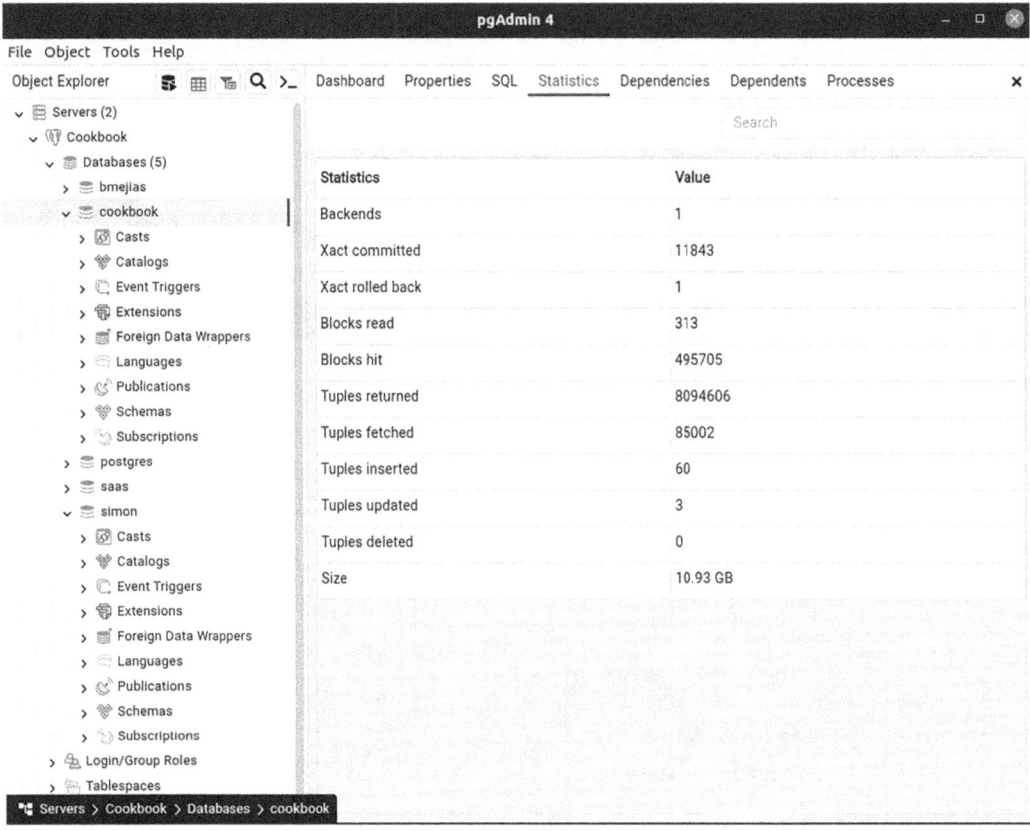

Figure 1.4: The pgAdmin tree view with the Statistics tab

pgAdmin easily displays much of the data that is available from PostgreSQL. The information is context-sensitive, allowing you to navigate and see everything quickly and easily. Except for the dashboard, the information is not dynamically updated; this will occur only when you navigate the application, where every click will refresh the data, so bear this in mind when using the application.

pgAdmin also provides **Grant Wizard**. This is useful for DBAs for review and immediate maintenance. In the example shown in the screenshot, the user first selected the **Sequences** on the navigation tree and then selected Tools → Grant Wizard. This will open a pop-up window to select the objects on which privileges will be granted to a selected role:

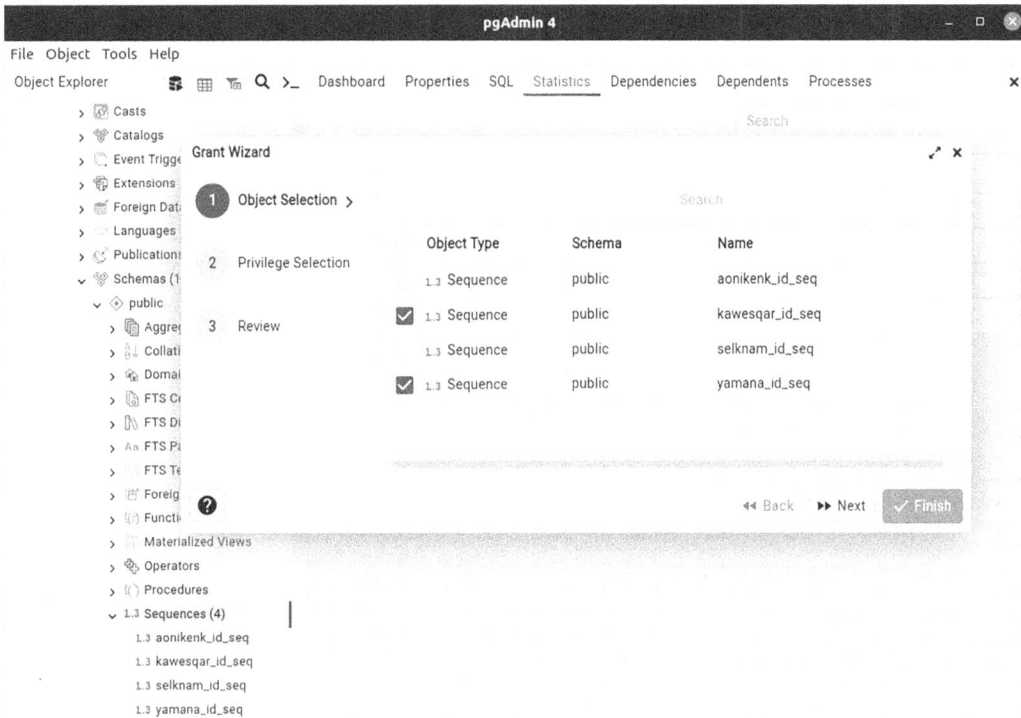

Figure 1.5: Grant Wizard: selecting sequences to grant privileges to a role

The pgAdmin query tool allows you to have multiple active sessions. The query tool has a good-looking visual **Explain** feature, which displays the EXPLAIN plan for your query. To do this, go to Tools → Query Tool. Write your query in the Query text box, and then click on the E (Explain) button, as shown in the following screenshot. The graphical execution tree is shown below the query:

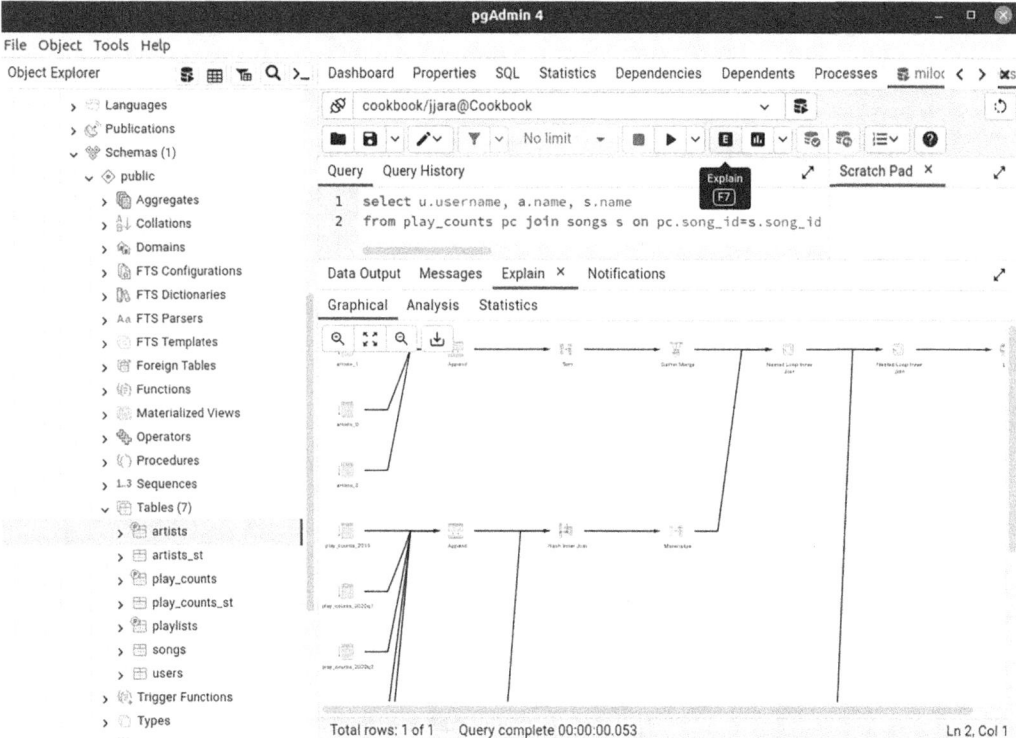

Figure 1.6: The visual Explain feature

How it works...

pgAdmin provides a wide range of features, many of which are provided by other tools as well. This gives us the opportunity to choose which of those tools we want. For many reasons, it is best to use the right tool for the right job, and that is always a matter of expertise, experience, and personal taste.

pgAdmin submits SQL to the PostgreSQL server and displays the results quickly and easily. As a database browser, it is fantastic. For performing small DBA tasks, it is ideal. As you might've guessed from these comments, I don't recommend GUI tools for every task.

Scripting is an important technique for DBAs. You keep an exact copy of the task executed, so you document all the actions in a way that is automatically repeatable, and you can edit and resubmit if problems occur. It's also easy to put all the tasks in a script into a single transaction, which isn't possible using the current GUI tools. For scripting, I strongly recommend the psql utility, which has many additional features that you'll increasingly appreciate over time.

Although I recommend psql as a scripting tool, many people find it convenient as a query tool. Some people may find this strange and assume that it is a choice for experts only. Two great features of psql as an interactive query tool are the online help for SQL and the tab completion feature, which allows you to build up SQL quickly without having to remember the syntax. This is why the *Using the psql query and scripting tool* recipe (which we recommend particularly for more information) is named that way.

pgAdmin provides the PSQL Tool in the Tools menu, which allows you to run psql alongside pgAdmin. This is a great innovation and allows you to get the power of a GUI alongside the power of psql.

pgAdmin also provides pgAgent, a job scheduler, which we will discuss in *Chapter 7, Database Administration*.

A quick warning! When you create an object in pgAdmin, the object will be created with a mixed-case name if you use capitals or spaces anywhere in the object name. If I ask for a table named MyTable, the only way to access that table is by referring to it in double quotes as "MyTable". See the *Handling objects with quoted names* recipe in *Chapter 5, Tables and Data*:

Figure 1.7: Table options

See also

You may also be interested in commercial tools of various kinds for PostgreSQL. A full listing is given in the PostgreSQL software catalog at http://www.postgresql.org/download/products/1.

Using the psql query and scripting tool

psql is the query tool supplied as a part of the core distribution of PostgreSQL, so it is available in all environments and works similarly in all of them. This makes it an ideal choice for developing portable applications and techniques.

psql provides features for use as both an interactive query tool and as a scripting tool.

Getting ready

From here on, we will assume that the psql command alone is enough to allow you access to the PostgreSQL server. This assumes that all your connection parameters are defaults, or that you have set environment variables appropriately, as previously explained in the *Enabling access for remote/network users* recipe.

Written in full, the connection parameters will be either of these options:

```
psql -h myhost -p 5432 -d mydb -U myuser
psql postgresql://myuser@myhost:5432/mydb
```

The default value for the port (-p) is 5432. By default, mydb and myuser are both identical to the operating system's username. The default myhost on Windows is localhost, while on Unix, we use the default directory for Unix socket connections. The location of such directories varies across distributions and is set at compile time. However, note that you don't actually need to know its value because, on local connections, both the server and the client are normally compiled together, so they use the same default.

How to do it...

The command that executes a single SQL command and prints the output is the easiest, as shown here:

```
$ psql -c "SELECT current_time"
     timetz
-----------------
 18:48:32.48101
(1 row)
```

The -c command is non-interactive. If we want to execute multiple commands, we can write those commands in a text file and then execute them using the -f option. This command loads a very small and simple set of examples:

```
$ psql -f examples.sql
```

Chapter 1

The contents of the examples.sql file are as follows:

```
SET client_encoding = 'UTF8';
SET standard_conforming_strings = on;
SET check_function_bodies = false;
SET xmloption = content;
SET client_min_messages = warning;
SET row_security = off;

DROP SCHEMA IF EXISTS myschema CASCADE;
CREATE SCHEMA myschema;

SET default_tablespace = '';
SET default_table_access_method = heap;
SET search_path = myschema;

CREATE TABLE mytable (
    id integer PRIMARY KEY,
    col1 text
);

CREATE TABLE mytable2 (
    id integer,
    fid integer REFERENCES mytable(id),
    col2 timestamp with time zone DEFAULT clock_timestamp(),
    PRIMARY KEY (id, fid)
);

COPY mytable (id, col1) FROM stdin;
1    Ananas
2    Banana
3    Cucumber
4    Dasheen
5    Endive
\.

COPY mytable2 (id, fid, col2) FROM stdin;
1001 1      2023-11-15 18:49:14.84806+01
```

```
1001  2       2023-11-15 18:49:14.848334+01
1002  5       2023-11-15 18:49:14.848344+01
\.
```

The above command produces the following output when successful, which is a list of command tags that show the command that was executed, and how many rows were affected:

```
SET
SET
SET
SET
SET
SET
DROP SCHEMA
CREATE SCHEMA
SET
SET
SET
CREATE TABLE
CREATE TABLE
COPY 5
COPY 3
```

The examples.sql script is very similar to a dump file produced by PostgreSQL backup tools, so this type of file and the output it produces are very common; in fact, we produced it by creating a dump file and then removing some parts that were not needed by this example.

When a command is executed successfully, PostgreSQL outputs a command tag equal to the name of that command; this is how the preceding output was produced.

The psql tool can also be used with both the -c and -f modes together; each one can be used multiple times. In this case, it will execute all the commands consecutively:

```
$ psql -c "SELECT current_time" -f examples.sql -c "SELECT current_time"
    timetz
-----------------
18:52:15.287+01
(1 row)
    ...output removed for clarity...
    timetz
```

Chapter 1

```
-----------------
18:58:23.554+01
(1 row)
```

The `psql` tool can also be used in interactive mode, which is the default, so it requires no option:

```
$ psql
postgres=#
```

The first interactive command you'll need is the following:

```
postgres=# help
```

You can then enter SQL or other commands. The following is the last interactive command you'll need:

```
postgres=# \quit
```

Unfortunately, you cannot type quit on its own, nor can you type \exit or other options. Sorry – it's just \quit, or \q for short!

How it works...

In psql, you can enter the following two types of command:

- psql meta-commands
- SQL

A meta-command is a command for the psql client, which may (or may not) send SQL to the database server, depending on what it actually does, whereas an SQL command is always sent to the database server. An example of a meta-command is \q, which tells the client to disconnect. All lines that begin with \ (a backslash) as the first non-blank character are presumed to be meta-commands of some kind.

If it isn't a meta-command, it's SQL, in which case psql keeps reading SQL until we find a semicolon, so we can spread SQL across many lines and format it any way we find convenient.

The help command is the only exception. We provide this for people who are completely lost, which is a good thought; so let's start from there ourselves.

There are two types of help commands, which are as follows:

- \?: This provides help on psql meta-commands.
- \h: This provides help on specific SQL commands.

Consider the following snippet as an example:

```
postgres=# \h DELETE
Command: DELETE
Description: delete rows of a table
Syntax:
[ WITH [ RECURSIVE ] with_query [, ...] ]
DELETE FROM [ ONLY ] table [ [ AS ] alias ]
    [ USING usinglist ]
    [ WHERE condition | WHERE CURRENT OF cursor_name ]
    [ RETURNING * | output_expression [ AS output_name ] [,]]
```

I find this a great way to discover and remember options and syntax. You'll also appreciate having the ability to scroll back through the previous command history if your terminal allows it.

You'll get a lot of benefits from tab completion, which will fill in the next part of the syntax when you press the *Tab* key. This also works for object names, so you can type in just the first few letters and then press *Tab*; all the options will be displayed. Thus, you can type in just enough letters to make the object name unique and then hit *Tab* to get the rest of the name.

Like most programming languages, SQL also supports comments. One-line comments begin with two dashes, as follows:

```
-- This is a single-line comment
```

Multiline comments are similar to those in C and Java:

```
/*
Multiline comment
Line 2
Line 3
*/
```

You'll probably agree that psql looks a little daunting at first, with strange backslash commands. I do hope you'll take a few moments to understand the interface and keep digging for more information. The psql tool is one of the most surprising parts of PostgreSQL, and it is incredibly useful for database administration tasks when used alongside other tools.

There's more...

psql works across releases and works well with older versions. It may not work at all with newer server versions, so use the latest client level of the server you are accessing.

See also

Check out some other useful features of psql, which are as follows:

- Informational metacommands, such as \d, \dn, and more
- Formatting, for output, such as \x
- Execution timing using the \timing command
- Input/output and editing commands, such as \copy, \i, and \o
- Automatic startup files, such as .psqlrc
- Substitutable parameters (variables), such as \set and \unset
- Access to the OS command line using \!
- Crosstab views with \crosstabview
- Conditional execution, such as \if, \elif, \else, and \endif

Changing your password securely

If you are using password authentication, then you may wish to change your password from time to time. This can be done from any interface. pgAdmin is a good choice, but here we will show how to do that from psql.

How to do it...

The most basic method is to use the psql tool. The \password command will prompt you once for a new password and again to confirm. Connect to the psql tool and type the following:

```
postgres=# SET password_encryption = 'scram-sha-256';
postgres=# \password
```

Enter a new password. This causes psql to send a SQL statement to the PostgreSQL server, which contains an already encrypted password string. An example of the SQL statement sent is as follows:

```
ALTER USER postgres PASSWORD 'SCRAM-SHA-256$4096:H4
5+UIZiJUcEXrB9SHlv5Q==$I0mc87UotsrnezRKv9Ijqn/
zjWMGPVdy1zHPARAGfVs=:nSjwT9LGDmAsMo+GqbmC2X/9LMgowTQBjUQsl45gZzA=';
```

Make sure you use the SCRAM-SHA-256 encryption, not the older and easily compromised MD5 encryption. Whatever you do, don't use postgres as your password. This will make you vulnerable to idle hackers, so make it a little more difficult than that!

Make sure you don't forget your password either. It may prove difficult to maintain your database if you can't access it.

How it works...

As changing the password is just an SQL statement, any interface can do this.

If you don't use one of the main routes to change the password, you can still do it yourself, using SQL from any interface. Note that you need to encrypt your password because if you do submit one in plain text, such as the following, it will be shipped to the server in plaintext:

```
ALTER USER myuser PASSWORD 'secret';
```

Luckily, the password in this case will still be stored in an encrypted form, but it will also be recorded in plaintext in the psql history file, as well as in any server and application logs, depending on the actual log-level settings.

PostgreSQL doesn't enforce a password change cycle, so you may wish to use more advanced authentication mechanisms, such as GSSAPI, SSPI, LDAP, or RADIUS.

Avoiding hardcoding your password

We can all agree that hardcoding your password is a bad idea. This recipe shows you how to keep your password in a secure password file.

Getting ready

Not all database users need passwords; some databases use other means of authentication. Don't perform this step unless you know you will be using password authentication and you know your password.

First, remove the hardcoded password from where you set it previously. Completely remove the password = xxxx text from the connection string in a program. Otherwise, when you test the password file, the hardcoded setting will override the details you are about to place in the file. Keeping the password hardcoded and in the password file is not any better. Using PGPASSWORD is not recommended either, so remove that as well.

If you think someone may have seen your password, change it before placing it in the secure password file.

How to do it...

A password file contains the usual five fields that we require when connecting, as shown here:

```
host:port:dbname:user:password
```

An example of how to set this is as follows:

```
myhost:5432:postgres:sriggs:moresecure
```

The password file is located using an environment variable named PGPASSFILE. If PGPASSFILE is not set, a default filename and location must be searched for, as follows:

- On *nix systems, look for ~/.pgpass.
- On Windows systems, look for %APPDATA%\postgresql\pgpass.conf, where %APPDATA% is the application data subdirectory in the path (for me, that would be C:\).

> **Note**
>
>
>
> Don't forget to set the file permissions on the file so that security is maintained. File permissions are not enforced on Windows, although the default location is secure. On *nix systems, you must issue the following command: chmod 0600 ~/.pgpass.
>
> If you forget to do this, the PostgreSQL client will ignore the .pgpass file. While the psql tool will issue a clear warning, many other clients will just fail silently, so don't forget!

How it works...

Many people name the password file .pgpass, whether or not they are on Windows, so don't get confused if they do this.

The password file can contain multiple lines. Each line is matched against the requested host:port:dbname:user combination until we find a line that matches. Then, we use that password.

Each item can be a literal value or *, a wildcard that matches anything. There is no support for partial matching. With appropriate permissions, a user can potentially connect to any database. Using the wildcard in the dbname and port fields makes sense, but it is less useful in other fields. The following are a few examples of wildcards:

- myhost:5432:*:sriggs:moresecurepw
- myhost:5432:perf:hannu:okpw
- myhost:*:perf:gianni:sicurissimo

There's more...

This looks like a good improvement if you have a few database servers. If you have many different database servers, you may want to think about using a connection service file instead (see the *Using a connection service file* recipe) or perhaps even storing details on a **Lightweight Directory Access Protocol (LDAP)** server.

Using a connection service file

As the number of connection options grows, you may want to consider using a connection service file.

The connection service file allows you to give a single name to a set of connection parameters. This can be accessed centrally to avoid the need for individual users to know the host and port of the database, and it is more resistant to future change.

You can set up a system-wide file as well as individual per-user files. The default file paths for these files are /etc/pg_service.conf and ~/.pg_service.conf respectively.

A system-wide connection file controls service names for all users from a single place, while a per-user file applies only to that particular user. Keep in mind that the per-user file overrides the system-wide file – if a service is defined in both files, then the definition in the per-user file will prevail.

How to do it...

First, create a file named pg_service.conf with the following content:

```
[dbservice1]
host=postgres1
port=5432
dbname=postgres
```

You can then copy it to either /etc/pg_service.conf or another agreed-upon central location. You can then set the PGSYSCONFDIR environment variable to that directory location.

Alternatively, you can copy it to ~/.pg_service.conf. If you want to use a different name, indicate it using PGSERVICEFILE. Either way, you can then specify the name of the service in a connection string, such as in the following example:

```
psql "service=dbservice1=cookbook user=gciolli"
```

The service can also be set using an environment variable named PGSERVICE.

How it works...

The connection service file can also be used to specify the user, although that means that the database username will be shared.

The pg_service.conf and .pgpass files can work together, or you can use just one of the two. Note that the pg_service.conf file is shared, so it is not a suitable place for passwords. The per-user connection service file is not shared, but in any case, it seems best to keep things separate and confine passwords to .pgpass.

There's more...

This feature applies to libpq connections only, so it does not apply to clients using other libraries, such as **Java database connectivity (JDBC)**.

Troubleshooting a failed connection

This recipe is all about what you should do when things go wrong.

Bear in mind that 90% of problems are just misunderstandings, and you'll quickly be on track again.

How to do it...

Here, we've made a checklist to be followed if a connection attempt fails:

- **Check whether the database name and the username are accurate**. You may be requesting a service on one system when the database you require is on another system. Recheck your credentials; ensure that you haven't mixed things up and that you are not using the database name as the username, or vice versa. If you receive an error for too many connections, then you may need to disconnect another session before you can connect or request the administrator to allow further connections.

- **Check for explicit rejections**. If you receive the pg_hba.conf rejects connection for host... error message, it means that your connection attempt has been explicitly rejected by the database administrator for that server. You will not be able to connect from the current client system using those credentials. There is little point in attempting to contact the administrator, as you are violating an explicit security policy with what you are attempting to do.

- **Check for implicit rejections.** If the error message you receive is no `pg_hba.conf entry for...`, it means there is no explicit rule that matches your credentials. This is likely an oversight on the part of the administrator and is common in very complex networks. Contact the administrator and request a ruling on whether your connection should be allowed (hopefully) or explicitly rejected in the future.
- **Check whether the connection works with** `psql`. If you're trying to connect to PostgreSQL from anything other than the `psql` command-line utility, switch to that now. If you can make `psql` connect successfully but cannot make your main connection work correctly, the problem may be in the local interface you are using.
- **Check the status of the database server** using the `pg_isready` utility, shipped with PostgreSQL. This tool checks the status of a database server, either local or remote, by establishing a minimal connection. Only the hostname and port are mandatory, which is great if you don't know the database name, username, or password. The following outcomes are possible:
    - The server is running and accepting connections.
    - The server is running but not accepting connections (because it is starting up, shutting down, or in recovery).
    - A connection attempt was made, but it failed.
    - No connection attempt was made because of a client problem (invalid parameters or out of memory).
- **Check whether the server is up.** If a server is shut down, you cannot connect. The typical problem here is simply mixing up the server to which you are connecting. You need to specify the hostname and port, so it's possible that you are mixing up those details.
- **Check whether the server is up and accepting new connections.** A server that is shutting down will not accept new connections, apart from superusers. Also, a standby server may not have the `hot_standby` parameter enabled, preventing you from connecting.
- **Check whether the server is listening correctly; also, check the port to which the server is actually listening.** Confirm that the incoming request is arriving on the interface listed in the `listen_addresses` parameter. Check whether it is set to * for remote connections and `localhost` for local connections.
- **Check whether the database name and username exist.** It's possible that the database or user no longer exists.

- **Check the connection request** – that is, check whether the connection request was successful and was somehow dropped following the connection. You can confirm this by looking at the server log when the following parameters are enabled:

    ```
    log_connections = on
    log_disconnections = on
    ```

- **Check for other reasons for disconnection**. If you are connecting to a standby server, it is possible that you have been disconnected because of hot standby conflicts. See *Chapter 12*, *Replication and Upgrades*, for more information.

There's more...

Client authentication and security are the rapidly changing areas in subsequent major PostgreSQL releases. You will also find differences between maintenance release levels. The PostgreSQL documents on this topic can be viewed at http://www.postgresql.org/docs/current/interactive/client-authentication.html.

Always check which release level you are using before consulting the manual or asking for support. Many problems are caused simply by confusing the capabilities between release levels.

PostgreSQL in the cloud

Like many other systems, PostgreSQL is available in the cloud as a **Database as a Service** (**DBaaS**). These services create and manage databases for you, with high availability and backup included. So it's less work, but not zero work, and you still have responsibilities…which you will see later.

Getting ready

We will select EDB's BigAnimal as an example of a PostgreSQL cloud service, since EDB has the largest number of contributors to open source PostgreSQL, over the longest period.

EDB's BigAnimal creates clusters within your own cloud account, allowing you to understand and control the costs you incur when running PostgreSQL. So, the first step is to log in to your host cloud account: https://www.biganimal.com/.

How to do it...

Using EDB's BigAnimal as a specific example, navigate through these steps:

1. If you don't have an account, you can sign in using the Free Trial at http://biganimal.com/; click **Try for free**, sign up, and sign in. This will take you to *Step 5* of this sequence. If you do already have an account, then you can start at *Step 2*.

2. Connect to the cloud portal – for example, Azure. If you have multiple accounts, as we do, then make sure you are connected to the right account. BigAnimal is then available as a marketplace subscription.

3. Go to https://portal.biganimal.com/:

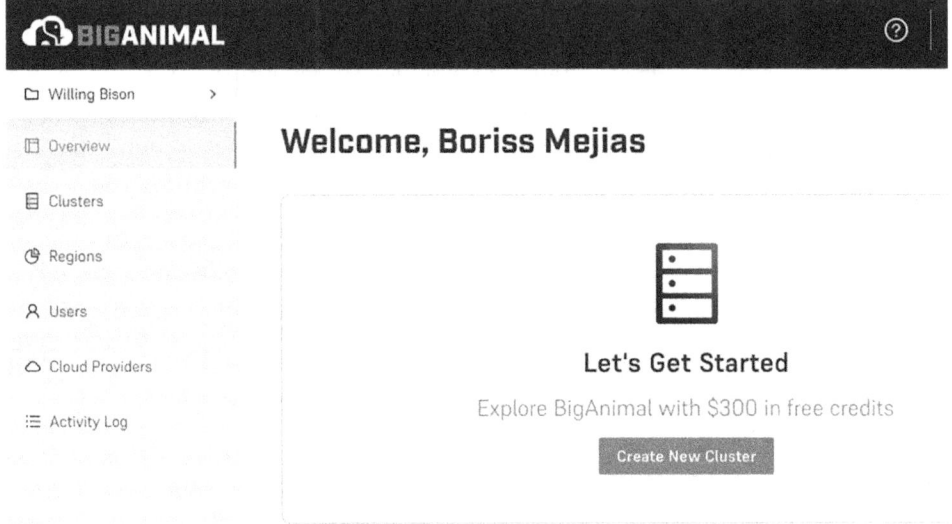

Figure 1.8: The portal welcome screen

4. Manage your cloud limits, if necessary.

5. Select **Create New Cluster**, and then set **Cluster Name** and **Password**:

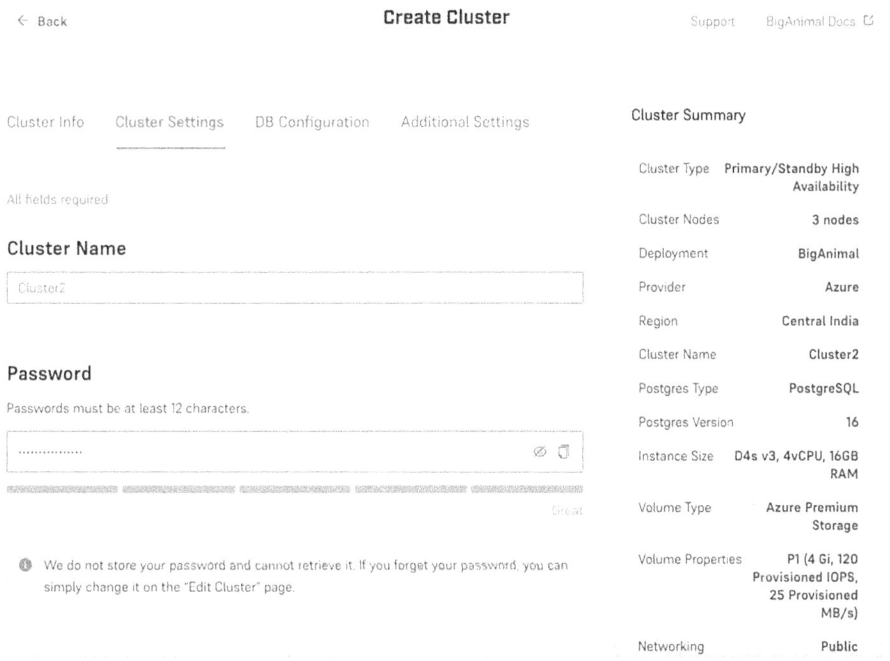

Figure 1.9: The portal main screen

6. In this example, we will create a cluster called Cluster2. Specify **Database Type**. Select the software type and version – for example, PostgreSQL 16. Select the cloud provider and distribution across region(s) – for example, **Azure** and **Central India**:

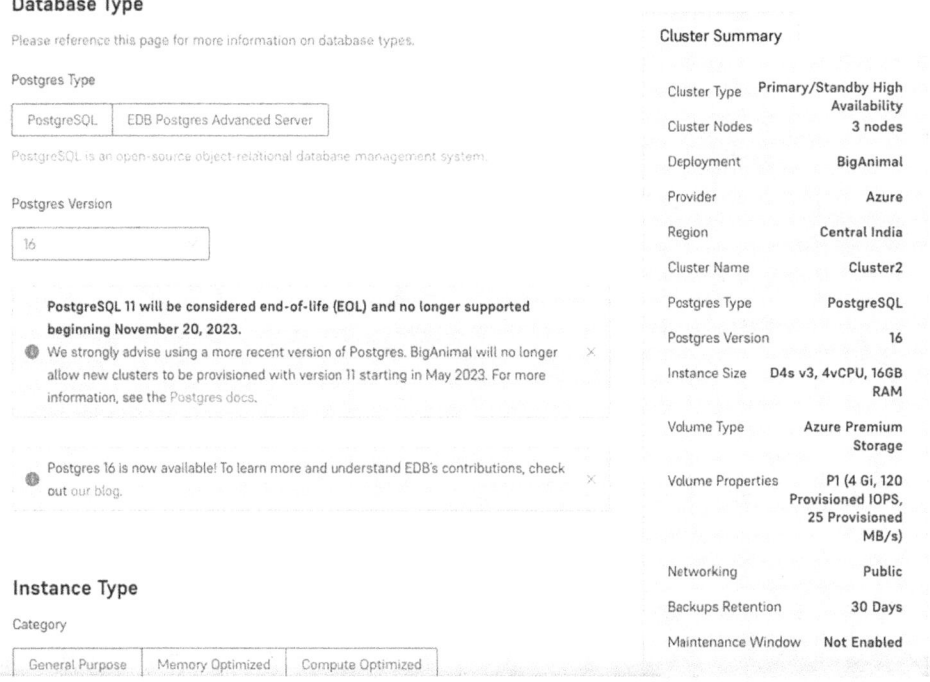

Figure 1.10: BigAnimal database type

7. Specify the instance type and key details, all of which will then be provisioned for you:

- Specify the instance type – for instance, **D4s v3**:
    - How many CPUs? (such as **4 vCPUs**)
    - How much RAM? (such as **16GB RAM**)
- Specify storage:
    - Volume type? (**Azure Premium Storage**)
    - Provisioned IOPS? (**4 Gi, 120 IOPS, 25 MB/s**)
- Specify other aspects:
    - Networking? (**Public**)
    - High availability? (**Yes**)

- HA clusters are configured with a single primary and two replica nodes using streaming physical replication. Clusters are configured across availability zones automatically. **synchronous_replication** is configured by default.

8. Create the cluster. Wait for the cluster to be built, which will usually be very quick, yet varies according to the options selected in the previous step. Assume it will take 1 hour to avoid sitting and watching it:

Figure 1.11: The BigAnimal progress bar

9. Set up **Connection Info** for our new Cluster2:

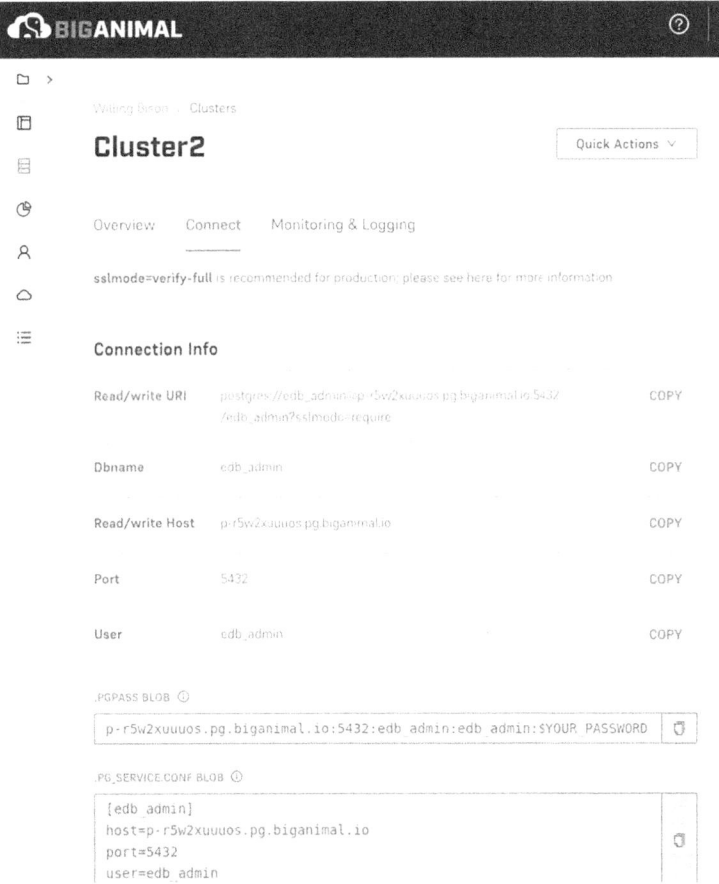

Figure 1.12: EDB's BigAnimal connection details

Test the connection and then set up the connection details, as discussed in earlier recipes. Assign the new instance a shortcut name, since remembering a node name such as `p-r5w2xuuuos.pg.biganimal.io` will not be easy!

How it works…

The cloud (or DBaaS) means that PostgreSQL is managed for you, so this is all you need to do.

EDB's BigAnimal provides a GUI to allow you to create PostgreSQL clusters manually on demand. One of the main themes in this cookbook is using repeatable, scriptable mechanisms where possible, so I recommend that you use either a **Command-Line Interface (CLI)** or an **Application Programming Interface (API)**. The API uses a RESTful interface to define and manage clusters.

Note that when you run a database service, you still have these and other responsibilities:

- You are responsible for contacting the support team if things are not as you think they should be.
- You are responsible for keeping your passwords to the cluster secure.
- You are responsible for creating users with appropriate access rights to your data.
- You are responsible for choosing whether to enable high availability and for noting the availability level offered by the service.
- You are responsible for data modeling, query performance, and scaling the cluster to meet your performance needs.
- You are responsible for choosing the appropriate resources for your workload, including instance type, storage, and connections. You are also responsible for managing your cloud resource limits to ensure the underlying infrastructure can scale.
- You are responsible for periodically restoring and verifying the restores to ensure that archives are completed frequently and successfully to meet your needs.
- You are responsible for paying!

So, the cloud is a good way forward, but not a way to avoid taking full responsibility for your overall application and database.

There's more…

Cloud services are also available from these and others:

- Aiven
- Amazon Web Services

- Crunchy
- Google
- Microsoft

PostgreSQL with Kubernetes

In this recipe, we discuss Kubernetes (K8s for short), the industry's most prominent solution for automated application deployment, scaling, and management. It is free software, vendor neutral, and maintained by the **Cloud Native Computing Foundation** (**CNCF**).

CloudNativePG (**CNPG**) is the newest and fastest-rising Kubernetes *operator* for PostgreSQL. In other words, it provides automation around the entire Postgres lifecycle, taking care of deployment, scaling, and the management of database clusters.

In this recipe, we'll use Minikube, a lightweight and fuss-free Kubernetes distribution for testing software deployment. It's not suitable for production usage, but whatever we do in Minikube also holds true for any Kubernetes cluster, so you can take what you learn here and apply it to production-ready clusters.

Getting ready

First off, we install Minikube to provide a minimal Kubernetes cluster. Install Docker (or Podman) from your OS's default package manager, then visit https://minikube.sigs.k8s.io/docs/start/ to find download and installation instructions for your operating system and architecture. For example, if you use Debian, then the installation is as simple as:

```
curl -LO \
https://storage.googleapis.com/minikube/releases/latest/minikube_latest_amd64.deb
sudo dpkg -i minikube_latest_amd64.deb
```

Next, assuming that your user has permission to use Docker, you can start Minikube with:

```
minikube start
```

At this point, you can install the kubectl utility, which lets you interact with the Kubernetes cluster:

```
minikube kubectl -- get pods -A
```

The above command is a bit verbose; you can wrap it in a shorter alias:

```
alias kubectl="minikube kubectl --"
```

Now everything should be ready; you can verify that by running:

```
kubectl get nodes
NAME       STATUS   ROLES           AGE   VERSION
minikube   Ready    control-plane   12m   v1.27.4
```

which means that you're ready to start your CloudNativePG journey.

How to do it...

In order to install the latest version (at the time of writing, v1.21.0) of the CloudNativePG operator into your Kubernetes cluster, run:

```
kubectl apply -f \
    https://raw.githubusercontent.com/cloudnative-pg/cloudnative-pg/
release-1.21/releases/cnpg-1.21.0.yaml
```

We verify the installation with:

```
kubectl get deployment -n cnpg-system cnpg-controller-manager
NAME                      READY   UP-TO-DATE   AVAILABLE   AGE
cnpg-controller-manager   0/1     1            0           15s
```

Let's deploy a sample PostgreSQL cluster.

Kubernetes works in a *declarative* way: you *declare* what the cluster should look like, and then CNPG (the *operator*) will perform all the necessary *operations* that will end up with the cluster in the exact state that you declared.

In practice, we create a YAML file called `sample-cluster.yaml` with the following content:

```
apiVersion: postgresql.cnpg.io/v1
kind: Cluster
metadata:
  name: sample-cluster
spec:
  instances: 3
  storage:
    size: 1Gi
```

And then we apply that file by running:

```
kubectl apply -f sample-cluster.yaml
```

We can check what is going on by seeing which Postgres pods are up and running:

```
kubectl get pods
NAME                             READY    STATUS            RESTARTS    AGE
sample-cluster-1-initdb-74xf7    0/1      PodInitializing   0           30s
```

Looks like we're not done yet. Give it a moment, and then you will see:

```
kubectl get pods
NAME               READY    STATUS     RESTARTS    AGE
sample-cluster-1   1/1      Running    0           2m19s
sample-cluster-2   1/1      Running    0           1m41s
sample-cluster-3   1/1      Running    0           1m12s
```

Our Postgres nodes are up! They are now ready to be accessed by applications running inside the Kubernetes cluster by connecting to the following Services created by CNPG:

```
kubectl get svc
NAME                TYPE        CLUSTER-IP      EXTERNAL-IP    PORT(S)
AGE
kubernetes          ClusterIP   10.96.0.1       <none>         443/TCP
77m
sample-cluster-r    ClusterIP   10.101.133.29   <none>         5432/TCP
42m
sample-cluster-ro   ClusterIP   10.100.24.250   <none>         5432/TCP
42m
sample-cluster-rw   ClusterIP   10.99.79.108    <none>         5432/TCP
42m
```

The `sample-cluster-rw` Service lets you connect to the primary node for read/write operations, `sample-cluster-ro` to standbys only for read-only operations, and `sample-cluster-r` to any node (including the primary) for read operations.

You can find more sample configurations with more features at `https://cloudnative-pg.io/documentation/current/samples/`.

How it works...

The operator defines a new Kubernetes resource called `Cluster`, representing a PostgreSQL cluster made up of a single primary and an optional number of physical replicas that co-exist in the chosen Kubernetes namespace for high availability and offloading of read-only queries.

Applications in the Kubernetes cluster can now access the Postgres database through the Service that the operator manages, without worrying about which node is primary and whether the primary changes due to a failover or switchover. For applications from outside the Kubernetes cluster, you need to expose Postgres via TCP by configuring a Service or Ingress object.

In our cluster, 1 GB of disk space was allocated for Postgres in the default Kubernetes storage. Be aware that we deployed Postgres with the default configuration, which is conservative and safe for testing on a laptop, but definitely not suitable for production usage.

You can find CNPG's extensive documentation, which describes all you can do with the operator, including detailed Prometheus monitoring, backup and recovery, upgrades, migration, scaling, etc., and how to configure it for production use, at `https://cloudnative-pg.io/documentation/current/`.

There's more...

CloudNativePG is able to react to the failure of a PostgreSQL instance by performing failover and/or creating new replicas, depending on what is needed to restore the desired state, which in our example is one primary node and two physical replicas.

We recommend this method for Kubernetes PostgreSQL deployments because it is not an attempt to shoehorn Postgres into Kubernetes with additional sidecar software to take care of the high availability aspect. It is built from the ground up with Postgres-specific resources, while respecting the cloud-native declarative conventions and using Kubernetes's built-in facilities and features.

High availability has historically been a complex subject for PostgreSQL, as for other database systems, because the most difficult part is to diagnose failures correctly. The various middleware tools – for which we refer you to *Chapter 12, Replication and Upgrades* – employ a number of techniques to reduce the risk of doing the wrong thing due to a mistaken diagnosis.

Kubernetes changes the way high availability is achieved because it provides a very reliable interface for detecting node failures. CNPG is called "native" because it follows this approach strictly, and as a result it is becoming very popular in the Kubernetes world, probably also because people who are experienced with Kubernetes will recognize this approach as familiar and reliable.

CloudNativePG is the first PostgreSQL-related project to aim for CNCF certification through the Sandbox/Incubation/Graduation process. You can find the CNPG repository at `https://github.com/cloudnative-pg/cloudnative-pg`.

PostgreSQL with TPA

Trusted Postgres Architect (**TPA**) is a software based on Ansible that can be used to deploy database clusters on a variety of platforms.

In this recipe, we will use TPA to configure and deploy a small cluster on our own Linux workstation.

This recipe uses TPA's `docker` platform, which is meant to be used only for test clusters. TPA currently supports two other platforms:

- The `aws` platform, to provision and use instances on AWS EC2
- The `bare` platform, to use existing instances (including bare-metal and already provisioned servers)

For more information on how to use these platforms, please refer to the corresponding TPA documentation pages:

- `https://www.enterprisedb.com/docs/tpa/latest/platform-aws/`
- `https://www.enterprisedb.com/docs/tpa/latest/platform-bare/`

Getting ready

First, we need to install TPA, which is free software, released under the GPL v3 license. Therefore, you can download it from public repositories, as explained in the installation instructions:

`https://www.enterprisedb.com/docs/tpa/latest/INSTALL/`

Make sure you have the latest version installed; you can check it by typing:

`tpaexec info`

At the time when this recipe was written, TPA version 23.23 was the latest release available. Given that TPA tries hard to keep compatibility with clusters installed using previous versions, you should definitely always use the latest version of TPA, and be able to repeat this recipe even with releases newer than 23.23.

Then, we need to install Docker. If you don't have it already on your laptop you can install it as described here: `https://www.enterprisedb.com/docs/tpa/latest/platform-docker/#installing-docker`.

In the common microservices approach, each container runs a specific service. The way TPA uses Docker is quite different because each container runs a miniature copy of a Linux OS. This approach is not meant for production use, but it is a great way to test the behavior of a cluster with minimal resource use.

How to do it...

This is our first TPA example, so we will deploy the smallest possible PostgreSQL cluster, composed of a single instance with a backup server. No replication, no high availability (which most of the time means no production!)

First, we create the cluster configuration using the `tpaexec configure` command as follows:

```
tpaexec configure myfirstcluster --architecture M1 \
    --platform docker --enable-repmgr --postgresql 16
```

This command creates a directory named `myfirstcluster` with the following contents:

```
myfirstcluster/
├── commands
│   ├── status.yml -> /opt/EDB/TPA/architectures/M1/commands/status.yml
│   ├── switchover.sh -> /opt/EDB/TPA/architectures/M1/commands/switchover.sh
│   ├── switchover.yml -> /opt/EDB/TPA/architectures/M1/commands/switchover.yml
│   └── upgrade.yml -> /opt/EDB/TPA/architectures/M1/commands/upgrade.yml
├── config.yml
└── deploy.yml -> /opt/EDB/TPA/architectures/M1/deploy.yml
```

The `commands` directory contains some **symlinks** to commands that are specific to the architecture that we have chosen, while `deploy.yml` is a **symlink** to the playbook used for the deploy command. As you can see, all these are files distributed together with TPA, which are linked to this cluster directory so they can easily be used.

The only new file that has been created by this invocation is `config.yml`, which describes the cluster. It is effectively a template that the user can modify if they want to fine-tune the cluster; in fact, editing that file is quite common because only some of the settings can be specified as options of the `tpaexec configure` command.

We created a configuration file specifying this architecture:

https://www.enterprisedb.com/docs/tpa/latest/architecture-M1/

As we want a smaller example, we will now edit `config.yml` to remove some of the instances because in this first example, we just want to deploy one PostgreSQL instance and one Barman instance instead of the full M1 architecture, which by default includes a three-node physical replication cluster plus a Barman node, which also acts as a log server and as a monitoring server.

Let's locate the instances section, at the end of the file:

```
instances:
- Name: kennel
  backup: karma
  location: main
  node: 1
  role:
  - primary
- Name: quintet
  location: main
  node: 2
  role:
  - replica
  upstream: kennel
- Name: karma
  location: main
  node: 3
  role:
  - barman
  - log-server
  - monitoring-server
- Name: kinship
  location: dr
  node: 4
  role:
  - replica
  upstream: quintet
```

The instance names in your example will likely be different every time you run tpaexec configure because TPA by default picks them at random from a built-in list of words; however, the structure will be the same.

From there, we can remove:

- The physical replicas – that is, instances 2 and 4 (here, quintet and kinship)
- The additional roles for the Barman instance – that is, log-server and monitoring-server from instance 3 (here, karma)

We end up with the following instances section:

```
instances:
- Name: kennel
  backup: karma
  location: main
  node: 1
  role:
  - primary
- Name: karma
  location: main
  node: 3
  role:
  - barman
```

After making these changes, we can deploy the cluster, which is as simple as issuing the following command:

```
tpaexec deploy myfirstcluster
```

This command will display copious output, ending like this after a few minutes:

```
PLAY RECAP *********************************************************
******
karma                      : ok=177    changed=40   unreachable=0
failed=0    skipped=163   rescued=0    ignored=0
kennel                     : ok=316    changed=97   unreachable=0
failed=0    skipped=222   rescued=0    ignored=1
localhost                  : ok=4      changed=0    unreachable=0
failed=0    skipped=0     rescued=0    ignored=0


real 5m35.687s
user 1m13.249s
sys  0m30.098s
```

The output is also collected in the ansible.log file, with millisecond timestamps, if you need to inspect the (many) steps afterward.

Now that we have a cluster, we can use it. Let's connect with SSH to the Postgres host:

```
$ cd myfirstcluster
$ ssh -F ssh_config kennel
[root@kennel ~]# su - postgres
postgres@kennel:~ $ psql
psql (15.4)
Type "help" for help.

postgres=#
We can also open another terminal and connect to the Barman host:
$ ssh -F ssh_config karma
Last login: Mon Sep 18 21:35:41 2023 from 172.17.0.1
[root@karma ~]# su - barman
[barman@karma ~]$ barman list-backup all
kennel 20230918T213317 - Mon Sep 18 21:33:19 2023 - Size: 22.2 MiB - WAL Size: 0 B
kennel 20230918T213310 - Mon Sep 18 21:33:11 2023 - Size: 22.2 MiB - WAL Size: 36.2 KiB
kennel 20230918T213303 - Mon Sep 18 21:33:05 2023 - Size: 22.2 MiB - WAL Size: 36.8 KiB
```

There's more

TPA reads the `config.yml` file, where the cluster is described in a declarative way, and then performs all the actions needed to deploy the cluster, or to modify an already-deployed cluster if `config.yml` has been changed since the last run of the `deploy` command.

The `tpaexec deploy` command automatically performs the preliminary `tpaexec provision`, which is the step where TPA populates the Ansible inventory based on the contents of `config.yml` and then creates the required resources, such as SSH keys, passwords, and instances. Here, "instances" means:

- Containers, when using the `docker` platform
- VMs, when using the `aws` platform
- Nothing, when using the bare platform (TPA will expect "bare metal" instances, in the sense that they exist already and TPA has `sudo` SSH access to them)

For more details, please refer to the TPA online documentation:

- The `configure` command: https://www.enterprisedb.com/docs/tpa/latest/tpaexec-configure/
- The `provision` command: https://www.enterprisedb.com/docs/tpa/latest/tpaexec-provision/
- The `deploy` command: https://www.enterprisedb.com/docs/tpa/latest/tpaexec-deploy/

Learn more on Discord

To join the Discord community for this book – where you can share feedback, ask questions to the author, and learn about new releases – follow the QR code below:

`https://discord.gg/pQkghgmgdG`

2
Exploring the Database

To understand PostgreSQL, you need to see it in use. An empty database is like a ghost town without houses.

For now, we will assume that you already have a database. There are over a thousand books on how to design your own database from nothing. So, here we aim to help people who are still learning to use the PostgreSQL database management system with handy routines to explore the database.

The best way to start the process of understanding is by asking some simple questions to orient yourself. Incidentally, these are also questions that you'll need to answer if you ask someone else for help.

In this chapter, we'll cover the following recipes:

- What type of server is this?
- What version is the server?
- What is the server uptime?
- Locating the database server files
- Locating the database server's message log
- Locating the database's system identifier
- Listing databases on the database server
- How many tables are there in a database?
- How much memory does a database currently use?
- How much disk space does a database use?

- How much disk space does a table use?
- Which are my biggest tables?
- How many rows are there in a table?
- Quickly estimating the number of rows in a table
- Listing extensions in this database
- Understanding object dependencies

What type of server is this?

PostgreSQL is an open source **object-relational database management system (ORDBMS)**, distributed under a very permissive license and developed by an active community.

There are a number of PostgreSQL-related services and software (`https://wiki.postgresql.org/wiki/PostgreSQL_derived_databases`), either open source or not, that are provided by other software companies. Here, we discuss how to recognize which one you are using.

It is not so easy to detect the variant of PostgreSQL from the name; many of the products and services involving PostgreSQL include the word **Postgres** or **PostgreSQL**.

However, if you need to check the documentation, or to buy services such as support and consulting, you need to find out exactly what type your server is, as the available options will vary.

If you are paying a license fee or a cloud service subscription, you will already know the name of the company you are paying, and the specific variant of PostgreSQL you are subscribed to. However, it's not rare to have multiple servers of different types, so it is still useful to be able to tell them apart.

How to do it...

Unfortunately, there isn't a single function or parameter that works on each *variant* of PostgreSQL and, at the same time, is able to answer that question. The closest you can get is the `version()` function, which is used in the next recipe, *What version is the server?*, and returns a textual description of the version you are running, including (but not limited to) the version number.

In some cases, this is enough, but if not, you have to determine the specific version from other clues, such as the following:

- The version number for stable releases of community PostgreSQL is either *X.Y* (with *X=10* or above) or *X.Y.Z* (up to *X=9*). An extra number usually indicates that you are running a variant of PostgreSQL.

- The presence of certain objects that are available only on a specific variant, for instance, an extension. More details on how to work with extensions can be found in the *Listing extensions in this database* recipe in this chapter.

There's more...

Some of the PostgreSQL-based services on the cloud will return the same value of version() as community PostgreSQL does. While this is correct, in the sense that they are indeed running that version of PostgreSQL, it doesn't mean that you have the same level of control. For instance, you might not be given a superuser account, and you will probably be unable to install extensions freely.

What version is the server?

PostgreSQL has internal version numbers for the data file format, database catalog layout, and crash recovery format. Each of these is checked as the server runs to ensure that the data doesn't become corrupt. PostgreSQL doesn't change these internal formats for a single minor release; they only change across major releases.

From a user's perspective, each release differs in terms of the way the server behaves. If you know your application well, then it should be possible to assess the differences simply by reading the release notes for each version. In many cases, a retest of the application is the safest thing to do.

If you experience any general problems related to setup and configuration with your database, then you'll need to double-check which version of the server you have. This will help you to report a fault or to consult the correct version of the manual.

How to do it...

We will find out the version by querying the database server directly:

1. Connect to the database and issue the following command:
   ```
   postgres=# SELECT version();
   ```

2. You'll get a response that looks something like this:
   ```
   PostgreSQL 16.0 (Debian 16.0-1.pgdg120+1) on x86_64-pc-linux-gnu,
   compiled by gcc (Debian 12.2.0-14) 12.2.0, 64-bit
   ```

That's probably too much information all at once!

The most important bit is the "16.0" at the beginning, which is the PostgreSQL version on the server. The remaining information is mainly useful when debugging because the ordinary user doesn't usually need to know the OS, the package version, the architecture, etc.

Another way of checking the version number in your programs is as follows:

```
postgres=# SHOW server_version;
       server_version
-------------------------------
 16.0 (Debian 16.0-1.pgdg120+1)
(1 row)
```

The preceding shows the version in text form, so you may also want a numerical value that is easier to compare using a greater than symbol, in which case you execute this command instead:

```
postgres=# SHOW server_version_num;
 server_version_num
--------------------
 160000
(1 row)
```

Another alternative is via command-line utilities, such as this:

```
$ psql --version
psql (PostgreSQL) 16.0 (Debian 16.0-1.pgdg120+1)
```

However, be wary that this shows the client software version number, which may differ from the server software version number. This will usually be reported to you so that you're aware, as in the following example:

```
psql
psql (16.0 (Debian 16.0-1.pgdg120+1), server 15.4 (Debian 15.4-2.
pgdg120+1))
Type "help" for help.
postgres=#
```

Here, you can see that the server version is 15.4, while 16.0 is the client version.

How it works...

The current PostgreSQL server version format is composed of two numbers; the first number indicates the major release, and the second one denotes subsequent maintenance releases for that major release. It is common to mention just the major release when discussing what features are supported, as they are unchanged on a maintenance release.

Chapter 2

16.0 is the first release of PostgreSQL 16, and subsequent maintenance releases will be 16.1, 16.2, 16.3, and so on. In the preceding example, we see that 16.0 is the version of that PostgreSQL server.

For each major release, there is a separate version of the manual, since the feature set is not the same. If something doesn't work exactly the way you think it should, make sure you are consulting the correct version of the manual.

There's more...

Prior to release 10, PostgreSQL used a three-part numbering series, meaning that the feature set and compatibility related to the first two numbers, while maintenance releases were denoted by the third number. For instance, version 9.6 contained more additional features and compatibility changes when compared to version 9.5; version 9.6.0 was the initial release of 9.6, and version 9.6.1 was a later maintenance release. But this is ancient history: release 9.6 ended its life in November 2021, and as of today, the oldest supported release is PostgreSQL 11, which is about to expire as well. So it is fair to say that from now on, you are unlikely to encounter many instances of PostgreSQL 9.6 or earlier.

The release support policy for PostgreSQL is available at http://www.postgresql.org/support/versioning/. This article explains that each release will be supported for a period of five years. Since we release one major version per year, this means that normally five major releases will be supported at any given time.

Support for all releases up to and including 10 ended in November 2022. So by the time you're reading this book, only PostgreSQL 11 and higher versions will be supported, with version 11 having reached its **EOL (End-of-Life)** date at the end of 2023. The earlier versions are still robust, although many performance and enterprise features are missing from those releases. The future end-of-support dates are as follows:

Version	Release Date	Last Supported Date
11	October 2018	October 2023
12	October 2019	October 2024
13	September 2020	September 2025
14	September 2021	September 2026
15	October 2022	October 2027
16	September 2023	September 2028

Table 2.1: A table showing PostgreSQL version release dates

What is the server uptime?

You may be wondering, how long has it been since the server started?

For instance, you might want to verify that there was no server crash if your server is not monitored, or to see when the server was last restarted, for instance, to change the configuration. We will find this out by asking the database server.

How to do it...

Issue the following SQL from any interface:

```
postgres=# SELECT date_trunc('second', current_timestamp
- pg_postmaster_start_time()) as uptime;
```

You should get the following output:

```
    uptime
---------------------------------------
 2 days 02:48:04
```

How it works...

Postgres stores the server start time, so we can access it directly, as follows:

```
postgres=# SELECT pg_postmaster_start_time();
pg_postmaster_start_time
----------------------------------------------
2023-09-26 09:24:44.292012+02
```

Then, we can write a SQL query to get the uptime, like this:

```
postgres=# SELECT current_timestamp - pg_postmaster_start_time();
?column?
------------------------
 6 days 06:02:16.655631
```

Finally, we can apply some formatting:

```
postgres=# SELECT date_trunc('second', current_timestamp
- pg_postmaster_start_time()) as uptime;
    uptime
-----------------
 6 days 06:03:04
```

See also

This is simple stuff. Further monitoring and statistics are covered in *Chapter 8, Monitoring and Diagnosis*.

Locating the database server files

Database server files are initially stored in a location referred to as the `data` directory. Additional data files may also be stored in tablespaces if any exist.

In this recipe, you will learn how to find the location of these directories on a given database server.

Getting ready

You'll need to get OS access to the **database system**, which is what we call the platform on which the database runs.

How to do it...

If you can connect using `psql`, then you can use this command:

```
postgres=# SHOW data_directory;
    data_directory
----------------------
/opt/postgres/data/
```

If not, the following are the system's default `data` directory locations:

- Debian or Ubuntu systems: `/var/lib/postgresql/MAJOR_RELEASE/main`
- Red Hat RHEL, CentOS, and Fedora: `/var/lib/pgsql/data/`
- Systems deployed with **Trusted Postgres Architect** (**TPA**): `/opt/postgres/data`
- Windows: `C:\Program Files\PostgreSQL\MAJOR_RELEASE\data`

`MAJOR_RELEASE` is composed of just one number (for release 10 and above) or two (for releases up to 9.6).

So far, we have discussed the location of the `data` directory; now, we will discuss the location of the **configuration files**, which varies widely across the various software distributions.

On Debian or Ubuntu systems, the configuration files are located in `/etc/postgresql/MAJOR_RELEASE/main/`, where `main` is just the name of a database server. Other names are also possible. For the sake of simplicity, we assume that you only have a single installation, although the point of including the release number and database server name as components of the directory path is to allow multiple database servers to coexist on the same host.

> **NOTE**
>
> Debian and Ubuntu systems have the pg_lsclusters utility, which displays a list of all the available database servers, including information for each server.

The information for each server includes the following:

- Major release number
- Port
- Status (for example, online and down)
- Data directory
- Log file

pg_lsclusters is part of the postgresql-common Debian/Ubuntu package, which provides a structure under which multiple versions of PostgreSQL can be installed, and multiple clusters can be maintained, at the same time.

In the packages distributed with Red Hat RHEL, CentOS, and Fedora, the default data directory location also contains the configuration files (*.conf) by default. However, note that the packages distributed by the PostgreSQL community use a different default location: /var/lib/pgsql/MAJOR_RELEASE/data/.

Again, that is just the default location. You can create additional data directories using the initdb utility.

Systems deployed with TPA use a data directory that is different from the OS defaults. Given that the various operating systems have their different ways of managing the PostgreSQL service, the TPA developers decided to use separate service definitions and data directories in order to avoid any possible conflicts when installing PostgreSQL packages.

The initdb utility populates the given data directory with the initial content. The directory will be created for convenience if it is missing, but for safety, the utility will stop if the data directory is not empty. The initdb utility will read the data directory name from the PGDATA environment variable unless the -D command-line option is used to specify an override.

How it works...

Even though the Debian/Ubuntu and Red Hat file layouts are different, they both follow the Linux **Filesystem Hierarchy Standard (FHS)**, so neither layout is wrong.

The Red Hat layout is simpler and easier to understand. The Debian/Ubuntu layout is more complex, but it has different and more adventurous goals. The Debian/Ubuntu layout is similar to the **Optimal Flexible Architecture (OFA)** of other database systems. As pointed out earlier, the goals are to provide a file layout that will allow you to have multiple PostgreSQL database servers on one system, and many versions of the software existing in the filesystem at once.

Again, the layouts for the Windows and OS X installers are different. Multiple database clusters are possible, but they are also more complex to manage than on Debian/Ubuntu, where the defaults are designed to allow that.

The Debian/Ubuntu layout is a good choice for application developers or database consultants who need to have multiple versions of PostgreSQL already installed on their laptop. However, now that containers have become common and easy to use, there are better solutions to that problem. For instance, you can use TPA to deploy an instance using the `docker` platform, as shown in the *PostgreSQL with TPA* recipe of *Chapter 1, First Steps*, and in just a few minutes, you get a container that behaves mostly like a separate virtual machine, which is ideal for developers. This is more flexible than running multiple instances without containers on different ports and data directories; for instance, with TPA you can create multiple instances and emulate a cluster with multiple nodes connected to each other.

Note that server applications such as `initdb` can only work with one major PostgreSQL version. On distributions that allow several major versions, such as Debian or Ubuntu, these applications are placed in dedicated directories, which are not put in the default command path. This means that if you just type `initdb`, the system will not find the executable, and you will get an error message.

This may look like a bug, but in fact, it is the desired behavior. Instead of accessing `initdb` directly, you are supposed to use the `pg_createcluster` utility from `postgresql-common`, which will select the right `initdb` utility, depending on the major version you specify.

There's more...

Once you've located the data directory, you can look for the files that comprise the PostgreSQL database server. The layout is as follows:

Subdirectory	Purpose
base	This is the main table storage. Beneath this directory, each database has its own directory, within which the files for each database table or index are located.
global	Tables that are shared across all databases, including the list of databases.

pg_commit_ts	Transaction commit timestamp data.
pg_dynshmem	Dynamic shared memory information.
pg_logical	Logical decoding status data.
pg_multixact	Files used for shared row-level locks.
pg_notify	LISTEN/NOTIFY status files.
pg_replslot	Information about replication slots.
pg_serial	Information on committed serializable transactions.
pg_snapshots	Exported snapshot files.
pg_stat	Permanent statistics data.
pg_stat_tmp	Transient statistics data.
pg_subtrans	Subtransaction status data.
pg_tblspc	Symbolic links to tablespace directories.
pg_twophase	State files for prepared transactions (a.k.a. two-phase commit).
pg_wal	Transaction log or, more appropriately, **Write-Ahead Log (WAL)**.
pg_xact	Transaction status files.

Table 2.2: Contents of the PostgreSQL data directory

None of the aforementioned directories contain user-modifiable files, nor should any of the files be manually deleted to save space, or for any other reason. *Don't touch it, because you'll break it, and you may not be able to fix it!* It's not even sensible to copy files in these directories without carefully following the procedures described in *Chapter 11*, *Backup and Recovery*. Keep off the grass!

We'll talk about tablespaces later in the book. We'll also discuss a performance enhancement that involves putting the transaction log on its own set of disk drives in *Chapter 10*, *Performance and Concurrency*.

The only things you are allowed to touch are configuration files, which are all *.conf files, and server message log files. Server message log files may or may not be in the data directory. For more details on this, refer to the next recipe, *Locating the database server's message log*.

Locating the database server's message log

The database server's message log is a record of all messages recorded by the database server. This is the first place to look if you have server problems and is a good place to check regularly.

This log will include messages that look something like the following:

```
2023-09-01 19:37:41 GMT [2507-1] LOG:  database system was shut down at
2021-09-01 19:37:38 GMT
```

```
2023-09-01 19:37:41 GMT [2506-1] LOG:   database system is ready to accept
connections
```

We'll explain some more about these logs once we've located the files.

Getting ready

You'll need to get operating system access to the **database system**, which is what we call the platform on which the database runs.

The **server log** can be in a few different places, so let's list all of them first so that we can locate the log or decide where we want it to be placed:

- It may be in a directory beneath the data directory.
- It may be in a directory elsewhere on the filesystem.
- It may be redirected to syslog.
- There may be no server log at all. In this case, it's time to add a log soon.

If not redirected to syslog, the server log consists of one or more files. You can change the names of these files, so it may not always be the same on every system.

How to do it…

The following are the default server log locations:

- Debian or Ubuntu systems: /var/log/postgresql
- Red Hat, RHEL, CentOS, and Fedora: /var/lib/pgsql/data/pg_log
- System deployed with TPA: Messages are sent to syslog, and to /var/log/postgres/postgres.log
- Windows systems: The messages are sent to the Windows Event Log

The current server log file is named postgresql-MAJOR_RELEASE-SERVER.log, where SERVER is the name of the server (by default, main), and MAJOR_RELEASE represents the major release of the server, for example, 9.6 or 11 (as we mentioned in a prior recipe, from release 10 onward, the major release is composed of just one number). An example is postgresql-14-main.log, while older log files are numbered as postgresql-14-main.log.1. The higher the final number, the older the file, since they are being rotated by the logrotate utility.

As already explained in this chapter, TPA intentionally uses a default location that is different from the OS defaults, so it avoids any conflict with the PostgreSQL configurations and services from the OS distribution.

How it works...

The server log is just a file that records messages from the server. Each message has a severity level, the most typical of them being `LOG`, although there are others, as shown in the following table:

PostgreSQL severity	Meaning	Syslog severity	Windows Event Log
`DEBUG 1` to `DEBUG 5`	This comprises the internal diagnostics.	DEBUG	INFORMATION
`INFO`	This is the command output for the user.	INFO	INFORMATION
`NOTICE`	This is helpful information.	NOTICE	INFORMATION
`WARNING`	This warns of likely problems.	NOTICE	WARNING
`ERROR`	This is the current command that is aborted.	WARNING	ERROR
`LOG`	This is useful for sysadmins.	INFO	INFORMATION
`FATAL`	This is the event that disconnects one session only.	ERR	ERROR
`PANIC`	This is the event that crashes the server.	CRIT	ERROR

Table 2.3: PostgreSQL message severity levels

Watch out for `FATAL` and `PANIC`. This shouldn't happen in most cases during normal server operation, apart from certain cases related to replication, for which we refer you to *Chapter 12, Replication and Upgrades*.

You can adjust the number of messages that appear in the log by changing the `log_min_messages` server parameter. You can also change the amount of information that is displayed for each event by changing the `log_error_verbosity` parameter. If the messages are sent to a standard log file, then each line in the log will have a prefix of useful information that can also be controlled by the system administrator, with a parameter named `log_line_prefix`.

You can also alter the *what* and the *how much* that goes into the logs by changing other settings, such as `log_statements`, `log_checkpoints`, `log_connections`/`log_disconnections`, `log_verbosity`, and `log_lock_waits`.

There's more...

The `log_destination` parameter controls where log messages are stored. The valid values are `stderr`, `csvlog`, `syslog`, and `eventlog` (the latter is only on Windows).

The logging collector is a background process that writes to a log file everything that the PostgreSQL server outputs to stderr. This is probably the most reliable way to log messages in case of problems, since it depends on fewer services.

Log rotation can be controlled with settings such as log_rotation_age and log_rotation_size if you are using the logging collector. Alternatively, it is possible to configure the logrotate utility to perform log rotation, which is the default on Debian and Ubuntu systems.

In general, monitoring activities are covered in *Chapter 8, Monitoring and Diagnosis*, and examining the message log is just one part of it. Refer to the *Monitoring the PostgreSQL message log* recipe in *Chapter 8, Monitoring and Diagnosis*, for more details.

Locating the database's system identifier

Each database server has a system identifier assigned when the database is initialized (created). The server identifier remains the same if the server is backed up, cloned, and so on.

Many actions on the server are keyed to the system identifier, and you may be asked to provide this information when you report a fault.

In this recipe, you will learn how to display the system identifier.

Getting ready

You need to connect as the Postgres OS user, or another user with execute privileges on the server software.

How to do it...

In order to display the system identifier, we just need to launch the following command:

```
pg_controldata <data-directory> | grep "system identifier"
Database system identifier:             7015545877453537036
```

Note that the preceding syntax will not work on Debian or Ubuntu systems, because of the choice of offering a multi-version directory structure. That choice implies that every PostgreSQL executable tool is to be accessed via a wrapper that selects the right path, based on the major version. These are the same reasons explained in relation to initdb in the *Locating the database server files* recipe. However, there is no postgresql-common wrapper for this command, as it is not deemed common enough, so if you must run pg_controldata, you need to specify the full path to the executable, as in this example:

```
/usr/lib/postgresql/16/bin/pg_controldata $PGDATA
```

How it works...

The pg_controldata utility is a PostgreSQL server application that shows global state information for a database cluster, by reading it from the server's control file. The control file is located within the data directory of a server, and it is created at database initialization time. Some of the information within it is updated regularly, and some is only updated when certain major events occur.

The output of pg_controldata looks like the following (some values may change over time as the server runs):

```
pg_control version number:            1300
Catalog version number:               202307071
Database system identifier:           7015545877453537036
Database cluster state:               in production
pg_control last modified:             Wed 27 Sep 2023 13:39:50 BST
Latest checkpoint location:           0/16F2EC0
... (not shown in full)
```

TIP

Never edit the PostgreSQL control file. If you do, the server probably won't start correctly, or you may mask other errors. And if you do that, people will be able to tell, so fess up as soon as possible!

Listing databases on the database server

When we connect to PostgreSQL, we always connect to just one specific database on any database server. If there are many databases on a single server, it can get confusing, so sometimes you may just want to find out which databases are parts of the database server.

This is also confusing because we can use the word database in two different, but related, contexts. Initially, we start off by thinking that PostgreSQL is a database in which we put data, referring to the whole database server by just the word *database*. In PostgreSQL, a **database server** (also known as a **cluster**) is potentially split into multiple, individual databases, so as you get more used to working with PostgreSQL, you'll start to separate the two concepts.

How to do it...

If you have access to psql, you can type the following command:

```
$ psql -l
                                                    List of databases
   Name    |  Owner   | Encoding | Locale Provider |  Collate   |
Ctype      | ICU Locale | ICU Rules |    Access privileges
-----------+----------+----------+-----------------+------------+--------
-----+------------+-----------+-----------------------
 postgres  | postgres | UTF8     | libc            | en_GB.UTF-8 | en_
GB.UTF-8  |            |          |
 template0 | postgres | UTF8     | libc            | en_GB.UTF-8 | en_
GB.UTF-8  |            |          | =c/postgres            +
           |            |          |                        |            |
           |            |          | postgres=CTc/postgres
 template1 | postgres | UTF8     | libc            | en_GB.UTF-8 | en_
GB.UTF-8  |            |          | =c/postgres            +
           |            |          |                        |            |
           |            |          | postgres=CTc/postgres
(3 rows)
```

You can also get the same information while running psql by simply typing \l.

The information that we just looked at is stored in a PostgreSQL catalog table named pg_database. We can issue a SQL query directly against that table from any connection to get a simpler result, as follows:

```
postgres=# select datname from pg_database;
 datname
-----------
 template1
 template0
 postgres
(3 rows)
```

How it works...

PostgreSQL starts with three databases: template0, template1, and postgres. The main user database is postgres.

You can create your own databases as well, like this:

```
CREATE DATABASE cookbook;
```

You can do the same from the command line, using the following expression:

```
bash $ createdb cookbook
```

After you've created your databases, be sure to secure them properly, as discussed in *Chapter 6, Security*.

From now on, we will run our examples in the cookbook database.

When you create another database, it actually takes a copy of an existing database, including all the objects such as schemas, tables, etc. By default, the source for the copy is the template1 database, although you could indicate a different source using the TEMPLATE keyword. Once the new database is created, there is no further link between the two databases. For instance, if I modified the cookbook database, then my changes would not affect the template1 database.

As you might guess, the template0 and template1 databases are known as **template databases**. Every database can also be a template, but those two are supposed to be used **only** as templates. The template1 database can be changed to allow you to create a custom template for any new databases that you create; for instance, if you want to place certain schemas and tables in every new database, you can place them in template1 so that you don't have to create them again every time you create a new database. The template0 database exists so that, when you alter template1, you still have a pristine copy to fall back on. In other words, if you break template1, then you can drop it and recreate it from template0.

You can drop the database named postgres. But don't, okay? Similarly, don't try to touch template0, because you won't be allowed to do anything with it, except use it as a template. On the other hand, the template1 database exists to be modified, so feel free to change it.

There's more...

The information that we just saw is stored in a PostgreSQL catalog table named pg_database. We can look at this directly to get some more information. In some ways, the output is less useful as well, as we need to look up some of the code in other tables. The output is also very long, which is the price you pay for the additional readability given by the \x command, so we added a limit 1 clause to display only one record:

```
cookbook=# \x
Expanded display is on.
cookbook=# select * from pg_database limit 1;
-[ RECORD 1 ]--+-------------------------------------
oid            | 5
datname        | postgres
datdba         | 10
encoding       | 6
datlocprovider | c
datistemplate  | f
datallowconn   | t
datconnlimit   | -1
datfrozenxid   | 722
datminmxid     | 1
dattablespace  | 1663
datcollate     | en_GB.UTF-8
datctype       | en_GB.UTF-8
daticulocale   |
daticurules    |
datcollversion | 2.36
datacl         |
```

First of all, look at the use of the \x command. It makes the output in psql appear as one column per line, rather than one row per line.

We've already discussed templates. The other interesting things are that we can turn connections on and off for a database, and we can set connection limits for them, as well.

Also, you can see that each database has a default tablespace. Therefore, data tables get created inside one specific database, and the data files for that table get placed in one tablespace.

You can also see that each database has a collation sequence, which is the way that various language features are defined. We'll cover more on that in the *Choosing good names for database objects* recipe in *Chapter 5*, *Tables and Data*.

How many tables are there in a database?

The number of tables in a relational database is a good measure of the complexity of a database, so it is a simple way to get to know any database. But what kind of complexity exactly? Well, a database may be complex because it has been designed to be deliberately flexible, in order to cover a variety of business situations. It could also be that a complex business process may have a limited portion of its details covered in the database. So a large number of tables might reveal a complex business process or just a complex piece of software.

In this recipe, we will show you how to compute the number of tables.

How to do it...

From any interface, type the following SQL command:

```
SELECT count(*) FROM information_schema.tables
WHERE table_schema NOT IN ('information_schema','pg_catalog');
```

You can also look at the list of tables directly and judge whether the list is small or large.

In psql, you can see your own tables by using the \d meta-command, which you can run either from the SQL prompt or by using the -c switch, as in the following example:

```
$ psql -c "\d"
         List of relations
 Schema |   Name   | Type  |  Owner
--------+----------+-------+----------
 public | accounts | table | postgres
 public | branches | table | postgres
```

In **pgAdmin 4**, you can see the tables in the tree view on the left-hand side, as shown in the following screenshot:

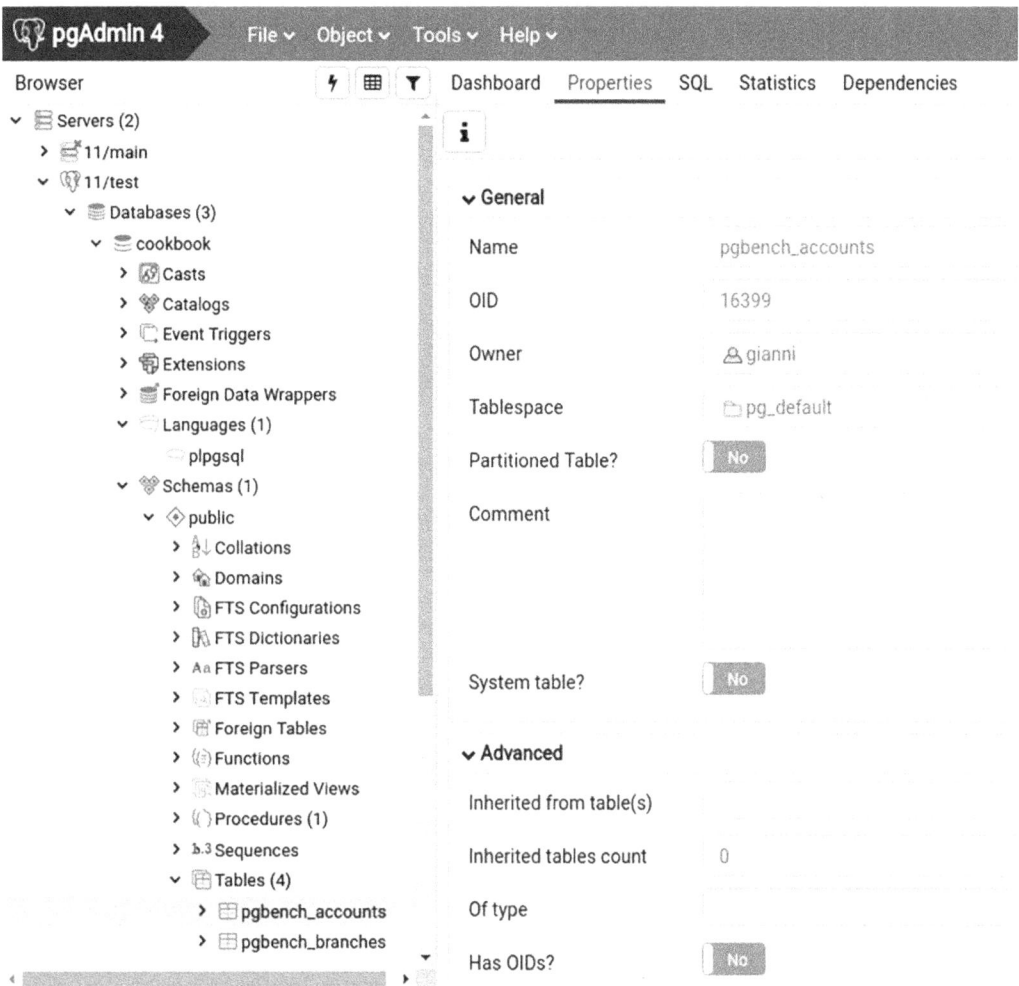

Figure 2.1: The tree view of database objects in pgAdmin 4

How it works...

PostgreSQL stores information about the database in catalog tables. They describe every aspect of the way the database has been defined. There is a main set of catalog tables stored in a schema, called pg_catalog. There is a second set of catalog objects called the **information schema**, which is the standard SQL way of accessing information in a relational database.

We want to exclude both of these schemas from our query, to avoid counting non-user objects. We excluded them in the preceding query, using the NOT IN phrase in the WHERE clause.

Table partitions are implemented as standard tables, which are then considered as part of a larger table, so you might want to exclude them from the total count of tables, i.e., you might prefer that a table with 100 partitions is counted as 1 table instead of 101 tables. However, doing that is more complicated. The information schema shows partitions as the same as tables, which is true for PostgreSQL, so it is somewhat misleading. So what we want to do is exclude tables that are also partitions. Partitions are marked in the pg_catalog.pg_class table, with a Boolean column of **relispartition**. If we use pg_class, we also need to exclude non-tables and ensure we don't include internal schemas, which leaves us with this much more complex query:

```
SELECT count(*) FROM pg_class
WHERE relkind = 'r'
AND not relispartition
AND relnamespace NOT IN (
    SELECT oid FROM pg_namespace
    WHERE nspname IN ('information_schema','pg_catalog', 'pg_toast')
    AND nspname NOT LIKE 'pg_temp%' AND nspname NOT LIKE 'pg_toast_temp%'
);
```

NOTE

Note that this query shows only the number of tables in one of the databases on the PostgreSQL server. You can only see the tables in the database to which you are currently connected, so you'll need to run the same query on each database in turn.

There's more...

The highest number of distinct, major tables I've ever seen in a database is 20,000, without counting partitions, views, and worktables. That clearly rates as a very complex system.

Number of distinct tables (entities – not partitions!)	Complexity rating
20,000	This is incredibly complex. You're either counting wrong or you have a big team to manage this.
2,000	This is a complex business database. Usually, few of these are seen.
200	This is a typical modern business database.
20	This is a simple business database.
2	This is a database with a single, clear purpose, strictly designed for performance or some other goal.
0	This tells you that you haven't loaded any data yet!

Table 2.4: Estimating database complexity based on the number of tables

Of course, you can't always easily tell which tables are entities, so we just need to count the tables. Some databases use a lot of partitions or similar tables, so the numbers can grow dramatically. I've seen databases with up to 200,000 tables (of any kind). That's not recommended, however, as some of the database catalog tables then begin to become awfully large, especially those that have multiple rows for each table. By this, we mean that most use cases and performance tests do not involve databases with so many tables; in practice, it means that some of the standard operating procedures will become painfully slow.

How much disk space does a database use?

It is very important to allocate sufficient disk space for your database. If the disk gets full, it will not corrupt the data, but it might lead to database server panic and then consequent shutdown.

For planning or space monitoring, we often need to know how big a database currently is, so that we get a hint of how fast it grows and are not caught by surprise later on.

How to do it...

We can do this in the following ways:

- Look at the size of the files that make up the database server.
- Run a SQL request to confirm the database size. If you look at the size of the actual files, you'll need to make sure that you include the data directory and all subdirectories, as well as all other directories that contain tablespaces. This can be tricky, and it is also difficult to break down all the different pieces.

The easiest way is to ask the database a simple query, like this:

```
SELECT pg_database_size(current_database());
 pg_database_size
------------------
          8147427
(1 row)
```

However, this is limited to just the current database. If you want to know the size of all the databases together, then you'll need a query such as the following:

```
SELECT sum(pg_database_size(datname)) from pg_database;
```

How it works...

The database server knows which tables it has loaded. It also knows how to calculate the size of each table, so the `pg_database_size()` function just looks at the file sizes.

How much memory does a database currently use?

This is one of those questions that are quick and simple when asked, but not so much when being answered.

PostgreSQL uses memory in a dynamic way across its databases, which is why the question includes the word "currently." By this, we mean that the shared buffers managed by a database server are used concurrently by all its databases. This is based on demand, so if a database is unused, then the amount of memory it uses will decrease, because the same memory will gradually be reused to cache data files from busier databases.

How to do it...

The largest amount of memory is usually consumed for the shared buffers. This is an area that is shared among all the sessions; e.g., if multiple sessions are working on the same table at the same time, there is only one copy of each table block in the shared buffers.

The `pg_buffercache` extension provides a view that has one row for each buffer page. It can be installed easily, as it is distributed with PostgreSQL:

```
cookbook=# CREATE EXTENSION pg_buffercache;
CREATE EXTENSION
```

Let's inspect the view created by this extension:

```
cookbook=# SELECT * FROM pg_buffercache LIMIT 1 \gx
-[ RECORD 1 ]----+-----
bufferid         | 1
relfilenode      | 1262
reltablespace    | 1664
reldatabase      | 0
relforknumber    | 0
relblocknumber   | 0
isdirty          | f
usagecount       | 5
pinning_backends | 0
```

This row represents one block of memory, which is being used for the database specified in the column **reldatabase**. The special value *0* means that this is actually used for data that does not belong to a specific database. In fact, the only time this happens is related to global tables, which only exist in the catalog. Finally, NULL indicates a block that is not used at all.

So we can count the number of blocks for each database and multiply them by the current setting of block_size, as in this example:

```
cookbook=# SELECT datname
, pg_size_pretty(CAST(current_setting('block_size') AS bigint)
    * count(*))
FROM pg_buffercache c
LEFT JOIN pg_database d
  ON c.reldatabase = d.oid
```

```
  GROUP BY datname ORDER BY datname;
  datname   | pg_size_pretty
------------+----------------
 cookbook   | 2464 kB
 postgres   | 568 kB
 template1  | 568 kB
            | 124 MB
(4 rows)
```

This result makes sense because, on the PostgreSQL instance that I am using to prepare the queries for this chapter, the shared_buffers parameter is set to the (extremely low) default value of 128 MB. In fact, it seems sufficient, since we can see that 124 MB is actually unused.

How it works...

The pg_buffercache view is just a wrapper around pg_buffercache_pages(), which is a C function reading the information in real time from the shared memory in PostgreSQL; it is presented in a view just because it is much easier to use.

The pg_size_pretty() function reads the number of bytes and returns a text string, which effectively rounds the result to the reasonable unit. We describe it in more detail in the next recipe.

By default, PostgreSQL uses blocks whose size is 8 KB = 8,192 bytes, but the users could have initialized their instance with a different block size, which is why it is safer to write current_setting('block_size') instead of just 8192.

How much disk space does a table use?

The maximum supported table size in the default configuration is 32 TB, and it does not require large file support from the operating system. The filesystem size limits do not impact the large tables, as they are stored in multiple 1 GB files.

Large tables can suffer performance issues. Indexes can take much longer to update, and query performance can degrade. In this recipe, we will see how to measure the size of a table.

How to do it...

We can see the size of a table by using this command:

```
cookbook=# select pg_relation_size('pgbench_accounts');
```

The output of this command is as follows:

```
 pg_relation_size
------------------
         13434880
(1 row)
```

We can also see the total size of a table, including indexes and other related spaces, as follows:

```
cookbook=# select pg_total_relation_size('pgbench_accounts');
```

The output is as follows:

```
 pg_total_relation_size
------------------------
               15728640
(1 row)
```

We can also use a `psql` command, like this:

```
cookbook=# \dt+ pgbench_accounts
                                    List of relations
  Schema |       Name       | Type  | Owner   | Persistence | Access method
 | Size  | Description
--------+------------------+-------+---------+-------------+---------------
-+-------+-------------
 public | pgbench_accounts | table | gciolli | permanent   | heap
 | 13 MB |
(1 row)
```

How it works...

In PostgreSQL, a table is made up of many relations. The main relation is the data table. In addition, there are a variety of additional data files. Each index created on a table is also a relation. Long data values are placed in a secondary table named TOAST, which is an acronym for **The Oversize Attribute Storage Technique**; in most cases, each table also has a TOAST table and a TOAST index.

Each relation consists of multiple data files. The main data files are broken into 1 GB pieces. The first file has no suffix; others have a numbered suffix (such as .2). There are also files marked _vm and _fsm, which represent the **Visibility Map** and **Free Space Map**, respectively. They are used as part of maintenance operations. They stay fairly small, even for very large tables.

There's more...

The preceding functions, which measure the size of a relation, output the number of bytes, which is normally too large to be immediately clear. You can apply the pg_size_pretty() function to format that number nicely, as shown in the following example:

```
SELECT pg_size_pretty(pg_relation_size('pgbench_accounts'));
```

This yields the following output:

```
pg_size_pretty
----------------
13 MB
(1 row)
```

As the name **TOAST** implies, this is a mechanism used to store long column values. PostgreSQL allows many data types to store values up to 1 GB in size. It transparently stores large data items in many smaller pieces, so the same data type can be used for data ranging from 1 byte to 1 GB. When appropriate, values are automatically compressed and decompressed before they are split and stored, so the actual limit will vary, depending on compressibility.

You may also see files ending in _init; they are used by unlogged tables and their indexes, to restore them after a crash. **Unlogged** objects are called this way because they do not produce any **write-ahead log** (**WAL** for short), the PostgreSQL mechanism to ensure consistent crash recovery. So they support faster writes, but in the event of a crash they must be truncated, that is, restored to an empty state.

Which are my biggest tables?

We've looked at getting the size of a specific table, so now it's time to widen the problem to related areas. Rather than having an absolute value for a specific table, let's look at the relative sizes.

How to do it...

The following basic query will tell us the 10 biggest tables:

```
SELECT quote_ident(table_schema)||'.'||quote_ident(table_name) as name
      ,pg_relation_size(quote_ident(table_schema)
             || '.' || quote_ident(table_name)) as size
 FROM information_schema.tables
 WHERE table_schema NOT IN ('information_schema', 'pg_catalog')
```

```
ORDER BY size DESC
LIMIT 10;
```

The tables are shown in descending order of size, with at the most 10 rows displayed, as in the following example:

```
          name            |   size
--------------------------+----------
 public.pgbench_accounts  | 13434880
 public.pgbench_branches  |     8192
 public.pgbench_tellers   |     8192
 public.pgbench_history   |        0
(4 rows)
```

In this case, we look at all the tables in all the schemas, apart from the tables in information_schema or pg_catalog, as we did in the *How many tables are in the database?* recipe.

How it works...

PostgreSQL provides a dedicated function, pg_relation_size, to compute the actual disk space used by a specific table or index. We just need to provide the table name. In addition to the main data files, there are other files (called **forks**) that can be measured by specifying an optional second argument. These include the **Visibility Map (VM)**, the **Free Space Map (FSM)**, and the **initialization fork** for unlogged objects.

How many rows are there in a table?

There is no limit on the number of rows in a table, but the table is limited to available disk space and memory/swap space. If you are storing rows that exceed an aggregated data size of 2 KB, then the maximum number of rows may be limited to 4 billion or fewer.

Counting is one of the easiest SQL statements, so it is also many people's first experience of a PostgreSQL query.

How to do it...

From any interface, the SQL command used to count rows is as follows:

```
SELECT count(*) FROM table;
```

This will return a single integer value as the result.

In psql, the command looks like the following:

```
cookbook=# select count(*) from orders;
 count
-------
   345
(1 row)
```

How it works...

PostgreSQL can choose between two techniques available to compute the SQL count(*) function. Both are available in all the currently supported versions:

- The first is called **sequential scan**. We access every data block in the table, one after the other, reading the number of rows in each block. If the table is on the disk, it will cause a beneficial disk access pattern, and the statement will be fairly fast.
- The other technique is known as an **index-only scan**. It requires an index on the table, and it covers a more general case than optimizing SQL queries with count(*).

PostgreSQL will choose whichever technique looks faster, based on table size and other information, as we will see in *Chapter 10, Performance and Concurrency*.

Some people think that the count SQL statement is a good test of the performance of a **Database Management System** (**DBMS**). Some DBMSs have specific tuning features for the count SQL statement, like Postgres, which can optimize this kind of query using index-only scans. While in the past PostgreSQL contributors have discussed optimizing the count() statement many times, few people actually thought we should optimize the specific query from the example above. Yes, the count function is frequently used within applications; however, without a WHERE clause, it is not that useful. Then, at some point, the **index-only scans** feature was implemented, and that includes the case described in this recipe, while also applying to many more real-world situations.

We need to scan every block of the table because of a major feature of Postgres, named **Multi-Version Concurrency Control** (**MVCC**). MVCC allows us to run the count SQL statement at the same time that we insert, update, or delete data from the table. That's a very cool feature, and we went through a lot of trouble in Postgres to provide it to users like you; however, it means that, in this case, counting rows quickly is a bit more complicated.

MVCC requires us to record information on each row of a table, stating when each change was made. If the changes were made after the SQL statement began to execute, then we just ignore those changes. This means that we need to carry out visibility checks on each row in the table to allow us to work out the results of the count SQL statement. The optimization provided by index-only scans is the ability to skip such checks on the table blocks that are already known to be visible to all current sessions. Rows in these blocks can be counted directly in the index, which is normally smaller than the table and is, therefore, faster.

If you think a little deeper about this, you'll see that the result of the count SQL statement is just the value at a moment in time. Depending on what happens to the table, that value could change a little or a lot while the count SQL statement executes. So once you've executed this, all you really know is that, at a particular point in the past, there were exactly x rows in the table.

Quickly estimating the number of rows in a table

We don't always need an accurate count of rows, especially on a large table that may take a long time to execute. Administrators often need to estimate how big a table is so that they can estimate how long other operations may take.

How to do it...

The Postgres optimizer can provide a quick estimate of the number of rows in a table, simply by using its statistics:

```
EXPLAIN SELECT * FROM mytable;
                       QUERY PLAN
-----------------------------------------------------------------
Seq Scan on mytable  (cost=0.00..2640.00 rows=100000 width=97)
(1 row)
```

You can see that the optimizer estimates 100,000 rows in output. We can directly compute a similar number using roughly the same calculation:

```
SELECT (CASE WHEN reltuples > 0 THEN pg_relation_size(oid)*reltuples/
(8192*relpages)
ELSE 0
END)::bigint AS estimated_row_count
FROM pg_class
WHERE oid = 'mytable'::regclass;
```

This gives us the following output:

```
estimated_row_count
-------------------
              99960
(1 row)
```

Both queries return a row count very quickly, no matter how large the table that we are examining is, because they use statistics that were collected in advance. You may want to create a SQL function for the preceding calculation so that you won't need to retype the SQL code repeatedly.

The following function estimates the total number of rows, using a mathematical procedure called **extrapolation**. In other words, we take the average number of bytes per row resulting from the last statistics collection, and then we apply it to the current table size:

```
CREATE OR REPLACE FUNCTION estimated_row_count(text)
RETURNS bigint
LANGUAGE sql
AS $$
SELECT (CASE WHEN reltuples > 0 THEN
               pg_relation_size($1)*reltuples/(8192*relpages)
             ELSE 0
             END)::bigint
FROM pg_class
WHERE oid = $1::regclass;
$$;
```

which we can use as follows:

```
select estimated_row_count('myschema.mytable');
 estimated_row_count
---------------------
             1048576
(1 row)
```

How it works...

We saw the pg_relation_size() function earlier, so we know that it brings back an accurate value for the current size of the table.

Later in this book, we will discuss extensively the **VACUUM** maintenance process. When that process runs against a table in Postgres, we record two pieces of information in the pg_class catalog entry for the table. These two items are the number of data blocks in the table (relpages) and the number of rows in the table (reltuples). Some people think they can use the value of reltuples in pg_class as an estimate, but it could be severely out of date. You will also be fooled if you use information in another table named pg_stat_user_tables, which is discussed in more detail in *Chapter 10, Performance and Concurrency*.

The Postgres optimizer uses the relpages and reltuples values to calculate the average rows per block, which is also known as the **average tuple density**.

If we assume that the average tuple density remains constant over time, then we can calculate the number of rows using this formula: *Row estimate = number of data blocks * rows per block*.

PostgreSQL includes some code to handle cases where the reltuples or relpages fields are zero. The Postgres optimizer actually works a little harder than we do in that case, so our estimate isn't very good.

The WHERE oid = 'mytable'::regclass; syntax is a shorthand that uses the cast syntax **value :: type**, converting the user-visible name of a table into the internal unique identifier number that every database object has.

There's more...

The good thing about the preceding recipe is that it returns a value in about the same time, no matter how big the table is. The bad thing about it is that pg_relation_size() requests a lock on the table, so if any other user has an AccessExclusiveLock lock on the table, then the table size estimate will wait for the lock to be released before returning a value.

Er... so what is an AccessExclusiveLock lock? It is a kind of lock that blocks any other access to the table, including our use of pg_relation_size(), until the lock is released, which happens only when the statement that originally took the lock terminates. In our example, it means that our estimate will have to wait if somebody else is performing maintenance actions, such as certain forms of ALTER TABLE.

The typical scenario is that you use pg_relation_size() to estimate the number of rows, as we did above, but then it doesn't return immediately as you would expect, because it has to wait for the other maintenance action to terminate and, thus, release the lock.

At that point, you think, *Oh, was that table bigger than I thought? How long will I be waiting?* Yes, it's better to calculate that beforehand, but hindsight doesn't get you out of the hole you are in right now. So we need a way to calculate the size of a table without needing the lock.

A solution is to look at the operating system files that Postgres uses to store data and figure out how large they are. This is not blocked by any Postgres lock, but at the same time, it requires a higher level of security than most people usually allow. In any case, looking at files without a lock could easily mislead the observer into a wrong conclusion if the table were dropped or changed at the exact same time.

Listing extensions in this database

Every PostgreSQL database contains some objects that are automatically brought in when the database is created. Every user will find a pg_database system catalog that lists databases, as shown in the *Listing databases on the database server* recipe. There is little point in checking whether these objects exist because even superusers are not allowed to drop them.

On the other hand, PostgreSQL comes with tens of collections of optional objects, called **modules**, or equivalently **extensions**. The database administrator can install or uninstall these objects, depending on the requirements. They are not automatically included in a newly created database because they might not be required by every use case. Users will install only the extensions they actually need, when they need them; an extension can be installed while a database is up and running.

In this recipe, we will explain how to list extensions that have been installed on the current database. This is important for getting to know the database better, and also because certain extensions affect the behavior of the database.

How to do it...

In PostgreSQL, there is a catalog table recording the list of installed extensions, so this recipe is quite simple. Issue the following command:

```
cookbook=> SELECT * FROM pg_extension;
```

This results in the following output:

```
-[ RECORD 1 ]--+---------
oid            | 13693
extname        | plpgsql
extowner       | 10
```

```
extnamespace  | 11
extrelocatable | f
extversion    | 1.0
extconfig     |
extcondition  |
```

NOTE

Note that we have chosen an expanded format, due to the high number of columns, which resembles the outcome of issuing the \x meta-command previously.

To get the same list with fewer technical details, you can use the \dx meta-command, as used when listing databases.

How it works...

A PostgreSQL extension is represented by a control file, <extension name>.control, located in the SHAREDIR/extension directory, plus one or more files containing the actual extension objects. The control file specifies the extension name, version, and other information that is useful for the extension infrastructure. Each time an extension is installed, uninstalled, or upgraded to a new version, the corresponding row in the pg_extension catalog table is inserted, deleted, or updated, respectively.

There's more...

In this recipe, we only mentioned extensions distributed with PostgreSQL, and solely for the purpose of listing which ones are being used in the current database. The infrastructure for extensions will be described in greater detail in *Chapter 3, Server Configuration*. We will talk about the version number of an extension, and we will show you how to install, uninstall, and upgrade extensions, including those distributed independently of PostgreSQL.

See also

To get an idea of which extensions are available, you can browse the list of additional modules shipped together with PostgreSQL, which are almost all extensions, at https://www.postgresql.org/docs/current/static/contrib.html.

Understanding object dependencies

In most databases, there will be dependencies between objects in a database. Sometimes, we need to understand these dependencies to figure out how to perform certain actions, such as modifying or deleting existing objects. Let's look at this in detail.

Getting ready

We'll use the following simple database to understand and investigate the dependencies:

1. Create two tables as follows:

    ```
    CREATE TABLE orders (
    orderid integer PRIMARY KEY
    );
    CREATE TABLE orderlines (
    orderid integer
    ,lineid smallint
    ,PRIMARY KEY (orderid, lineid)
    );
    ```

2. Now, we add a link between them to enforce what is known as **referential integrity**, as follows:

    ```
    ALTER TABLE orderlines ADD FOREIGN KEY (orderid)
    REFERENCES orders (orderid);
    ```

3. If we try to drop the referenced table, we get the following message:

    ```
    DROP TABLE orders;
    ERROR: cannot drop table orders because other objects depend on it
    DETAIL: constraint orderlines_orderid_fkey on table orderlines
    depends on table orders
    HINT: Use DROP ... CASCADE to drop the dependent objects too.
    ```

Be very careful! If you follow the hint, you may accidentally remove all the objects that have any dependency on the orders table. You might think that this would be a great idea, but it is not the right thing to do. It might work, but we need to ensure that it will work.

Therefore, you need to know what dependencies are present on the orders table, and then review them. Then, you can decide whether it is okay to issue the CASCADE version of the command, or whether you should reconcile the situation manually.

How to do it...

You can use the following command from `psql` to display full information about a table, the constraints that are defined upon it, and the constraints that reference it:

```
\d+ orders
                                      Table "public.orders"
 Column |  Type  | Collation | Nullable | Default | Storage | Compression
 | Stats target | Description
--------+--------+-----------+----------+---------+---------+-------------
--------------+-------------
 id     | bigint |           | not null |         | plain   |
 |              |
 aid    | bigint |           |          |         | plain   |
 |              |
Indexes:
    "orders_pkey" PRIMARY KEY, btree (id)
Foreign-key constraints:
    "orders_aid_fkey" FOREIGN KEY (aid) REFERENCES accounts(aid)
Access method: heap
```

You can also get specific details of the constraints by using the following query:

```
SELECT * FROM pg_constraint
WHERE confrelid = 'orders'::regclass;
-[ RECORD 1 ]--+----------------
oid            | 17227
conname        | orders_aid_fkey
connamespace   | 2200
contype        | f
condeferrable  | f
condeferred    | f
convalidated   | t
conrelid       | 17222
contypid       | 0
conindid       | 17215
conparentid    | 0
confrelid      | 17211
confupdtype    | a
confdeltype    | a
```

```
confmatchtype   | s
conislocal      | t
coninhcount     | 0
connoinherit    | t
conkey          | {2}
confkey         | {1}
conpfeqop       | {410}
conppeqop       | {410}
conffeqop       | {410}
confdelsetcols  |
conexclop       |
conbin          |
```

We included the entire output for completeness, but most of those rows are difficult to interpret without joins to many other catalog tables. That's why, for most users, the information provided by the \d+ syntax is sufficient.

The aforementioned queries only covered constraints between tables. This is not the end of the story, so read the *There's more...* section.

How it works...

When we create a foreign key, we add a constraint to the catalog table, known as pg_constraint. Therefore, the query shows us how to find all the constraints that depend upon the orders table.

There's more...

With Postgres, there's always a little more when you look beneath the surface. In this case, there's a lot more, and it's important.

We didn't discuss dependencies with other kinds of objects. Two important types of objects that might have dependencies on tables are **views** and **functions**.

Consider the following command:

```
DROP TABLE orders;
```

If you issue this, the dependency on any of the **views** will prevent the table from being dropped. So, you need to remove those views and then drop the table.

The story with function dependencies is not as useful, because PostgreSQL only tracks a portion of the dependencies, so you break a function by dropping a table or another function, without being stopped. Precisely, relationships between functions and tables are not recorded in the catalog, nor is the dependency information between functions. This is partly due to the fact that most PostgreSQL procedural languages allow dynamic query execution, so you wouldn't be able to tell which tables or functions a function would access until it executes. That's only partly the reason because most functions clearly reference other tables and functions, so it should be possible to identify and store those dependencies. However, right now, we won't do that. So make a note that you need to record the dependency information for your functions manually, ensuring that you'll know if and when it's okay to remove or alter a table or other objects that functions depend on.

Learn more on Discord

To join the Discord community for this book – where you can share feedback, ask questions to the author, and learn about new releases – follow the QR code below:

`https://discord.gg/pQkghgmgdG`

3
Server Configuration

We get asked many questions about parameter settings in PostgreSQL. Everybody's busy, and most people want a 5-minute tour of how things work. That's exactly what a cookbook is for, so we'll do our best.

Some people believe that some magical parameter settings will improve their performance, spending hours combing the pages of books to glean insights. Others feel comfortable because they have found a website somewhere that *explains everything*, and they *know* they have their database configured OK.

For the most part, the settings are easy to understand. However, finding the best setting can be difficult, and the optimal setting may change over time. This chapter is mostly about knowing how, when, and where to change parameter settings.

In this chapter, we will cover the following recipes:

- Read the fine manual (RTFM)
- Planning a new database
- Setting the configuration parameters for the database server
- Setting the configuration parameters in your programs
- Finding the configuration settings for your session
- Finding parameters with non-default settings
- Setting parameters for particular groups of users
- A basic server configuration checklist
- Adding an external module to PostgreSQL

- Using an installed module/extension
- Managing installed extensions

Read the fine manual (RTFM)

RTFM is often (rudely) used to mean *don't bother me; I'm busy*, or it is used as a stronger form of abuse. The strange thing is that asking you to read a manual is most often very good advice. Take the advice! The most important point to remember is that you should refer to a manual whose release version matches that of the server on which you are operating.

The PostgreSQL manual is very well-written and comprehensive in its coverage of specific topics. However, one of its main failings is that the documents aren't organized in a way that helps somebody who is trying to learn PostgreSQL. They are organized from the perspective of people checking specific technical points so that they can decide whether their problem is a user error or something else. It sometimes answers *what?* but it seldom answers *why?* or *how?*

How to do it...

The main documents for each PostgreSQL release are available at http://www.postgresql.org/docs/manuals/.

These are the most frequently accessed parts of the documents:

- The **Structured Query Language (SQL)** command reference, as well as the client and server tools' reference: http://www.postgresql.org/docs/current/interactive/reference.html
- Configuration: http://www.postgresql.org/docs/current/interactive/runtime-config.html
- Functions: http://www.postgresql.org/docs/current/interactive/functions.html

You can also grab yourself a **PDF** version of the manual, which can allow for easier searching in some cases. Don't print it! The documents are about 2,800 pages of A4-sized sheets.

How it works...

PostgreSQL documents are written in **Standard Generalized Markup Language (SGML)**, which is similar to, but not the same as, **Extensible Markup Language (XML)**. These files are then processed to generate **HyperText Markup Language (HTML)** files, PDFs, and so on. This ensures that all the formats have exactly the same content. Then, you can choose the format you prefer, and you can even compile it in other formats, such as **Electronic Publication (EPUB)**, **Interchange File Format (INFO)**, and so on.

Moreover, the PostgreSQL manual is actually a subset of the PostgreSQL source code, so it evolves together with the software. It is written by the same people who make PostgreSQL, which gives you even more reasons to read it!

There's more...

More information is also available at `http://wiki.postgresql.org`.

Many distributions offer packages that install static versions of the HTML documentation. For example, on Debian and Ubuntu, the documentation for the most recent stable PostgreSQL version is named `postgresql-doc-16`.

Planning a new database

Planning a new database can be a daunting task as there are so many things to consider, as we will see in this recipe. It's easy to get overwhelmed, so here, we will present some planning ideas. It's also easy to charge headlong at the task without planning, thinking that whatever you know is all you'll ever need to consider, but planning is an important step.

Getting ready

Don't wait to be told what to do. If you haven't been told what the requirements are, then write down what you think they are, clearly labeling them as *assumptions* rather than *requirements*; you must not confuse the two.

Iterate until you get some agreement, and then build a prototype.

How to do it...

Write a document that covers the following actions:

- Database design – plan your database design.
- Calculate the initial database sizing.
- Transaction analysis – how will we access the database?
- Look at the most frequent access paths (for example, queries).
- What are the requirements for the response times?
- Hardware specification (which is still needed in the cloud).
- Initial performance thoughts – will all of the data fit into the available **Random-Access Memory (RAM)**?
- Choose the **Operating System (OS)** and filesystem types.

- Create a localization plan.
- Decide the server encoding, locale, and time zone.
- Access and security administration plan.
- Identify client systems and specify the required drivers.
- Create roles according to a plan for access control.
- Specify connection routes and authentication for the server in `pg_hba.conf`.
- Monitoring – are there PostgreSQL plugins for the monitoring solution you are already using (usually, the answer is yes)? What are the business-specific metrics we need to monitor?
- Maintenance plan – who will keep it working? How?
- Availability plan – consider the availability requirements.
- If you are working with a cloud database cluster, you should also consider the following:
    - Plan your backup mechanism and test it.
    - Create a **High-Availability** (**HA**) plan.
    - Decide which form of replication you'll need – if any.
- If you are going to deploy on Kubernetes, plenty of these things will be managed by the operator, but you still need to think about:
    - RAM and CPU specs.
    - Sizing for the PVC.
    - Availability, e.g., will the Kubernetes workers be spread across data centers?

How it works...

One of the most important reasons to plan your database ahead of time is that retrofitting some things is difficult. This is especially true of server encoding and locale, which can cause a lot of downtime and exertion if we need to change them later. Security is also much more difficult to set up when a system is nearly live.

There's more...

Planning always helps. You may know what you're doing, but others may not. Tell everybody what you're going to do before you do it to avoid wasting time. If you're not sure yet, then build a prototype to help you decide. **Trusted Postgres Architect (TPA)** deployed with Docker containers and CloudNativePG are great for prototyping, as we covered in the recipes in *Chapter 1, First Steps*. Approach the administration framework as if it were a development task. Make a list of things you don't know yet, and work through them one by one. Collaboration with the development team is crucial to reduce uncertainty and increase the success of a project.

This is deliberately a short recipe. Everybody has their own way of doing things, and it's very important not to be too prescriptive about how to do things. If you already have a plan, great! If you don't, think about what you need to do, make a checklist, and then do it.

Setting the configuration parameters for the database server

The parameter file, which is known as postgresql.conf, is the main location that's used to define parameter values for the PostgreSQL server. All the parameters can be set in the parameter file. There are also two other parameter files: pg_hba.conf and pg_ident.conf. Both of these relate to connections and security, so we'll cover them in later chapters. Other locations can be added with the include directive inside postgresql.conf, as we will explain in this recipe.

Getting ready

In the pg_settings view of the pg_catalog, the context defines when each parameter can be set. The following table categorizes this so that we can see what action is needed for changes to take effect. SET is a command, but RELOAD and RESTART are actions, not specific commands. What is RESTART ALL? Some parameters marked POSTMASTER are marked as exceptions in the following table. These parameters must be set to a value less than or equal to their setting on standby. As a result, to increase them on the primary node, we must first increase them on ALL standby nodes and then restart them before restarting the primary node – for example, max_connections.

To simplify the table, we have avoided mentioning two more complex contexts. The word "procedures" has been shortened to "procs", "transaction" to "xact", and "timestamp" to "ts":

Context (Number of parameters of this type in PG16, excluding any added by extensions)	SET Set by user via SET command in session, ALTER USER, and/or connection option	RELOAD Altered in config file; reload implements change(s)	RESTART Altered in config; restart of primary implements changes(s)	RESTART ALL Altered in config; restart of primary and all standbys implements change(s)
USER (141)	✔ for local session only	✔ for all sessions not already SET	Not needed, but restart will apply	Not applicable
SUPERUSER (45)	✔ but only superuser	✔ (as above)	(as above)	Not applicable
SIGHUP (94)	✗	✔	(as above)	Not applicable
POSTMASTER (54)	✗	✗	✔	Not applicable
Exceptions: max_connections max_wal_senders max_worker_procs max_prepared_xacts max_locks_per_xact wal_level wal_log_hints track_commit_ts	✗	✗	✗	✔
INTERNAL (18)	✗	✗	✗	✗

Table 3.1: Based on the context of a parameter, we can observe when the value change will take effect

How to do it...

Let's start by looking at this in the cloud, and then move on to a discussion of command-line actions for on-prem users. Using EDB BigAnimal (the specific example we presented in the *PostgreSQL in the cloud* recipe in *Chapter 1, First Steps*), follow these steps:

1. Navigate to https://portal.biganimal.com and log in.
2. On the left navigation bar, select **Clusters** and locate your specific cluster.

3. On the right-hand side, select **Edit Cluster**:

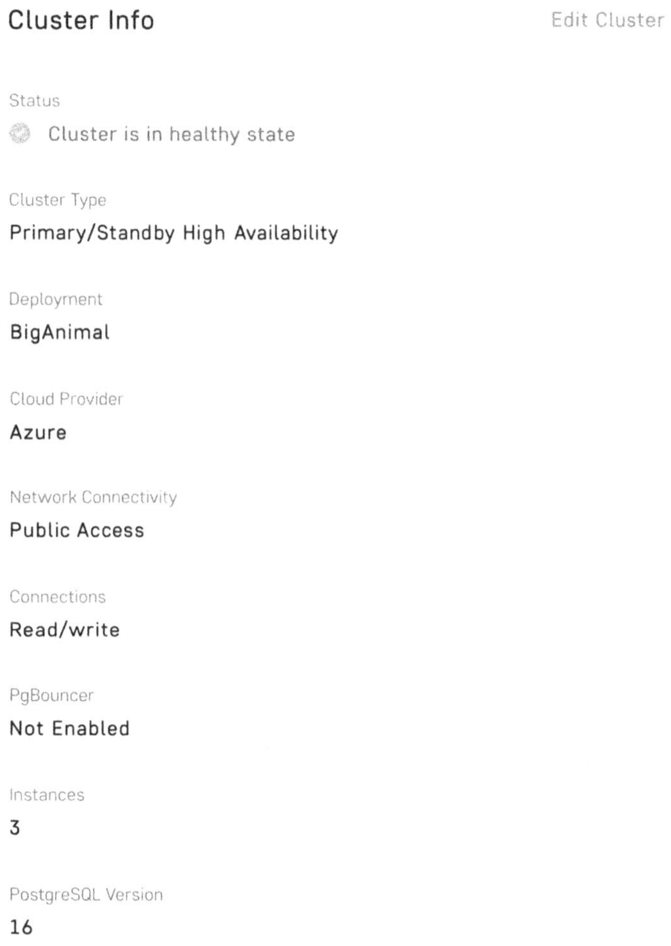

Cluster Info   Edit Cluster

Status
Cluster is in healthy state

Cluster Type
Primary/Standby High Availability

Deployment
BigAnimal

Cloud Provider
Azure

Network Connectivity
Public Access

Connections
Read/write

PgBouncer
Not Enabled

Instances
3

PostgreSQL Version
16

Figure 3.1: Cluster Info offering the Edit Cluster option

4. On the **Edit Cluster** page, select the **DB Configuration** tab:

![Edit Cluster DB Configuration tab screenshot]

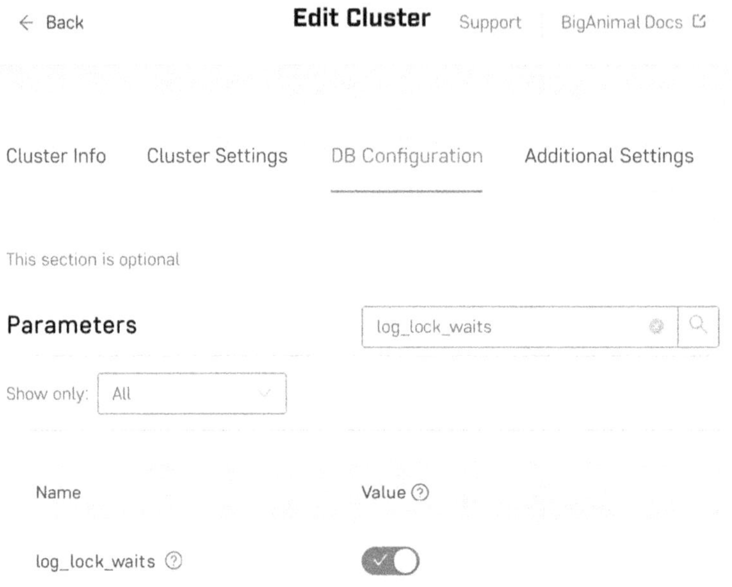

Figure 3.2: Edit Cluster | the DB Configuration tab with a search for one parameter

The preceding example shows how to find the `log_lock_waits` parameter. Flip the radio button to *on* to enable the logging of locks, and then, if you are happy, save the results.

Once parameters have been changed, the server will reload automatically. If any changed parameters are marked with a yellow exclamation mark (**!**), as shown in the next screenshot, the server will restart; however, if the parameter is on the exceptions list, this will cause the standbys to be restarted before restarting the primary. Make sure you understand the behavior of the managed database so that you don't cause unexpected downtime.

Figure 3.3: The max_connections parameter belongs to the postmaster context, and therefore, it will require a restart for the change to take effect

How it works...

On cloud-based PostgreSQL deployments, it is typical that access to the superuser is restricted, so a **Graphical User Interface** (**GUI**) is often available to allow configuration changes.

With more than 300 settable parameters, it is better to search for them individually and change the values. Hit **Save** to keep the changed settings before you leave the screen.

There's more...

In a self-hosted database environment, you may have direct access to the parameter files as your team manages it, in which case the postgresql.conf file would be located as described in the *Finding the configuration settings for your session* recipe.

After changing parameters marked SIGHUP, we issue a reload command to the server, forcing PostgreSQL to re-read the postgresql.conf file (and all other configuration files). There are a number of ways to do that, depending on your distribution and OS. The most common way is to use pg_ctl with the same OS user that runs the PostgreSQL server process, as follows:

```
pg_ctl reload
```

This assumes the default data directory; otherwise, you have to specify the correct data directory with the -D option.

As we previously noted, Debian and Ubuntu have a different multi-version architecture, so you should issue the following command instead:

```
pg_ctlcluster 16 main reload
```

On modern distributions, you should use systemd, as follows:

```
sudo systemctl reload postgresql@16-main
```

NOTE

See the *Starting the database server manually* recipe in *Chapter 4, Server Control*, for more details on how to manage PostgreSQL via systemd; the *Reloading server configuration files* recipe, also in *Chapter 4, Server Control*, shows more ways to reload configuration files.

After changing parameters marked POSTMASTER, we issue a restart of the server for changes to take effect – for instance, listen_addresses. The syntax is very similar to a reload operation, as shown here:

```
pg_ctl restart
```

For Debian and Ubuntu, use this command:

```
pg_ctlcluster 16 main restart
```

With system, use this command:

```
sudo systemctl restart postgresql@16-main
```

Of course, a restart will affect all existing connections, and it will mean downtime while the restart happens. Therefore, it is important to understand the impact of this action and take preventative measures. See the *Restarting the server quickly* recipe in *Chapter 4, Server Control*, for further details. Furthermore, consider the usage of a connection pooler such as pgbouncer to minimize the impact at the application level, as explained in the *Using repmgr and pgbouncer* recipe in *Chapter 12, Replication and Upgrades*.

The postgresql.conf file is a normal text file that can be simply edited. Most of the parameters are listed in the file, so you can just search for them and then insert the desired value in the right place.

If you set the same parameter twice in different parts of the file, the last setting is what applies. This can cause lots of confusion if you add settings to the bottom of the file, so you are advised against doing this.

A longstanding and good practice is to version-control configuration files by using Git alongside any other code or configuration changes. An even better alternative is to use configuration management software such as Ansible, Chef, or Puppet, rather than editing configuration files directly.

The postgresql.conf file also supports an include directive. This allows the postgresql.conf file to reference other files, which can then reference other files, and so on. That may help you organize your parameter settings better if you don't make it too complicated. TPA, as discussed in *Chapter 1, First Steps*, in the *PostgreSQL with TPA* recipe, can also manage configuration parameters. It organizes them in coherent files inside the conf.d directory, which gets included at the end of postgresql.conf.

For more on reloading, see the *Reloading server configuration files* recipe in *Chapter 4, Server Control*.

Furthermore, you can change the values stored in the parameter files directly from your session with syntax such as the following, if you have superuser access:

```
ALTER SYSTEM SET shared_buffers = '1GB';
```

The behavior of this syntax is quite different compared to the other setting-related commands: you run it from within your session, and it changes the default value but not the value in the current session.

This command will not actually edit postgresql.conf. Instead, it writes the new setting to another file named postgresql.auto.conf. The effect is equivalent, albeit in a crash-safe way. The original configuration is never written, so it cannot be damaged in the event of a crash. If you mess up with too many ALTER SYSTEM commands, you can always delete postgresql.auto.conf manually and reload the configuration or restart PostgreSQL, depending on which parameters you changed. However, there are no serious checks on values passed to ALTER SYSTEM, so it's relatively easy to break the configuration of the system and have the server fail to start when the server does actually restart – which might be via a different person – sometime later.

Setting the configuration parameters in your programs

PostgreSQL allows you to override some parameter settings for each session or transaction using SQL commands. Here are some examples of parameters that are designed to be user-modifiable:

- application_name – to help identify the session for monitoring
- synchronous_commit – to set the level of durability desired
- Various timeouts and check intervals:
    - client_connection_check_interval
    - idle_in_transaction_session_timeout
    - idle_session_timeout
    - lock_timeout
    - statement_timeout
- Client-tuning parameters:
    - commit_siblings
    - cursor_tuple_fraction
    - maintenance_work_mem

- vacuum_cost_delay
- work_mem

- Data type-specific settings:

  - bytea_output
  - DateStyle
  - xmlbinary
  - xmloption

- Optimization settings (too many to list, but not normally changed)

How to do it...

Execute the following steps to set custom parameters:

1. You can change the value of a setting during your session, like this:

   ```
   SET work_mem = '16MB';
   ```

2. This value will then be used for every future transaction. You can also change it only for the duration of the current transaction:

   ```
   SET LOCAL work_mem = '16MB';
   ```

3. The setting will last until you issue this command:

   ```
   RESET work_mem;
   ```

4. Alternatively, you can issue the following command:

   ```
   RESET ALL;
   ```

The SET and RESET commands are SQL commands that can be issued from any interface. They apply only to PostgreSQL server parameters, but this does not mean that they affect the entire server. In fact, the parameters you can change with SET and RESET apply only to the current session. Also, note that there may be other parameters, such as **Java Database Connectivity (JDBC)** driver parameters, that cannot be set in this way. Refer to the *Connecting to the PostgreSQL server* recipe in *Chapter 1, First Steps*, for help with those parameters.

How it works...

Suppose you change the value of a setting during your session – for example, by issuing this command:

```
SET work_mem = '16MB';
```

Then, the following will show up in the pg_settings catalog view:

```
postgres=# SELECT name, setting, reset_val, source FROM pg_settings WHERE
source = 'session';
   name    | setting | reset_val | source
-----------+---------+-----------+---------
 work_mem  | 16384   | 4096      | session
```

This will show until you issue this command:

```
RESET work_mem;
```

After issuing it, the setting returns to reset_val, and source returns to the default. To see it, you will need to change the query to:

```
postgres=# SELECT name, setting, reset_val, source FROM pg_settings WHERE
name = 'work_mem';
   name    | setting | reset_val | source
-----------+---------+-----------+---------
 work_mem  | 4096    | 4096      | default
```

There's more...

You can change the value of a setting during your transaction as well, like this:

```
SET LOCAL work_mem = '16MB';
```

This results in the following output:

```
WARNING: SET LOCAL can only be used in transaction blocks
SET
```

In order to understand what the warning means, we can look up that setting in the pg_settings catalog view:

```
postgres=# SELECT name, setting, reset_val, source FROM pg_settings WHERE
source = 'session';
   name   | setting | reset_val | source
----------+---------+-----------+--------

(0 rows)
```

Huh? What happened to your parameter setting? The SET LOCAL command takes effect only for the transaction in which it was executed, which was just the SET LOCAL command in our case. We need to execute it inside a transaction block to be able to see the setting take hold, as follows:

```
BEGIN;
SET LOCAL work_mem = '16MB';
```

Here is what shows up in the pg_settings catalog view:

```
postgres=# SELECT name, setting, reset_val, source FROM pg_settings WHERE
source = 'session';
    name    | setting | reset_val | source
------------+---------+-----------+---------
 work_mem   |  16384  |   4096    | session
```

You should also note that the value of source is session rather than transaction, as you might have been expecting. Note that the global value of work_mem remains unchanged.

Finding the configuration settings for your session

At some point, it will occur to you to ask: *What are the current configuration settings?*

Most settings can be changed in more than one way, and some ways do not affect all users or all sessions, so it is quite possible to get confused.

How to do it...

Your first thought is probably to look in postgresql.conf, which is the configuration file and is described in detail in the *Setting the configuration parameters for the database server* recipe, in this chapter. That works, but only as long as there is only one parameter file. If there are two, then maybe you're reading the wrong file! How would you know? So the cautious and accurate way is to not trust a text file but to trust the server itself.

Moreover, you learned in the previous recipe, *Setting configuration parameters in your programs*, that each parameter has a scope that determines when it can be set. Some parameters can be set through `postgresql.conf`, but others can be changed afterward. So the current values of the configuration settings may have subsequently changed.

We can use the `SHOW` command like this:

```
postgres=# SHOW work_mem;
```

This is its output:

```
work_mem
----------
4MB
(1 row)
```

However, remember that it reports the current setting at the time it is run, and that can be changed in many places, as explained in different parts of this chapter.

Another way of finding the current settings is to access a PostgreSQL catalog view named `pg_settings`:

```
postgres=#  \x
Expanded display is on.
postgres=# SELECT * FROM pg_settings WHERE name = 'work_mem';
[ RECORD 1 ] -------------------------------------------
name            | work_mem
setting         | 4096
unit            | kB
category        | Resource Usage / Memory
short_desc      | Sets the maximum memory to be used for query workspaces.
extra_desc      | This much memory can be used by each internal sort
operation and hash table before switching to temporary disk files.
context         | user
vartype         | integer
source          | default
min_val         | 64
max_val         | 2147483647
enumvals        |
boot_val        | 4096
reset_val       | 4096
sourcefile      |
sourceline      |
pending_restart | f
```

Thus, you can use the `SHOW` command to retrieve the value for a setting, or you can access full details using the `catalog` table.

The actual location of each configuration file can be queried directly to the PostgreSQL server, as shown in this example:

```
postgres=# SHOW config_file;
```

This returns the following output, depending on the OS distribution:

```
          config_file
-----------------------------------------
/etc/postgresql/16/main/postgresql.conf
(1 row)
```

This shows the top-level file, which may include directives to other files.

The other configuration files can be located by querying similar variables – that is, `hba_file` and `ident_file`.

How it works...

Each parameter setting is cached within each session so that we can get quick access to the parameter settings. This allows us to access the parameter settings with ease.

Remember that the values displayed are not necessarily settings for the server as a whole. Many of those parameters will be specific to the current session, as we observed when we studied the **context** of the parameter in the *Setting configuration parameters for the database server* recipe, in this chapter. That's different from what you experience with many other types of database software, and it is also very useful.

Finding parameters with non-default settings

Often, we need to check which parameters have been changed, or whether our changes have taken effect correctly.

In the previous two recipes, we have seen that parameters can be changed in several ways and with different scopes. You learned how to inspect the value of one parameter or get a full list of parameters.

In this recipe, we will show you how to use SQL capabilities to list only those parameters whose value in the *current session* differs from the system-wide default value.

This list is valuable for several reasons. First, it includes only a few of the 200+ available parameters, so it is more immediate. Also, it is difficult to remember all our past actions, especially in the middle of a long or complicated session.

How to do it...

We write an SQL query that lists all parameter values, excluding those whose current value is either the default or set from a configuration file:

```
postgres=# SELECT name, source, setting, reset_val
                 FROM pg_settings
                 WHERE source != 'default'
                 AND source != 'override'
                 AND setting != reset_val
                 ORDER by 2, 1;
```

The output is displayed here, where reset_val shows what happens if you issue RESET:

```
   name   | source  |   setting   | reset_val
----------+---------+-------------+-----------
 TimeZone | session | Europe/Rome | Etc/UTC
```

How it works...

From pg_settings, you can see which parameters have non-default values, and what the source of the current value is. The SHOW command only tells you the current value but doesn't tell you whether a parameter is set at a non-default value. If the source is a configuration file, then the sourcefile and sourceline columns are also set. These can be useful in understanding where the configuration came from.

There's more...

The setting column of pg_settings shows the current value, but you can also look at the boot_val parameter. boot_val shows the value that was assigned when the PostgreSQL database cluster was initialized (initdb). On a typical configuration, you will see more than 40 parameters that differ from their boot_val parameter, but this is simply because many parameters are configured after initialization. Unfortunately, there is no way to view whether those values are "normal" (or what you would expect them to be) purely by looking at pg_settings.

Setting parameters for particular groups of users

PostgreSQL supports a variety of ways of defining parameter settings for various user groups. This is very convenient, especially for managing user groups that have different requirements. For instance, some users would constantly require a higher value of work_mem to execute larger analytical queries.

How to do it...

Follow these steps to set parameters at various levels as per the requirements:

1. For all users in the saas database, use the following commands:

    ```
    ALTER DATABASE saas
    SET configuration_parameter = value1;
    ```

2. For a user named simon connected to any database, use the following commands:

    ```
    ALTER ROLE simon
    SET configuration_parameter = value2;
    ```

3. Alternatively, you can set a parameter for a user only when they're connected to a specific database, as follows:

    ```
    ALTER ROLE simon
    IN DATABASE saas
    SET configuration_parameter = value3;
    ```

The user won't know that these have been executed specifically for them. The goal is to make the change transparent to the user for its convenience. An advanced user will be able to find the change in the PostgreSQL catalog. Note that these are default settings, and in most cases, they can be overridden if the user requires non-default values.

How it works...

You can set parameters for each of the following:

- A database
- A user (also called a role by PostgreSQL)
- A database and user combination

Each of the parameter defaults is overridden by the one following it.

As an example, we may wish to set the value of the work_mem configuration parameter. In the preceding three SQL statements, the following applies:

- If gianni connects to the saas database, then value1 will apply.
- If simon connects to a database other than saas, then value2 will apply.
- If simon connects to the saas database, then value3 will apply.

PostgreSQL implements this in exactly the same way as if the user had manually issued the equivalent SET statements immediately after connecting.

A basic server configuration checklist

PostgreSQL arrives configured for use on a shared system, though many people want to run dedicated database systems. The PostgreSQL project wishes to ensure that PostgreSQL will play nicely with other server software and will not assume that it has access to full server resources. If you, as the system administrator, know that there is no other important server software running on the system, then you can crank the values up much higher. The default values are conservative, so for many workloads, you will want a much larger value.

Getting ready

Before we start, we need to know two sets of information:

- The size of the physical RAM that will be dedicated to PostgreSQL
- The types of applications for which you will use PostgreSQL

How to do it...

If your database is larger than 128 **Megabytes** (**MB**), then you'll probably benefit from increasing shared_buffers, the physical cache size. You can increase this to a much larger value, but remember that running out of memory induces many problems.

For instance, PostgreSQL is able to store information on disk when the available memory is too small, and it employs sophisticated algorithms to treat each case differently and to place each piece of data on the disk or in memory, depending on each use case.

On the other hand, overstating the amount of available memory confuses such abilities and results in suboptimal behavior. For instance, if the memory is swapped to disk, then PostgreSQL will inefficiently treat all data as if it were the RAM. Another unfortunate circumstance is when the Linux **Out-Of-Memory** (**OOM**) killer terminates one of the various processes spawned by the PostgreSQL server. So it's better to be conservative.

It is good practice to set a low value in your postgresql.conf file and increment slowly to ensure that you get the benefits from each change. See *Chapter 8, Monitoring and Diagnosis*, for monitoring recipes and *Chapter 10, Performance and Concurrency*, for recipes on improving performance.

There's more...

Don't worry about setting effective_cache_size. It is much less important a parameter than you might think. Do not confuse this with the physical database cache, which is the shared_buffers mentioned before in the chapter.

If there is heavy write activity, you may want to set wal_buffers to a much higher value than the default. In fact, wal_buffers is automatically set from the value of shared_buffers, following a rule that fits most cases. However, it is always possible to specify an explicit value that overrides the computation for the very few cases where the rule is not good enough.

If you're doing heavy write activity and/or large data loads, you may want to set max_wal_size and min_wal_size higher than the default to avoid wasting input/output (I/O) in excessively frequent checkpoints. You may also wish to set checkpoint_timeout and checkpoint_completion_target.

PostgreSQL tries its best to decouple query latency from storage performance: synchronous writes are limited to the **Write-Ahead Logging** (**WAL**) directory, and most calculations are carried out in memory buffers. However, there are cases where a query will need to use the disk before returning (for example, to read data that was not already cached), meaning that fewer checkpoints will actually improve query latency.

If your database has many large queries, you may wish to set work_mem to a value higher than the default. However, remember that such a limit applies to *each* node separately in the query plan, so there is a real risk of over-allocating memory, with all the problems we discussed earlier.

Ensure that autovacuum is turned *on* unless you have a very good reason to turn it off; most people don't, because vacuuming needs to happen regardless. See later chapters for more information on autovacuum; in particular, see *Chapter 9, Regular Maintenance*.

Leave the settings as they are for now. Don't fuss too much about getting the settings right. You can change most of them later, so you can take an iterative approach to improving things.

And remember – don't turn off the fsync parameter. It's keeping you safe.

Adding an external module to PostgreSQL

Another strength of PostgreSQL is its **extensibility**. Extensibility was one of the original design goals, going back to the late 1980s. Now, in PostgreSQL 16, there are many additional modules that plug into the core PostgreSQL server.

There are many kinds of additional module offerings, such as the following:

- Additional functions
- Additional data types
- Additional operators
- Additional index types

Some extensions come preloaded with cloud services such as EDB BigAnimal, which preloads pg_stat_statements and pgaudit when selecting a PostgreSQL database. Other extensions are available from a pre-selected list, so move directly to the *Using an installed module/extension* recipe if using PostgreSQL alongside a cloud service.

Many tools and client interfaces work with PostgreSQL without any special installation. Here, we are discussing modules that extend and alter the behavior of the server beyond its normal range of SQL standard syntax, functions, and behavior.

The procedure that makes a module usable is actually a two-step process. First, you install the module's files on your system so that they become available to the database server. Then, you connect to the database (or databases) where you want to use the module and create objects as required. The first step is discussed in this recipe. For the second step, refer to the next recipe, *Using an installed module/extension*.

In this book, we will use the words *extension* and *module* as synonyms, as we did in the PostgreSQL documentation. Note, however, that these are the SQL commands to manage extensions, which we'll describe in the next recipe:

- `CREATE EXTENSION myext;`
- `ALTER EXTENSION myext UPDATE;`

Getting ready

If you want an extension that is not pre-installed, you can choose from a wide range of options from a number of sources, such as the following:

- **Contrib**: The PostgreSQL core includes many functions. There is also an official section for add-in modules, known as contrib modules. They are always available for your database server but are not automatically enabled in every database, because not all users might need them. In PostgreSQL 16, we have 50 such modules. These are documented at http://www.postgresql.org/docs/current/static/contrib.html.
- **PGXN**: This is the **PostgreSQL Extension Network**, a central distribution system dedicated to sharing PostgreSQL extensions. The website started in 2010 as a repository dedicated to the sharing of extension files. As of October 2023, there were 371 extensions from 410 different authors. You can learn more about it at http://pgxn.org/.
- **Separate projects:** These are large external projects, such as PostGIS, offering extensive and complex PostgreSQL modules. For more information, take a look at http://www.postgis.org/.

How to do it...

There are several ways to make additional modules available for your database server, as follows:

- Using a software installer
- Installing from PGXN
- Installing from a manually downloaded package
- Installing from source code

Often, a particular module will be available in more than one way, and users are free to choose their favorite, exactly as with PostgreSQL itself, which can be downloaded and installed through many different procedures.

Installing modules using a software installer

Certain modules are available exactly like any other software packages that you may want to install on your server, such as those provided by the **PostgreSQL Global Development Group (PGDG)** repositories. All main Linux distributions provide packages for the most popular modules such as PostGIS, procedural languages other than those distributed with the core, and so on.

Modules can sometimes be added during installation if you're using a standalone installer application – for example, the OneClick installer, or tools such as rpm, apt-get, and YaST on Linux distributions. The same procedure can also be followed after the PostgreSQL installation when a need for a certain module arises. We will actually describe this case, which is very common.

For example, let's say that you need to manage a collection of Debian package files and that one of your tasks is being able to pick the latest version of one of them. You start by building a database that records all package files. Clearly, you need to store the version number of each package. However, Debian version numbers are much more complex than what we usually call numbers. For instance, Debian may use something such as 16.0-1.pgdg120+1 for a version of the PostgreSQL client package. Despite being complicated, that string follows a clearly defined specification that includes many bits of information, including how to compare two versions to establish which of them is older.

Since this recipe discusses extending PostgreSQL with custom data types and operators, you might have already guessed that we will now consider a custom data type for Debian version numbers that is capable of tasks such as understanding the Debian version number format, sorting version numbers, choosing the latest version number in a given group, and so on. It turns out that somebody else already did the work of creating a required PostgreSQL data type, endowed with all the useful accessories: comparison operators, I/O functions, support for indexes, and maximum/minimum aggregates. All of this has been packaged as a PostgreSQL extension as well as a Debian package (which is not a big surprise), so it is just a matter of installing the postgresql-16 package with a Debian tool such as apt, aptitude, or synaptic. On my laptop, that boils down to the following command:

```
sudo apt install postgresql-16
```

This will download the required package and unpack all the files in the right locations, making them available to my PostgreSQL server.

Installing modules from PGXN

PGXN is a website (http://pgxn.org) that was launched in late 2010 with the purpose of providing a central distribution system for open source PostgreSQL extension libraries. Anybody can register and upload their own module, packaged as an extension archive. The website allows you to browse the available extensions and their versions, either via a search interface or from a directory of packages and usernames.

The simple way is to use a command-line utility called pgxnclient. It can be easily installed in most systems; see the PGXN website for how to do this. Its purpose is to interact with PGXN and take care of administrative tasks, such as browsing available extensions, downloading the package, compiling the source code, installing files in the proper places, and removing installed package files. Alternatively, you can download the extension files from the website and place them in the right location by following the installation instructions.

PGXN is different compared with the official repositories because it serves another purpose. Official repositories usually contain only seasoned extensions because they accept new software only after a certain amount of evaluation and testing. On the other hand, anybody can ask for a PGXN account and upload their own extensions, so there is no filter except requiring that the extension has an open source license and a few files that any extension must have.

Installing modules from source code

In many cases, useful modules may not have full packaging. In these cases, you may need to install the module manually. This isn't very hard, and it's a useful exercise that will help you understand what happens.

Each module will have different installation requirements. There are generally two aspects to installing a module, as follows:

- Building the libraries (only for modules that have libraries)
- Installing the module files in the appropriate locations

You need to follow the instructions for the specific module to build the libraries if any are required. The installation will then be straightforward, and there will usually be a suitably prepared configuration file for the make utility, so you just need to type the following command:

```
make install
```

Each file will be copied to the right directory. Remember that you normally need to be a system superuser in order to install files on the system's directories.

Once a library file is in the directory expected by the PostgreSQL server, it will be loaded automatically as soon as requested by a function.

How it works...

PostgreSQL can dynamically load libraries in the following ways:

- Using the explicit LOAD command in a session

- Using the shared_preload_libraries parameter in postgresql.conf at the server start
- At the session start, using the local_preload_libraries parameter for a specific user, as set using ALTER ROLE

PostgreSQL functions and objects can reference code in these libraries, allowing extensions to be bound tightly to the running server process. The tight binding makes this method suitable for use in even very high-performance applications, and there's no significant difference between additionally supplied features and native features.

Using an installed module/extension

In this recipe, we will explain how to enable an installed module so that it can be used in a particular database. The additional types, functions, and so on will exist only in those databases where we have carried out this step.

As we mentioned in the previous recipe, *Adding an external module to PostgreSQL*, specially packaged modules are called **extensions** in PostgreSQL. They can be managed with dedicated SQL commands.

Getting ready

The pg_available_extensions system view shows one row for each extension that can be installed. All you need to know is the extension name.

How to do it...

Each extension has a unique name, so it is just a matter of issuing the following command:

```
CREATE EXTENSION myextname;
```

This will automatically create all required objects inside the current database.

For security reasons, you need to do this as a database superuser, unless the extension is marked as trusted. In such a case, the user must have CREATEDB privileges in the current database. For instance, if you want to install the dblink extension, type this:

```
CREATE EXTENSION dblink;
```

How it works...

When you issue a CREATE EXTENSION command, the database server looks for a file named EXTNAME.control in the SHAREDIR/extension directory. That file tells PostgreSQL some properties of the extension, including a description, some installation information, and the default version number of the extension (which is unrelated to the PostgreSQL version number). Then, a creation script is executed in a single transaction; thus, if it fails, the database is unchanged. The database server also notes down the extension name and all the objects that belong to it in a catalog table.

Managing installed extensions

In the previous two recipes, we showed you how to install external modules in PostgreSQL to augment its capabilities.

In this recipe, we will show you some more capabilities that are offered by the extension infrastructure.

How to do it...

Here are the steps to manage extensions:

1. First, we list all the available extensions:

    ```
    postgres=# \x on
    Expanded display is on.
    postgres=# SELECT *
    postgres-# FROM pg_available_extensions
    postgres-# ORDER BY name;
    -[ RECORD 1 ]-----+------------------------------------
    name              | adminpack
    default_version   | 2.1
    installed_version |
    comment           | administrative functions for PostgreSQL
    (...)
    -[ RECORD 28 ]----+------------------------------------
    name              | pg_stat_statements
    default_version   | 1.10
    installed_version |
    comment           | track planning and execution statistics of all
    SQL statements executed
    (...)
    ```

In particular, if the dblink extension is installed, we see a record such as this:

```
-[ RECORD 10 ]----+-----------------------------------
name              | dblink
default_version   | 1.2
installed_version | 1.2
comment           | connect to other PostgreSQL databases from
within a database
```

2. Now, we can list all objects in the dblink extension, as follows:

```
postgres=# \x off
Expanded display is off.
postgres=# \dx+ dblink
                  Objects in extension "dblink"
                       Object Description
-------------------------------------------------------
foreign-data wrapper dblink_fdw
function dblink_build_sql_delete(text,int2vector,integer,text[])
function dblink_build_sql_
insert(text,int2vector,integer,text[],text[])
function dblink_build_sql_
update(text,int2vector,integer,text[],text[])
function dblink_cancel_query(text)
function dblink_close(text)
function dblink_close(text,boolean)
function dblink_close(text,text)
(...)
```

3. Objects created as parts of extensions are not special in any way, except that you can't drop them individually. This is done to protect you from mistakes:

```
postgres=# DROP FUNCTION dblink_close(text);
ERROR:  cannot drop function dblink_close(text) because extension dblink requires it
HINT:  You can drop extension dblink instead.
```

4. Extensions might have dependencies, too. The cube and earthdistance contrib extensions are a good example, since the latter depends on the former:

   ```
   postgres=# CREATE EXTENSION earthdistance;
   ERROR:  required extension "cube" is not installed
   HINT: Use CREATE EXTENSION ... CASCADE to install required extensions too.
   postgres=# CREATE EXTENSION earthdistance CASCADE;
   NOTICE:  installing required extension "cube"
   CREATE EXTENSION
   ```

   Note how the CASCADE keyword was used to automatically create all the other extensions that the extension being created depends on, as clearly indicated by the HINT message.

5. As you can reasonably expect, dependencies are considered when dropping objects, just as for other objects:

   ```
   postgres=# DROP EXTENSION cube;
   ERROR:  cannot drop extension cube because other objects depend on it
   DETAIL:  extension earthdistance depends on extension cube
   HINT:  Use DROP ... CASCADE to drop the dependent objects too.
   postgres=# DROP EXTENSION cube CASCADE;
   NOTICE:  drop cascades to extension earthdistance
   DROP EXTENSION
   ```

How it works...

The pg_available_extensions system view shows one row for each extension control file in the SHAREDIR/extension directory (see the *Using an installed module/extension* recipe). The pg_extension catalog table records only extensions that have already been created.

The psql command-line utility provides the \dx meta-command to examine the extensions. It supports an optional plus sign (+) to control verbosity, and an optional pattern for the extension name to restrict its range. Consider the following command:

```
\dx+ db*
```

This will list all extensions whose names start with db, together with all their objects.

The CREATE EXTENSION command creates all objects belonging to a given extension and then records the dependency of each object on the extension in pg_depend. That's how PostgreSQL can ensure that you cannot drop one such object without dropping its extension.

The extension control file admits an optional line, requires, that names one or more extensions on which the current one depends. The implementation of dependencies is still quite simple; for instance, there is no way to specify a dependency on a specific version number of other extensions.

As a general PostgreSQL rule, the CASCADE keyword tells the DROP command to delete all objects that depend on cube, which is the earthdistance extension in this example.

There's more...

Another system view, pg_available_extension_versions, shows all the versions that are available for each extension. It can be valuable when there are multiple versions of the same extension available at the same time – for example, when preparing for an extension upgrade.

When a more recent version of an already installed extension becomes available to the database server – for instance, because of a distribution upgrade that installs updated package files – the superuser can perform an upgrade by issuing the following command:

```
ALTER EXTENSION mytext UPDATE TO '1.1';
```

This assumes that the author of the extension taught it how to perform the upgrade.

Extensions interact with logical backup and restore nicely, a topic that will be fully discussed in *Chapter 11*, *Backup and Recovery*. As an example, if your database contains the cube extension, then you will surely want a single line (CREATE EXTENSION cube) in the dump file instead of lots of lines recreating each object individually, which is inefficient and also dangerous.

The use of CASCADE in a CREATE statement only applies to extensions because, for other object types, the dependency is not predefined in the object metadata and only exists after creating a specific object (for example, a **Foreign Key** (**FK**)).

Remember that CREATE EXTENSION ... CASCADE will only work if all the extensions it tries to install have already been placed in the appropriate location.

Learn more on Discord

To join the Discord community for this book – where you can share feedback, ask questions to the author, and learn about new releases – follow the QR code below:

https://discord.gg/pQkghgmgdG

4
Server Control

The recipes in this chapter will show you how to control the database server directly. Database servers in the cloud do not give access to the privileges that are required to perform many of the actions listed in this chapter, but there are things worth considering if you want to understand what is happening under the hood.

This chapter includes the following recipes:

- An overview of controlling the database server
- Starting the database server manually
- Stopping the server safely and quickly
- Stopping the server in an emergency
- Reloading server configuration files
- Restarting the server quickly
- Preventing new connections
- Restricting users to only one session each
- Pushing users off the system
- Deciding on a design for multitenancy
- Using multiple schemas
- Giving users their own private databases
- Running multiple servers on one system
- Setting up a connection pool
- Accessing multiple servers using the same host and port
- Running multiple PgBouncers on the same port to leverage multiple cores

An overview of controlling the database server

PostgreSQL consists of a set of server processes, the group leader of which is named the **postmaster**, though that name is not visible as a process title in later versions. Starting the server is the act of creating these processes, and stopping the server is the act of terminating those processes.

Each postmaster listens for client connection requests on a defined port number. Multiple concurrently running postmasters cannot share that port number. The port number is often used to uniquely identify a particular postmaster and, hence, the database server that it leads.

When we start a database server, we refer to a data directory, which contains the heart and soul – or at least the data – of our database. Subsidiary tablespaces may contain some data outside the main data directory, so the data directory is just the main central location and not the only place where data for that database server is held. Each running server has, at a minimum, one data directory; one data directory can have, at most, one running server (or instance).

To perform any action on a database server, we must know the data directory for that server. The basic actions we can perform on the database server are start and stop. We can also perform a restart, although that is just a stop followed by a start. In addition, we can reload the server, which means that we can reread the server's configuration files.

We should also mention a few other points.

The default port number for PostgreSQL is 5432. This has been registered with the **Internet Assigned Numbers Authority** (**IANA**), so it should already be reserved for PostgreSQL's use in most places. Because each PostgreSQL server requires a distinct port number, the normal convention is to use subsequent numbers for any additional server – for example, 5433 and 5434. Subsequent port numbers may not be as easily recognized by the network infrastructure, which may, in some cases, make life more difficult for you in large enterprises, especially in more security-conscious ones.

Port number 6432 has been registered with IANA for **PgBouncer**, the connection pooler that we will describe in the *Setting up a connection pool* recipe. This only happened recently, and many installations use non-standard port numbers such as 6543 because they were deployed earlier.

A database server is also sometimes referred to as a **database cluster**. I don't recommend this term for normal usage, as it makes people think about multiple nodes and not one database server on one system.

Starting the database server manually

Typically, the PostgreSQL server will start automatically when the system boots. You may opt to stop and start the server manually, or you may need to start it or shut it down for various operational reasons.

Getting ready

First, you need to understand the difference between the *service* and the *server*. The word *server* refers to the database server and its processes. The word *service* refers to the operating system wrapper that the server gets called by. The server works in essentially the same way on every platform, whereas each operating system and distribution has its own concept of a service.

Moreover, the way services are managed has changed recently; for instance, at the time of writing, most Linux distributions have adopted the `systemd` service manager. This means that you need to know which *distribution* and *release* you are using to find the correct variant of this recipe.

With `systemd`, a PostgreSQL server process is represented by a **service unit**, which is managed via the `systemctl` command. The systemd command syntax is the same on all distributions, but the name of the service unit isn't. For example, it will have to be adjusted depending on your distribution.

In other cases, you need to type the actual **data directory** path as part of the command line to start the server. More information on how to find out what is in the data directory path can be found in the *Locating the database server files* recipe of *Chapter 2, Exploring the Database*.

How to do it...

On each platform, there is a specific command to start the server.

If you are using a modern Linux distribution, then you are probably using `systemd`. In this case, PostgreSQL can be started with the following command:

```
sudo systemctl start SERVICEUNIT
```

This must be issued with operating system superuser privileges, after replacing `SERVICEUNIT` with the appropriate `systemd` service unit name.

The `systemctl` command must always be issued with operating system superuser privileges. Remember that, throughout this book, we will always prepend `systemctl` invocations with `sudo`.

There are a couple of things to keep in mind:

- The above command will only work if the user executing the command has been previously granted the appropriate sudo privileges by the system administrator.
- If the command is executed from a superuser account, then the sudo keyword is unnecessary, although not harmful.

As we mentioned previously, the service unit name depends on what distribution you are using, as follows:

- On Ubuntu and Debian, the service unit's name is as follows:

```
postgresql@RELEASE-CLUSTERNAME
```

- For each database server instance, there is another service unit called postgresql, which can be used to manage all the database servers at once. Therefore, you can issue the following command:

```
sudo systemctl start postgresql
```

- To start all the available instances, and to start only the default version *16* instance, use the following command:

```
sudo systemctl start postgresql@16-main
```

- Default Red Hat/Fedora packages call the service unit simply postgresql, so the syntax is as follows:

```
sudo systemctl start postgresql
```

- Red Hat/Fedora packages from the PostgreSQL Yum repository create a service unit called postgresql-RELEASE, so we can start version 16 as follows:

```
sudo systemctl start postgresql-16
```

As we noted previously, systemctl is part of systemd, which is only available on Linux and is normally used by most of the recent distributions.

The following commands can be used where systemd is not available.

On Debian and Ubuntu releases, you must invoke the PostgreSQL-specific pg_ctlcluster utility, as follows:

```
pg_ctlcluster 16 main start
```

This command will also work when systemd is available; it will just redirect the start request to systemctl and print a message on the screen so that next time, you will remember to use systemctl directly.

For older versions of Red Hat/Fedora/Rocky/AlmaLinux, you can use the following command:

```
sudo service postgresql-16 start
```

For Windows, the command is as follows:

```
net start postgresql-x64-16
```

For Red Hat/Fedora, you can also use the following command:

```
pg_ctl -D $PGDATA start
```

Here, PGDATA is set to the data directory path.

This command works on most distributions, including **macOS**, **Solaris**, and **FreeBSD**, although bear the following points in mind:

- It is recommended that you use, whenever possible, the distribution-specific syntax we described previously.
- You may have to specify the full path to the pg_ctl executable if it's not in your path already. This is normally the case with multi-version directory schemes such as Debian/Ubuntu, where distribution-specific scripts pick the appropriate executable for your version.

How it works...

On Ubuntu/Debian, the pg_ctlcluster wrapper is a convenient utility that allows multiple servers to coexist more easily, which is especially good when you have servers with different versions. This was invented by Debian and is not found on other PostgreSQL distributions. This capability is very useful and is transposed on systemd, as shown in the examples using @ in the name of the service unit, where @ denotes the usage of a service file template.

Another interesting systemd feature is the capability to enable/disable a service unit to specify whether it will be started automatically on the next boot, with syntax such as the following:

```
sudo systemctl enable postgresql@16-main
```

This can be very useful for setting the appropriate behavior based on the purpose of each instance.

A similar feature is implemented on Ubuntu and Debian via the start.conf file, which is located next to the other configuration files (that is, in the same directory). Apart from the informational comments, it contains only a single word, each with its specific meaning. The start.conf file can contain any one of the following words:

- auto: The server will be started automatically on boot. This is the default when you create a new server. It is suitable for frequently used servers, such as those powering live services or those used for everyday development activities.
- manual: The server will not be started automatically on boot, but it can be started with pg_ctlcluster. This is suitable for custom servers that are seldom used.
- disabled: The server is not supposed to be started. This setting only acts as protection from starting the server accidentally. The pg_ctlcluster wrapper won't let you start it, but a skilled user can easily bypass this protection.

If you need to reserve a port for a server that's not managed by pg_ctlcluster, such as when you're compiling directly from the source code, then you can create a cluster with start.conf set to disabled and then use its port. Any new servers will be assigned different ports.

If PostgreSQL's startup is taking time and you wish to check its progress, you can refer to the log files. The frequency of these log updates is based on the log_startup_progress_interval setting. PostgreSQL will log a message in the log file at intervals defined by this setting. By default, this parameter is set to 10 seconds, which means if PostgreSQL's startup takes longer than 10 seconds, it will begin logging information about each stage of the startup process in the log files.

Stopping the server safely and quickly

There are several modes you can use to stop the server, depending on the level of urgency. We'll compare the effects in each mode.

How to do it...

There are two variants: with and without systemd. This is similar to the previous recipe, *Starting the database server manually*, which we'll refer to for further information. For example, what is the exact name of the systemd service unit for a given database server on a given GNU/Linux distribution?

When using systemd, you can stop PostgreSQL using *fast* mode by issuing the following, after replacing SERVICEUNIT with the appropriate system service unit name:

```
sudo systemctl stop SERVICEUNIT
```

If systemd is not available and you are using Debian or Ubuntu the command is as follows, which applies to the default version 16 instance:

```
pg_ctlcluster 16 main stop -m fast
```

Fast mode has been the default since PostgreSQL 9.5 the previous default was to use smart mode, meaning we have to *wait for all users to finish before we exit*. This can take a very long time, and all while new connections are refused.

On other Linux/Unix distributions, you can issue a database server stop command using fast mode, as follows:

```
pg_ctl -D datadir -m fast stop
```

How it works...

When you do a fast stop, all the users have their transactions aborted and all the connections are disconnected. This is not very polite to users, but it still treats the server and its data with care, which is good.

PostgreSQL is similar to other database systems in that it creates a shutdown checkpoint before it closes. This means that the startup that follows will be quick and clean. The more work the checkpoint has to do, the longer it will take to shut down.

One difference between PostgreSQL and some other **RDBMSs (Relational Database Management Systems)**, such as Oracle, DB2, and SQL Server, is that the transaction rollback is very quick. On those other systems, if you shut down the server in a mode that rolls back transactions, it can cause the shutdown to take a while, possibly a very long time. This difference is for internal reasons and isn't in any way unsafe. Debian and Ubuntu's pg_ctlcluster supports the --force option, which is nice because it attempts a fast shutdown first; if that fails, it performs an immediate shutdown. After that, it kills the postmaster.

See also

The technology that provides immediate rollback for PostgreSQL is called **Multiversion Concurrency Control (MVCC)**. More information on this is provided in the *Identifying and fixing bloated tables and indexes* recipe in *Chapter 9, Regular Maintenance*.

Stopping the server in an emergency

If nothing else is working, we may need to stop the server quickly, without caring about disconnecting the clients gently.

Break the glass in case of emergency!

How to do it...

Follow these steps to stop the server:

1. The basic command to perform an emergency stop on the server is as follows:

   ```
   pg_ctl -D $PGDATA stop -m immediate
   ```

2. On Debian/Ubuntu, you can also use the following command:

   ```
   pg_ctlcluster 16 main stop -m immediate
   ```

As we mentioned in the previous recipe, this is just a wrapper around `pg_ctl`. From this example, we can see that it can pass through the `-m immediate` option.

In the previous recipe, we saw examples where the `systemctl` command was used to stop a server safely; however, this command cannot be used to perform an emergency stop.

How it works...

When you do an immediate stop, all the users have their transactions aborted, and all their connections are disconnected. There is no clean shutdown, nor is there polite warning of any kind.

An immediate mode stop is similar to a database crash. Some cached files will need to be rebuilt, and the database itself will need to undergo crash recovery when it comes back up.

Note that for **DBAs** (**Database Administrators**) with Oracle experience, immediate mode is the same thing as a `shutdown abort`. The PostgreSQL immediate mode stop is *not* the same thing as `shutdown immediate` on Oracle.

Reloading server configuration files

Some PostgreSQL configuration parameters can only be changed by reloading the entire configuration files. Note that in some cloud-based database services, reload occurs automatically when parameters are changed, so this recipe is not relevant.

How to do it...

There are two variants of this recipe, depending on whether you are using `systemd`. This is similar to the previous recipes in this chapter, especially the *Starting the database server manually* recipe. More details are provided there, such as the exact names of the `systemd` service units, depending on which database server you want to reload, and which GNU/Linux distribution you are working on.

With `systemd`, configuration files can be reloaded with the following syntax:

```
sudo systemctl reload SERVICEUNIT
```

Here, SERVICEUNIT must be replaced with the exact name of the `systemd` service unit for the server(s) that you want to reload.

Otherwise, on each platform, there is a specific command you can use to reload the server without using `systemd`. These commands are as follows:

- On Ubuntu and Debian, you can issue the following command:

    ```
    pg_ctlcluster 16 main reload
    ```

- On older versions of Red Hat/Fedora, you must use the following command:

    ```
    service postgresql-16 reload
    ```

- You can also use the following command on Red Hat/Fedora:

    ```
    pg_ctl -D /var/lib/pgsql-16/data reload
    ```

The above command also works on macOS, Solaris, and FreeBSD, where you must replace /var/lib/pgsql-16/data with your actual data directory if it's different.

On all platforms, you can also reload the configuration files while still connected to PostgreSQL. If you are a superuser, or the privilege for this function has been granted to you, this can be done with the following command:

```
postgres=# SELECT pg_reload_conf();
```

The output is rather short:

```
pg_reload_conf
----------------
t
```

This function is also often executed from an admin tool, such as **pgAdmin**.

If you do this, you should realize that it's possible to implement a new authentication rule that is violated by the current session. It won't force you to disconnect, but when you do disconnect, you may not be able to reconnect.

Any error in a configuration file will be reported in the message log, so we recommend that you look there immediately after reloading. You will quickly notice (and fix!) syntax errors in the parameter file because they prevent any logins from occurring before reloading. Other errors, such as typos in parameter names or wrong units, will only be reported in the log; moreover, only some non-syntax errors will prevent you from reloading the whole file, so it's best to always check the log.

How it works...

To reload the configuration files, we must send the SIGHUP signal to the postmaster, which then passes the files to all the connected backends. That's why some people call reloading the server *sigh-up-ing*.

If you look at the pg_settings catalog table, you'll see that there is a column named context. Each setting has a time and a place where it can be changed. Some parameters can only be reset by a server reload, so the value of context for those parameters will be sighup. Here are a few of the parameters you may want to change during server operation (there are others, however; please refer to *Chapter 3, Server Configuration*):

```
postgres=# SELECT name, setting, unit
                  ,(source = 'default') as is_default
           FROM pg_settings
           WHERE context = 'sighup'
           AND (name like '%delay' or name like '%timeout')
           AND setting != '0';
           name                       | setting | unit | is_default
-------------------------------+---------+------+------------
 authentication_timeout        | 60      | s    | t
 autovacuum_vacuum_cost_delay  | 20      | ms   | t
 bgwriter_delay                | 200     | ms   | f
 checkpoint_timeout            | 300     | s    | f
 max_standby_archive_delay     | 30000   | ms   | t
 max_standby_streaming_delay   | 30000   | ms   | t
 wal_receiver_timeout          | 60000   | ms   | t
 wal_sender_timeout            | 60000   | ms   | t
 wal_writer_delay              | 200     | ms   | t
(9 rows)
```

There's more...

Since reloading the configuration file is achieved by sending the SIGHUP signal, we can only reload the configuration file for a single backend using the `kill` command. As you may expect, you may get some strange results from doing this, so don't try this at home.

First, find the **PID (Process ID)** of the backend using `pg_stat_activity`. Then, from the operating system prompt, issue the following command:

```
kill -SIGHUP <pid>
```

Alternatively, we can do both at once, as shown in the following command:

```
kill -SIGHUP <pid>\
   && psql -t -c "select pid from pg_stat_activity limit 1";
```

This is only useful with a sensible WHERE clause.

Restarting the server quickly

Some of the database server parameters require you to stop and start the server again fully. Doing this as quickly as possible can be very important in some cases. The best time to do this is usually a quiet time, with lots of planning, testing, and forethought.

How to do it...

Many of the recipes in this chapter are presented in two forms: one with systemd and one without. This may look repetitive or boring, but it's unavoidable because introducing a new system does not automatically eliminate all existing alternatives or migrate old installations to new ones.

As we mentioned previously, you can find further systemd details, including details on service unit names, in the previous recipe, *Starting the database server manually*.

A PostgreSQL server that's managed by systemd can be restarted in *fast* mode by issuing the following command:

```
sudo systemctl restart SERVICEUNIT
```

As we mentioned previously, change SERVICEUNIT to the appropriate service unit name – for example, postgresql@16-main for a PostgreSQL 16 cluster running in Debian or Ubuntu.

If systemd is not available, then you can use the following syntax:

```
pg_ctlcluster 16 main restart -m fast
```

The basic command to restart the server is as follows:

```
pg_ctl -D $PGDATA restart -m fast
```

A `restart` is just a stop that's followed by a start, so it sounds very simple. In many cases, it will be simple, but there are times when you'll need to restart the server while it is fairly busy. That's when we need to start performing some tricks to make that restart happen quicker.

First, the stop that's performed needs to be a `fast` stop. If we do a default or a `smart` stop, then the server will just wait for everyone to finish. If we do an immediate stop, then the server will crash, and we will need to crash-recover the data, which will be slower overall.

The running database server has a cache full of data blocks, many of which are dirty. PostgreSQL is similar to other database systems in that it creates a shutdown checkpoint before it closes. This means that the startup that follows will be quick and clean. The more work the checkpoint has to do, the longer it will take to shut down.

The actual shutdown will happen much quicker if we issue a normal checkpoint first, as the shutdown checkpoint will have much less work to do. So, flush all the dirty shared buffers to disk with the following command, issued by a database superuser:

```
psql -c "CHECKPOINT"
```

The next consideration is that once we restart, the database cache will be empty again and will need to refresh itself. The larger the database cache, the longer it will take for the cache to get warm again, and 30 to 60 minutes is not uncommon before returning to full speed. So, what was a simple restart can have a large business impact if handled badly.

There's more...

There is an extension called `pgfincore` that implements a set of functions to manage PostgreSQL data pages in the operating system's file cache. One possible use is to preload some tables so that PostgreSQL will load them quicker when requested. The general idea is that you can provide more detailed information for the operating system cache so that it can behave more efficiently.

Some distributions include a prebuilt `pgfincore` package, which makes installation easier.

There is also a `contrib` module called `pg_prewarm`, which addresses a similar problem. While there is some overlap with `pgfincore`, the feature sets are not the same; for instance, `pgfincore` can operate on files that aren't in the shared buffer cache, and it can also preload full relations with only a few system calls while taking the existing cache into account. Conversely, `pg_prewarm` can operate on the PostgreSQL shared buffer cache, and it also works on Windows.

Preventing new connections

In certain emergencies, you may need to lock down a server completely, or just prevent specific users from accessing a database. It's hard to foresee all the situations where you may need to do this, so we will present a range of options.

How to do it...

Connections can be prevented in several ways, as follows:

1. Pause and resume the session pool. See the *Setting up a connection pool* recipe, later in this chapter, on controlling connection pools.
2. Stop the server (although this is not recommended)! See the *Stopping the server safely and quickly* and the *Stopping the server in an emergency* recipes.
3. Restrict the connections for a specific database to 0 by setting the connection limit to 0:

   ```
   ALTER DATABASE foo_db CONNECTION LIMIT 0;
   ```

   This will limit normal users from connecting to that database, although it will still allow superuser connections.

4. Restrict the connections for a specific user to 0 by setting the connection limit to 0 (see the *Restricting users to only one session each* recipe):

   ```
   ALTER USER foo CONNECTION LIMIT 0;
   ```

   This will prevent normal users from connecting to that database, but it will still allow connections if the user is a superuser; so, luckily, you cannot shut yourself out accidentally.

5. Change the **Host-Based Authentication** (**HBA**) file to refuse all incoming connections, and then reload the server.
6. Create a new file called pg_hba_lockdown.conf and add the following two lines to it. This puts rules in place that will completely lock down the server, including superusers. Note that this is a serious and drastic action:

   ```
   # TYPE  DATABASE  USER  ADDRESS     METHOD
     local all       all               reject
     host  all       all   0.0.0.0/0   reject
   ```

If you still want superuser access, then try something such as the following:

```
# TYPE    DATABASE    USER        ADDRESS      METHOD
  local   all         postgres                 peer
  local   all         all                      reject
  host    all         all         0.0.0.0/0    reject
```

This will prevent connections to the database by any user except the `postgres` operating system user ID, which connects locally to any database. It's worth keeping the header line just for that reason. The `peer` method should be replaced with other authentication methods if a more complex configuration is in use.

7. Copy the existing `pg_hba.conf` file to `pg_hba_access.conf` so that it can be replaced later if required.
8. Copy `pg_hba_lockdown.conf` to `pg_hba.conf`.
9. Reload the server by following the *Restarting the server quickly* recipe.

How it works...

The `pg_hba.conf` file is where we specify the host-based authentication rules. We do not specify the authentications themselves; we just specify which authentication mechanisms will be used. This is the top-level set of rules for PostgreSQL authentication. These rules are specified in a file and applied by the postmaster process when connections are attempted. To prevent denial-of-service attacks, the HBA rules never involve database access, so we do not know whether a user is a superuser. As a result, you can lock out all users, but note that you can always re-enable access by editing the file and reloading it.

Restricting users to only one session each

If resources need to be closely controlled, you may wish to restrict users so that they can only connect to the server once, at most. The same technique can be used to prevent connections entirely for that user.

How to do it...

We can restrict users to only one connection using the following command:

```
postgres=# ALTER ROLE fred CONNECTION LIMIT 1;
ALTER ROLE
```

Chapter 4

This will then cause any additional connections to receive the following error message:

```
FATAL: too many connections for role "fred"
```

You can eliminate this restriction by setting the value to -1.

It's possible to set the limit to 0 or any positive integer. You can set this to a number other than max_connections, although it is up to you to make sense of that if you do.

Setting the value to 0 will completely restrict normal connections. Note that even if you set the connection limit to 0 for superusers, they will still be able to connect.

How it works...

The connection limit is applied during the session connection. Raising this limit will never affect any connected users. Lowering the limit doesn't have any effect either unless they try to disconnect and reconnect.

So, if you lower the limit, you should immediately check whether there are more sessions connected than the new limit you just set. Otherwise, you may come across some surprises if there is a crash:

```
postgres=> SELECT rolconnlimit
             FROM pg_roles
             WHERE rolname = 'fred';
rolconnlimit
--------------
            1
(1 row)
postgres=> SELECT count(*)
             FROM pg_stat_activity
             WHERE usename = 'fred';
count
-------
     2
(1 row)
```

If you have more connected sessions than the new limit, you can ask users to politely disconnect, or you can apply the next recipe, *Pushing users off the system*.

Users can't raise or lower their connection limit, just in case you are worried that they might be able to override this somehow.

Pushing users off the system

Sometimes, we may need to remove groups of users from the database server for various operational reasons. Let's learn how to do this.

How to do it...

You can terminate a user's session with the pg_terminate_backend() function, which is included with PostgreSQL. This function takes the PID, or the process ID, of the user's session on the server. This process is known as the **backend**, and it is a different system process from the program that runs the client.

To find the PID of a user, we can look at the pg_stat_activity view. We can use it in a query, like this:

```
SELECT pg_terminate_backend(pid)
FROM pg_stat_activity
WHERE ...
```

There are a couple of things to note if you run this query. If the WHERE clause doesn't match any sessions, then you won't get any output from the query. Similarly, if it matches multiple rows, you will get a fairly useless result – that is, a list of Boolean true values. Unless you are careful enough to exclude your session from the query, you will disconnect yourself! What's even funnier is that you'll disconnect yourself halfway through disconnecting the other users. This is because the query will run pg_terminate_backend() in the order in which sessions are returned from the outer query.

Therefore, I suggest a safer and more useful query that gives a useful response in all cases, which is as follows:

```
postgres=# SELECT count(pg_terminate_backend(pid))
FROM pg_stat_activity
WHERE usename NOT IN
(SELECT usename
FROM pg_user
WHERE usesuper);
 count
-------
     1
```

The preceding code assumes that superusers are performing administrative tasks.

Some other good filters are as follows:

- `WHERE application_name = 'myappname'`
- `WHERE wait_event_type IS NOT NULL AND wait_event_type != 'Activity'`
- `WHERE state = 'idle in transaction'`
- `WHERE state = 'idle'`

How it works...

The `pg_terminate_backend()` function sends a signal (`SIGINT` or `SIGTERM`) directly to the operating system process for that session.

The session may have closed by the time `pg_terminate_backend()` is called. As PID numbers are assigned by the operating system, you may try to terminate a given session (let's call it *session A*), but you terminate another session while doing so (let's call it *session B*).

Here is how it could happen. Suppose you take note of the PID of session A and decide to disconnect it. Before you issue `pg_terminate_backend()`, session A disconnects, and right after, a new session, session B, is given the same PID. So, when you terminate that PID, you hit session B instead.

On the one hand, you need to be careful. On the other hand, this case is really unlikely and is only mentioned for completeness. For this to happen, the following events must occur as well:

- One of the sessions you are trying to close must terminate independently in the very short interval between the moment `pg_stat_activity` is read and the moment `pg_terminate_backend()` is executed.
- Another session on the same database server must be started in the even shorter interval between the old session closing and the execution of `pg_terminate_backend()`.
- The new session must get the same PID value as the old session, which is less than a 1 in 32,000 chance on a 32-bit Linux machine.

Nonetheless, probability theory is tricky, even for experts. Therefore, it's better to be aware that there is a tiny risk, especially if you use the query many times per day over a long period, in which case the probability of getting caught at least once builds up.

It's also possible that new sessions could start after we get the list of active sessions. There's no way to prevent this, other than by following the *Preventing new connections* recipe.

Finally, remember that superusers can terminate any session, while a non-superuser can only terminate a session that belongs to the same user.

Deciding on a design for multitenancy

There are many reasons why we may want to split groups of tables or applications: security, resource control, convenience, and so on. Whatever the reason, we often need to separate groups of tables (I avoid saying the word *database*, just to avoid various kinds of confusion).

This topic is frequently referred to as **multitenancy**, although this is not a fully accepted term yet.

The purpose of this recipe is to discuss the various options or strategies for implementing multitenancy in a database environment so that we can move on to other, more detailed recipes.

How to do it...

If you want to run multiple physical databases on one server, then you have five main options, which are as follows:

- **Option 0 (default)**: Run separate PostgreSQL instances in separate virtual machines on the same physical server. This is the default option in cloud systems such as EDB BigAnimal, as well as in on-premises deployments such as VMware or Kubernetes-based services.
- **Option 1**: Run multiple sets of tables in different schemas in one database of a PostgreSQL instance (covered in the *Using multiple schemas* recipe).
- **Option 2**: Run multiple databases in the same PostgreSQL instance (covered in the *Giving users their own private databases* recipe).
- **Option 3**: Run multiple PostgreSQL instances on the same virtual/physical system (covered in the *Running multiple servers on one system* recipe).
- **Option 4**: Place all the data in one schema and one database but use **row-level security** (**RLS**) to ensure that users only have access to some subset of the data. This provides security but not resource control or convenience.

Option 0 can be applied using virtualization technology, which is outside the scope of this book. Having said that, this is the "default" mode.

Which is best? Well, that's certainly a question many people ask and an area where many views exist. The answer lies in looking at the specific requirements, which are as follows:

- If our goal is to separate physical resources, then option 0 works best, although option 3 is also viable. Separate database servers can easily be assigned different disks, individual memory allocations can be assigned, and we can take the servers up or down without impacting the others.

- If our goal is security, then option 2 is sufficient.
- If our goal is merely to separate tables for administrative clarity, then option 1 or option 2 can be useful.

Option 2 allows complete separation for security purposes. However, this does prevent someone with privileges on both groups of tables from performing a join between those tables. So, if there is a possibility of future cross-analytics, it might be worth considering option 1. However, it may also be argued that such analytics should be carried out on a separate data warehouse, not by co-locating production systems.

Option 3 is difficult to implement in many of the PostgreSQL distributions; the default installation uses a single location for a database, making it a little harder to configure that option. Ubuntu/Debian handles this aspect particularly well, making it more attractive in that environment.

Option 4 is covered in the *Granting user access to specific rows* recipe in *Chapter 6, Security*.

How it works...

I've seen people use PostgreSQL with thousands of databases, but it is my opinion that the majority of people use only one database, such as `postgres` (or at least, only a few databases). I've also seen people use a great many schemas.

One thing you will find is that almost all admin GUI tools become significantly less useful if there are hundreds or thousands of items to display. In most cases, administration tools use a tree view, which doesn't cope gracefully with a large number of items.

Using multiple schemas

We can separate groups of tables into namespaces, referred to as **schemas** by PostgreSQL. In many ways, they can be thought of as being similar to directories, although that is not a precise description, and schemas are not arranged in a hierarchy.

Getting ready

Make sure you've read the *Deciding on a design for multitenancy* recipe so that you're certain that this is the route you wish to take. Other options exist, and they may be preferable in some cases.

How to do it...

Follow these steps:

1. Schemas can easily be created using the following commands:

   ```
   CREATE SCHEMA finance;
   CREATE SCHEMA sales;
   ```

2. Then, we can create objects directly within those schemas using *fully qualified* names, like this:

   ```
   CREATE TABLE finance.month_end_snapshot (.....)
   ```

   The default schema where an object is created is known as current_schema. We can find out what our current schema is by using the following query:

   ```
   postgres=# select current_schema;
   ```

   This returns an output similar to the following:

   ```
   current_schema
   ---------------
   public
   (1 row)
   ```

3. When we access database objects, we use the user-set table search_path parameter to identify the schemas to search for. current_schema is the first schema in the search_path parameter. There is no separate parameter for current_schema.

   So, if we only want to let a specific user look at certain sets of tables, we can modify their search_path parameter. This parameter can be set for each user so that the value will be set when they connect. The SQL queries for this would be something like this:

   ```
   ALTER ROLE fiona SET search_path = 'finance';
   ALTER ROLE sally SET search_path = 'sales';
   ```

   The public schema is not mentioned on search_path, so it will not be searched. All the tables that are created by fiona will go into the finance schema by default, whereas all the tables that are created by sally will go into the sales schema by default.

4. The users for finance and sales will be able to see that the other schema exists and change search_path to use it, but we will be able to GRANT or REVOKE privileges so that they can neither create objects nor read data in other people's schemas:

```
GRANT ALL ON SCHEMA finance TO fiona;
GRANT ALL ON SCHEMA sales TO sally;
```

An alternative technique is to grant user-create privileges to only one schema but grant usage rights to all other schemas. We can set up this arrangement like this:

```
GRANT USAGE ON SCHEMA finance TO fiona;
GRANT CREATE ON SCHEMA finance TO fiona;
GRANT USAGE ON SCHEMA sales TO sally;
GRANT CREATE ON SCHEMA sales TO sally;
GRANT USAGE ON SCHEMA sales TO fiona;
GRANT USAGE ON SCHEMA finance TO sally
```

In PostgreSQL, there is another technique to create a schema using the AUTHORIZATION option. This option designates a role as the owner of the schema and grants usage privileges on other schemas. Below are the commands:

```
CREATE SCHEMA finance AUTHORIZATION fiona;
GRANT USAGE ON SCHEMA sales TO fiona;
CREATE SCHEMA sales AUTHORIZATION sally;
GRANT USAGE ON SCHEMA finance TO sally;
```

5. Note that you need to grant the privileges for usage on the schema, as well as specific rights on the objects in the schema. So you will also need to issue specific grants for objects, as shown here:

```
GRANT SELECT ON month_end_snapshot TO public;
```

You can also set default privileges so that they are picked up when objects are created, by using the following command:

```
ALTER DEFAULT PRIVILEGES FOR USER fiona IN SCHEMA finance
GRANT SELECT ON TABLES TO PUBLIC;
```

Please note if you have created a schema with the AUTHORIZATION option for a specific user, then the above command won't be needed.

How it works...

Earlier, I mentioned that schemas work like directories, or at least a little.

The PostgreSQL concept of search_path is similar to the concept of a PATH environment variable.

The PostgreSQL concept of the current schema is similar to the concept of the current working directory. There is no cd command to change the directory. The current working directory is changed by altering search_path.

A few other differences exist; for example, PostgreSQL schemas are not arranged in a hierarchy like filesystem directories are.

Many people create a user with the same name as the schema to make this work in a way similar to other RDBMSs, such as Oracle.

Both the finance and sales schemas exist within the same PostgreSQL database, and they run on the same database server. They use a common buffer pool, and many global settings tie the two schemas fairly close together.

Giving users their own private databases

Separating data and users is a key part of administration. There will always be a need to give users a private, secure, or simply risk-free area (sandbox) to use the database. Here's how.

Getting ready

Again, make sure you've read the *Deciding on a design for multitenancy* recipe so that you're certain this is the route you wish to take. Other options exist, and they may be preferable in some cases.

How to do it...

Follow these steps to create a database with restricted access for a specific user:

1. We can create a database for a specific user with some ease. From the command line, as a superuser, we can do the following:

   ```
   postgres=# create user fred;
   CREATE ROLE
   postgres=# create database fred owner fred;
   CREATE DATABASE
   ```

2. As database owners, users have login privileges, so they can connect to any database by default. There is a command named ALTER DEFAULT PRIVILEGES for this; however, this does not currently apply to databases, tablespaces, or languages. The ALTER DEFAULT PRIVILEGES command also only currently applies to roles (that is, users) that already exist.

So, we need to revoke the privilege to connect to our new database from everybody except the designated user. There isn't a REVOKE ... FROM PUBLIC EXCEPT command. Therefore, we need to revoke everything and then just re-grant everything we need, all in one transaction, as shown in the following code:

```
postgres=# BEGIN;
BEGIN
postgres=# REVOKE connect ON DATABASE fred FROM public;
REVOKE
postgres=# GRANT connect ON DATABASE fred TO fred;
GRANT
postgres=# COMMIT;
COMMIT
postgres=# create user bob;
CREATE ROLE
```

3. Then, try to connect as bob to the fred database:

```
os $ psql -U bob fred
psql: FATAL:  permission denied for database "fred"
DETAIL:  User does not have CONNECT privilege.
```

This is exactly what we wanted.

How it works...

If you didn't catch it before, PostgreSQL allows transactional DDL (Data Definition Language) in most places, so either the REVOKE and GRANT commands in the preceding section work or neither works. This means that the fred user never loses the ability to connect to the database. Note that CREATE DATABASE cannot be performed as part of a transaction, although nothing serious happens as a result.

There's more...

Superusers can still connect to the new database, and there is no way of preventing them from doing so. No other users can see the tables that were created in the new database, nor can they know the names of any of the objects. The new database can be seen to exist by other users, and they can also see the name of the user who owns the database.

See also

See *Chapter 6, Security*, for more details on these issues.

Running multiple servers on one system

Running multiple PostgreSQL servers on one physical system is possible if it is convenient for your needs.

Getting ready

Once again, make sure that you've read the *Deciding on a design for multitenancy* recipe so that you're certain this is the route you wish to take. Other options exist, and they may be preferable in some cases.

How to do it...

The core version of PostgreSQL easily allows multiple servers to run on the same system, but there are a few wrinkles to be aware of.

Some installer versions create a PostgreSQL data directory named data. When this happens, it gets a little difficult to have more than one data directory without using different directory structures and names.

Debian/Ubuntu packagers chose a layout specifically designed to allow multiple servers to potentially run with different software release levels. You may remember this from the *Locating the database server files* recipe in *Chapter 2, Exploring the Database*.

Starting from /var/lib/postgresql, which is the home directory of the postgres user, there are subdirectories for each major version, such as 16, 15, 14, or 13, inside which the individual data directories are placed. When you install PostgreSQL server packages, a data directory is created with the default name of main. Configuration files are placed separately in /etc/postgresql/<version>/<name>, and log files are created in /var/log/postgresql/postgresql-<version>-<name>.log.

Thus, not all the files will be found in the data directory. As an example, let's create an additional data directory:

1. We start by running the following command:

    ```
    sudo -u postgres pg_createcluster 16 main2
    ```

2. Then, the new database server can be started using the following command:

    ```
    sudo -u postgres pg_ctlcluster 16 main2 start
    ```

This is sufficient to create and start an additional database cluster in version 16, named main2. The data and configuration files are stored inside the /var/lib/postgresql/16/main2/ and /etc/postgresql/16/main2/ directories, respectively, giving the new database the next unused port number, such as 5433, if this is the second PostgreSQL server on that machine.

Local access to multiple PostgreSQL servers has been simplified as well. PostgreSQL client programs, such as psql, are wrapped by a special script that takes the cluster name as an additional parameter and automatically uses the corresponding port number. Hence, you don't need the following command:

```
psql --port 5433 -h /var/run/postgresql ...
```

Instead, you can refer to the database server by name, as shown here:

```
psql --cluster 16/main2 ...
```

This has its advantages, especially if you wish (or need) to change the port in the future. I find this extremely convenient, and it works with other utilities, such as pg_dump and pg_restore.

With Red Hat systems, you will need to run initdb directly, selecting your directories carefully:

1. First, initialize your data directory with something such as the following:

   ```
   sudo -u postgres initdb -D /var/lib/pgsql/datadir2
   ```

2. Then, modify the port parameter in the postgresql.conf file and start using the following command:

   ```
   sudo -u postgres pg_ctl -D /var/lib/pgsql/datadir2 start
   ```

This will create an additional database server at the default server version, with the files stored in /var/lib/pgsql/datadir2.

You can also set up the server with the chkconfig utility to ensure it starts on boot if your distribution supports it.

How it works...

PostgreSQL servers are controlled using pg_ctl. Everything else is a wrapper of some kind around this utility. The only constraints of running multiple versions of PostgreSQL come from file locations and naming conventions, assuming (of course) that you have enough resources, such as disk space and memory. Everything else is straightforward. Having said that, the Debian/Ubuntu design is currently the only design that makes it easy to run multiple servers.

Setting up a connection pool

A **connection pool** is a term that's used for a collection of already-connected sessions that can be used to reduce the overhead of connection and reconnection.

There are various ways by which connection pools can be provided, depending on the software stack in use. The best option is to look at the server-side connection pool software because that works for all connection types, not just within a single software stack.

In this recipe, we're going to look at **PgBouncer**, which is designed as a very lightweight connection pool. Its name comes from the idea that the pool can be paused and resumed, allowing the server to be restarted or *bounced*.

Getting ready

First of all, decide where you're going to store the **PgBouncer** parameter files, log files, and PID files. PgBouncer can manage more than one database server's connections at the same time, although that probably isn't wise for simple architectures. If you keep the PgBouncer files associated with the database server, then it should be easy to manage.

How to do it...

Follow these steps to configure PgBouncer:

1. Create a pgbouncer.ini file, as follows:

    ```
    ;
    ; pgbouncer configuration example
    ;
    [databases]
    postgres = port=5432 dbname=postgres
    [pgbouncer]
    listen_addr = 127.0.0.1
    listen_port = 6432
    admin_users = postgres
    ;stats_users = monitoring userid
    auth_type = scram-sha-256
    ; put these files somewhere sensible:
    auth_file = users.txt
    ```

```
            logfile = pgbouncer.log
            pidfile = pgbouncer.pid
            server_reset_query = DISCARD ALL;
            ; default values
            pool_mode = session
            default_pool_size = 20
            log_pooler_errors = 0
```

2. Create a users.txt file. This must contain the minimum users mentioned in admin_users and stats_users. Its format is very simple – it's a collection of lines with a username and a password. Consider the following as an example:

    ```
    "postgres"      ""
    ```

3. PgBouncer also supports **SCRAM (Salt Challenge Response Authentication Mechanism)** authentication. If the PgBouncer to server connection requires SCRAM authentication, then you must also connect from the client to PgBouncer using SCRAM authentication. To use that effectively, you need to copy the SCRAM secrets from the database server into the users.txt file.

4. You may wish to create the users.txt file by directly copying the details from the server. This can be done by using the following psql script (this is the same one that was used for md5 authentication, back when that was recommended):

    ```
    postgres=> \o users.txt
    postgres=> \t
    postgres=> SELECT '"'||rolname||'" "'||rolpassword||'"'
    postgres-> FROM pg_authid;
    postgres=> \q
    ```

5. Launch pgbouncer:

    ```
    pgbouncer -d pgbouncer.ini
    ```

6. Test the connection; it should respond to reload:

    ```
    psql -p 6432 -h 127.0.0.1 -U postgres pgbouncer -c "reload"
    ```

7. Finally, verify that PgBouncer's max_client_conn parameter does not exceed the max_connections parameter on PostgreSQL.

How it works...

PgBouncer is a great piece of software. Its feature set is carefully defined to ensure that it is simple, robust, and very quick. PgBouncer is not multithreaded, so it runs in a single process and, thus, on a single CPU. It is very efficient, but very large data transfers will take more time and reduce concurrency, so create those data dumps using a direct connection.

PgBouncer provides connection pooling. If you set pool_mode = transaction, then PgBouncer will also provide connection concentration. This allows hundreds or even thousands of incoming connections to be managed, while only a few server connections are made.

As new connections, transactions, or statements arrive, the pool will increase in size up to the user-defined maximum values. Those connections will stay around until the server_idle_timeout value before the pool releases them.

PgBouncer also releases sessions every server_lifetime. This allows the server to free backends in rotation, avoiding issues with very long-lived session connections.

The query that creates users.txt only includes database users that have a password. All other users will have a null rolpassword field, so the whole string evaluates to NULL, and the line is omitted from the password file. This is intentional; users without a password represent a security risk unless they are closely guarded. An example of this is the postgres system user connecting from the same machine, which bypasses PgBouncer, and is used only for maintenance by responsible and trusted people.

It is possible to use an HBA file with the same syntax as pg_hba.conf. This allows for more flexibility when enabling TLS encryption (which includes SSL) for connections to remote servers, while using more efficient peer authentication for local servers.

There's more...

Instead of retrieving passwords from the userlist.txt file, PgBouncer can retrieve them directly from PostgreSQL, using the optional auth_user and auth_query parameters. If auth_user is set, PgBouncer will connect to the database using that user and run auth_query every time it needs to retrieve the password of some user trying to log in. The default value of auth_query is as follows:

```
SELECT usename, passwd FROM pg_shadow WHERE usename=$1
```

This default is just a minimal functioning example, which illustrates the idea of auth_query; however, it requires giving PgBouncer superuser access to PostgreSQL.

Hence, it is good practice to use the more sophisticated approach of creating a SECURITY DEFINER function that can retrieve the username and password, possibly making some checks on the username to allow only applicative connections. This is a good restriction because database administration connections should not go through a connection pooler.

It's also possible to connect to PgBouncer itself to issue commands. This can be done interactively, as if you were entering psql, or using single commands or scripts.

To shut down PgBouncer, we can just type SHUTDOWN or enter a single command, as follows:

```
psql -p 6432 pgbouncer -c "SHUTDOWN"
```

You can also use the RELOAD command to make PgBouncer reload (which means reread) the parameter files, as we did to test that everything is working.

If you are doing a switchover, you can use the WAIT_CLOSE command, followed by RELOAD or RECONNECT, to wait until the respective configuration change has been fully activated.

If you are using pool_mode = transaction or pool_mode = statement, then you can use the PAUSE command. This waits for the current transaction to complete before holding further work on that session. Thus, it allows you to perform DDL more easily or restart the server.

PgBouncer also allows you to use SUSPEND mode, which waits for all server-side buffers to flush.

The PAUSE or SUSPEND mode should eventually be followed by RESUME when the work is done.

In addition to the PgBouncer control commands, there are many varieties of SHOW commands, as shown here:

SHOW Command	Result Set
SHOW STATS	Show traffic stats, total and average requests, query duration, bytes sent/received, and so on. Also, take a look at SHOW STATS_TOTAL and SHOW STATS_AVERAGES.
SHOW SERVERS	One row per connection to the database server.
SHOW CLIENTS	One row per connection from the client.
SHOW POOLS	One row per pool of users.
SHOW LISTS	Gives a good summary of resource totals.
SHOW USERS	Lists users in users.txt.
SHOW DATABASES	Lists databases in pgbouncer.ini.
SHOW CONFIG	Lists configuration parameters.

SHOW FDS	Shows file descriptors.
SHOW SOCKETS	Show file sockets.
SHOW VERSION	Shows the pgBouncer version.

Table 4.1: PgBouncer SHOW commands

Accessing multiple servers using the same host and port

We will now show you one simple, yet important, application of the previous recipe, *Setting up a connection pool*. In that recipe, you learned how to reuse connections with PgBouncer, thus reducing the cost of disconnecting and reconnecting.

Here, we will demonstrate another way to use PgBouncer – one instance can connect to databases hosted by different database servers at the same time. These databases can be on separate hosts and can even have different major versions of PostgreSQL!

Getting ready

Suppose we have three database servers, each one hosting one database. All you need to know beforehand is the connection string for each database server.

More complex arrangements are possible, but those are left to you as an exercise.

Before you try this recipe, you should have already gone through the previous recipe. These two recipes have many steps in common, but we've kept them separate because they have different goals.

How to do it...

Each database is identified by its connection string. PgBouncer will read this information from its configuration file. Follow these steps:

1. All you need to do is set up PgBouncer, as you did in the previous recipe, by replacing the `databases` section of `pgbouncer.ini` with the following:

    ```
    [databases]
    myfirstdb = port=5432 host=localhost
    anotherdb = port=5437 host=localhost
    sparedb = port=5435 host=localhost
    ```

2. Once you have started PgBouncer, you can connect to the first database:

```
$ psql -p 6432 -h 127.0.0.1 -U postgres myfirstdb
psql (14.1)
Type "help" for help.
myfirstdb=# show port;
port
------
5432
(1 row)
myfirstdb=# show server_version;
server_version
----------------
14.1
(1 row)
```

3. Now, you can connect to the anotherdb database as if it were on the same server:

```
myfirstdb=# \c anotherdb
psql (14.1, server 9.5.15)
You are now connected to database "anotherdb" as user "postgres".
```

4. The server's greeting message suggests that we have landed on a different server, so we must check the port and the version (wow! This server needs an upgrade soon!):

```
anotherdb=# show port;
port
------
  5437
(1 row)
anotherdb=# show server_version;
server_version
----------------
     9.5.15
(1 row)
```

There's more...

The *Listing databases on the database server* recipe in *Chapter 2, Exploring the Database*, shows you how to list the available databases on the current database server, using either the \l meta-command or a couple of equivalent variations. Unfortunately, this doesn't work when you're using PgBouncer, for the very good reason that the current database server cannot know the answer.

We need to ask PgBouncer instead, which we can do using the SHOW command when connected to the pgbouncer special administrative database:

```
myfirstdb=# \c pgbouncer
psql (14.1, server 1.8.1/bouncer)
You are now connected to database "pgbouncer" as user "postgres".
pgbouncer=# show databases;
    name    |    host    | port | database  | force_user | pool_size | reserve_pool
------------+------------+------+-----------+------------+-----------+-------------
 anotherdb  | localhost  | 5437 | anotherdb |            |        20 |           0
 myfirstdb  | localhost  | 5432 | myfirstdb |            |        20 |           0
 pgbouncer  |            | 6432 | pgbouncer | pgbouncer  |         2 |           0
 sparedb    | localhost  | 5435 | sparedb   |            |        20 |           0
(4 rows)
```

Running multiple PgBouncer on the same port to leverage multiple cores

PgBouncer is a single process; therefore, it can only leverage one core on the system. Sometimes, leveraging a single thread becomes a bottleneck for performance. Therefore, PgBouncer also has specific parameters that allow users to run multiple instances of PgBouncer on the same port and host.

Getting ready

Suppose a user wants to run two instances of PgBouncer on the same port and hosts – you need three Unix directories and three different pgbouncer.ini files for the instances.

Before you try this recipe, you should have gone through the previous recipe on *Setting up a connection pool*.

How to do it...

Follow the steps given below:

1. Use the pgbouncer.ini mentioned in the recipe *Setting up a connection pool*, and create a separate configuration file for each PgBouncer instance, like pgbouncer1.ini and pgbouncer2.ini. For the first PgBouncer instance, add the following lines under the [pgbouncer] section for PgBouncer's first instance (pgbouncer1.ini):

    ```
    so_reuseport=1
    unix_socket_dir=/tmp/pgbouncer1
    peer_id=1 # id first PgBouncer instance
    ```

    Add the following section after the [pgbouncer] section:

    ```
    [peers]
    1 = host=/tmp/pgbouncer1
    2 = host=/tmp/pgbouncer2
    ```

2. For the second PgBouncer, add the following files in pgbouncer2.ini under [pgbouncer]:

    ```
    so_reuseport=1
    unix_socket_dir=/tmp/pgbouncer2
    peer_id=2 # id first PgBouncer instance
    ```

    Add the following section after the [pgbouncer] section:

    ```
    1 = host=/tmp/pgbouncer1
    2 = host=/tmp/pgbouncer2
    ```

3. Create the following socket directories on the server:

    ```
    mkdir /tmp/pgbouncer1
    mkdir /tmp/pgbouncer2
    ```

4. Launch PgBouncer instances using the following command:

    ```
    pgbouncer -d pgbouncer1.ini
    pgbouncer -d pgbouncer2.ini
    ```

How it works...

Once the PgBouncer instances are started, they will all listen on the same port and will share the connections. This will allow you to leverage multiple cores on your server to improve the performance of PgBouncer.

Here are some additional things to keep in mind when running multiple PgBouncer instances:

- Each PgBouncer instance will need its own unique configuration file.
- The `pool_size` setting in each configuration file should be set to the number of connections that you want each PgBouncer instance to handle.
- The `max_client_conn` setting in each configuration file should be set to the maximum number of connections that each PgBouncer instance can handle.
- You can use the `log_connections` setting in each configuration file to log all connections to PgBouncer. This can be helpful to troubleshoot problems.

Learn more on Discord

To join the Discord community for this book – where you can share feedback, ask questions to the author, and learn about new releases – follow the QR code below:

https://discord.gg/pQkghgmgdG

5
Tables and Data

This chapter covers a range of general recipes for your tables and for working with the data they contain. Many of the recipes contain general advice, with specific PostgreSQL examples.

Some system administrators that we've met work only on the external aspects of a database server. They see the data that is actually in the database, its meaning and quality, as someone else's problem.

Look after your data, and your database will look after you. Keep your data clean, and your queries will run faster and cause fewer application errors. You'll also gain many friends in the business. Getting called in the middle of the night to fix data problems just isn't cool.

In this chapter, we will cover the following recipes:

- Choosing good names for database objects
- Handling objects with quoted names
- Identifying and removing duplicates
- Preventing duplicate rows
- Finding a unique key for a set of data
- Generating test data
- Randomly sampling data
- Loading data from a spreadsheet
- Loading data from flat files
- Making bulk data changes using server-side procedures with transactions
- Dealing with large tables with table partitioning
- Finding good candidates for partition keys

- Consolidating data with MERGE
- Deciding when to use JSON data types

Choosing good names for database objects

The easiest way to help other people understand a database is to ensure that all the objects have a meaningful name.

Getting ready

What makes a name meaningful?

Take some time to reflect on your database to make sure you have a clear view of its purpose and main use cases. This is because all the items in this recipe describe certain naming choices that you need to consider carefully, given your specific circumstances.

How to do it...

Here are the points you should consider when naming your database objects:

- The name follows the existing standards and practices in place within your organization. Inventing new standards isn't helpful; enforcing existing standards is.
- The name clearly describes the role or table contents.
- For major tables, use short, powerful names.
- Name lookup tables after the table to which they are linked, such as account_status.
- For associative or linked tables, use all the names of the major tables to which they relate, such as customer_account.
- Make sure that the name is clearly distinct from other similar names.
- Use consistent abbreviations.
- Use underscores. Casing is not preserved by default, so using camel case names, such as customerAccount, as used in Java, will just leave them unreadable. See the *Handling objects with quoted names* recipe. Avoid names that include spaces and semicolons so that we can more easily tell names that have been deliberately crafted by attackers to defeat security.
- Use consistent plurals, or don't use them at all.
- Use suffixes to identify the content type or domain of an object. PostgreSQL already uses suffixes for the PostgreSQL catalog.
- Think ahead. Don't pick names that refer to the current role or location of an object. So don't name a table London because it exists on a server in London. That server might get moved to Los Angeles.

- Think ahead. Don't pick names that imply that an entity is the only one of its kind, such as a table named TEST or a table named BACKUP_DATA. On the other hand, such information can be put in the database name, which is not normally used from within the database.
- Avoid using acronyms in place of long table names. For example, money_allocation_decision is much better than MAD. This is especially important, as PostgreSQL translates names into lowercase, so the fact that it is an acronym may not be clear.
- The table name is commonly used as the root for other objects that are created, so don't add the table suffix or similar ideas.

There's more...

The standard names for the indexes in PostgreSQL are as follows:

{tablename}_{columnname(s)}_{suffix}

Here, the suffix is one of the following:

- pkey: This is used for a primary key constraint.
- key: This is used for a unique constraint.
- excl: This is used for an exclusion constraint.
- idx: This is used for any other kind of index.

The standard suffix for all sequences is seq.

Tables can have multiple triggers fired on each event. Triggers are executed in alphabetical order, so trigger names should have some kind of action name to differentiate them and to allow the order to be specified. It might seem like a good idea to put INSERT, UPDATE, or DELETE in the trigger name, but that can get confusing if you have triggers that work on both UPDATE and DELETE, and all of this may end up as a mess.

The alphabetical order for trigger names always follows the C locale, regardless of your actual locale settings. If your trigger names use non-ASCII characters, then the actual ordering might not be what you expect.

The following example shows how the è and é characters are ordered in the C locale. You can change the locale and/or the list of strings to explore how different locales affect ordering:

```
postgres=# WITH a(x) AS (
  VALUES ('è'),('é')
) SELECT *
FROM a
```

```
ORDER BY x
COLLATE "C";
 x
---
 è
 é
(2 rows)
```

A useful naming convention for triggers is as follows:

```
{tablename}_{actionname}_{after|before}_trig
```

If you do find yourself with strange or irregular object names, it might be a good idea to use the RENAME subcommands to tidy things up again. Here is an example of this:

```
ALTER INDEX badly_named_index RENAME TO tablename_status_idx;
```

You can enforce a naming convention using an event trigger. Event triggers can only be created by super users and will be called for all DDL statements, executed by any user. To enforce naming, run something like this:

```
CREATE EVENT TRIGGER enforce_naming_conventions
ON ddl_command_end
EXECUTE FUNCTION check_object_names();
```

The check_object_names() function can access the details of newly created objects using a query like this so that you can write programs to enforce naming:

```
SELECT object_identity
FROM pg_event_trigger_ddl_command()
WHERE NOT in_extension
   AND command_tag LIKE 'CREATE%';
```

Handling objects with quoted names

PostgreSQL object names can contain spaces and mixed-case characters if we enclose the table names in double quotes. This can cause some difficulties and security issues, so this recipe is designed to help you if you get stuck with this kind of problem.

Case-sensitivity issues can often be a problem for people more used to working with other database systems, such as MySQL, or for people who are facing the challenge of migrating code away from MySQL.

Getting ready

First, let's create a table that uses a quoted name with mixed cases, such as the following:

```
CREATE TABLE "MyCust"
AS
SELECT * FROM cust;
```

How to do it...

If we try to access these tables without the proper case, we get this error:

```
postgres=# SELECT count(*) FROM MyCust;
ERROR:  relation "mycust" does not exist
LINE 1: SELECT * FROM MyCust;
```

So we write it in the correct case:

```
postgres=# SELECT count(*) FROM MyCust;
ERROR:  relation "mycust" does not exist
LINE 1: SELECT * FROM MyCust;
```

This still fails and, in fact, gives the same error.

If you want to access a table that was created with quoted names, then you must use quoted names, such as the following:

```
postgres=# SELECT count(*) FROM "MyCust";
```

The output is as follows:

```
 count
-------
     5
(1 row)
```

The usage rule is that if you create your tables using quoted names, then you need to write your SQL using quoted names. Alternatively, if your SQL uses quoted names, then you will probably have to create the tables using quoted names as well.

How it works...

PostgreSQL folds all names to lowercase when used within a SQL statement. Consider this command:

```
SELECT * FROM mycust;
```

This is exactly the same as the following command:

```
SELECT * FROM MYCUST;
```

It is also exactly the same as this command:

```
SELECT * FROM MyCust;
```

However, it is not the same thing as the following command:

```
SELECT * FROM "MyCust";
```

There's more...

If you are handling object names in SQL, then you should use quote_ident() to ensure users don't call their objects a name that could cause security issues. quote_ident() puts double quotes around a value if PostgreSQL requires that for an object name, as shown here:

```
postgres=# SELECT quote_ident('MyCust');
quote_ident
-------------
"MyCust"
(1 row)
postgres=# SELECT quote_ident('mycust');
quote_ident
-------------
mycust
(1 row)
```

For a longer explanation of why this is necessary, see the *Performing actions on many tables* recipe in *Chapter 7, Database Administration*.

The quote_ident() function may be especially useful if you are creating a table based on a variable name in a PL/pgSQL function, as follows:

```
EXECUTE 'CREATE TEMP TABLE ' || quote_ident(tablename) ||
        '(col1              INTEGER);'
```

Identifying and removing duplicates

Relational databases work on the idea that items of data can be uniquely identified. However hard we try, there will always be bad data arriving from somewhere. This recipe shows you how to diagnose that and clean up the mess.

Getting ready

Let's start by looking at an example table, cust. It has a duplicate value in customerid:

```
CREATE TABLE cust (
customerid BIGINT NOT NULL
,firstname TEXT NOT NULL
,lastname  TEXT NOT NULL
,age       INTEGER NOT NULL);
INSERT INTO cust VALUES (1, 'Philip', 'Marlowe', 33);
INSERT INTO cust VALUES (2, 'Richard', 'Hannay', 37);
INSERT INTO cust VALUES (3, 'Harry', 'Palmer', 36);
INSERT INTO cust VALUES (4, 'Rick', 'Deckard', 4);
INSERT INTO cust VALUES (4, 'Roy', 'Batty', 41);
postgres=# SELECT * FROM cust ORDER BY 1;
customerid | firstname | lastname | age
------------+-----------+----------+-----
          1 | Philip    | Marlowe  |  33
          2 | Richard   | Hannay   |  37
          3 | Harry     | Palmer   |  36
          4 | Rick      | Deckard  |   4
          4 | Roy       | Batty    |  41
(5 rows)
```

Before you delete duplicate data, remember that sometimes it isn't the data that is wrong – it is your understanding of it. In those cases, it may be that you haven't properly normalized your database model and that you need to include additional tables to account for the shape of the data. You might also find that duplicate rows are caused because you decide to exclude a column somewhere earlier in a data load process. Check twice, and cut once.

How to do it...

First, identify the duplicates using a query, such as the following:

```
CREATE UNLOGGED TABLE dup_cust AS
SELECT *
FROM cust
WHERE customerid IN
(SELECT customerid
   FROM cust
   GROUP BY customerid
   HAVING count(*) > 1);
```

We save the list of duplicates in a separate table because the query can be very slow if the table is big, so we don't want to run it more than once.

An `UNLOGGED` table can be created with less I/O because it does not produce **Write Ahead Log (WAL)**. It is better than a temporary table because it doesn't disappear if you disconnect and then reconnect. The other side of the coin is that you lose its contents after a crash, but this is not too bad because if you choose to use an unlogged table, then you tell PostgreSQL that you can recreate the contents of that table in the (unlikely) event of a crash.

The results can be used to identify the bad data manually, and you can resolve the problem by carrying out the following steps:

1. Merge the two rows to give the best picture of the data, if required. This might use values from one row to update the row you decide to keep, as shown here:

   ```
   UPDATE cust
   SET age = 41
   WHERE customerid = 4
   AND lastname = 'Deckard';
   ```

2. Delete the remaining undesirable rows:

   ```
   DELETE FROM cust
   WHERE customerid = 4
   AND lastname = 'Batty';
   ```

In some cases, the data rows might be completely identical, so let's create an example:

```
CREATE TABLE new_cust (customerid BIGINT NOT NULL);
INSERT INTO new_cust VALUES (1), (1), (2), (3), (4), (4);
```

The new_cust table looks like the following:

```
postgres=# SELECT * FROM new_cust ORDER BY 1;
 customerid
------------
          1
          1
          2
          3
          4
          4
(5 rows)
```

Unlike the preceding case, we can't tell the data apart at all, so we cannot remove duplicate rows without any manual process. SQL is a set-based language, so picking only one row out of a set is slightly harder than most people want it to be.

In these circumstances, we should use a slightly different procedure to detect duplicates. We will use a hidden column named ctid. It denotes the physical location of the row you are observing, and rows will all have different ctid values. The steps are as follows:

1. First, we start a transaction:

   ```
   BEGIN;
   ```

2. Then, we lock the table in order to prevent any INSERT, UPDATE, or DELETE operations, which would alter the list of duplicates and/or change their ctid values:

   ```
   LOCK TABLE new_cust IN SHARE ROW EXCLUSIVE MODE;
   ```

3. Now, we locate all duplicates, keeping track of the minimum ctid value so that we don't delete it:

   ```
   CREATE TEMPORARY TABLE dups_cust AS
   SELECT customerid, min(ctid) AS min_ctid
   FROM new_cust
   GROUP BY customerid
   HAVING count(*) > 1;
   ```

4. Then, we can delete each duplicate, with the exception of the duplicate with the minimum ctid value:

```
DELETE FROM new_cust
USING dups_cust
WHERE new_cust.customerid = dups_cust.customerid
AND new_cust.ctid != dups_cust.min_ctid;
```

5. We commit the transaction, which also releases the lock we previously took:

```
COMMIT;
```

6. Finally, we clean up the table after the deletions:

```
VACUUM new_cust;
```

How it works...

The first query works by grouping together the rows on the unique column and counting rows. Anything with more than one row must be caused by duplicate values. If we're looking for duplicates of more than one column (or even all columns), then we have to use a SQL query of the following form:

```
SELECT *
FROM mytable
WHERE  (col1, col2, ... ,colN) IN
(SELECT col1, col2, ... ,colN
FROM mytable
GROUP BY  col1, col2, ... ,colN
HAVING count(*) > 1);
```

Here, col1, col2, and so on up until colN are the columns of the key.

Note that this type of query may need to sort the complete table due to the GROUP BY on all the key columns. That will require sort space equal to the size of the table, so you'd better think first before running that SQL on very large tables. You'll probably benefit from a large work_mem setting for this query, probably 128 MB or more.

The DELETE FROM ... USING query that we showed only works with PostgreSQL because it uses the ctid value, which is the internal identifier of each row in the table. If you wanted to run that query against more than one column, as we did earlier in the chapter, you'd need to extend the queries in *step 3*, as follows:

```
SELECT customerid, customer_name, ..., min(ctid) AS min_ctid
FROM ...
GROUP BY customerid, customer_name, ...
...;
```

Then, extend the query in *step 4*, like this:

```
DELETE FROM new_cust
...
WHERE new_cust.customerid = dups_cust.customerid
AND new_cust.customer_name = dups_cust.customer_name
AND ...
AND new_cust.ctid != dups_cust.min_ctid;
```

The preceding query works by grouping together all the rows with similar values and then finding the row with the lowest ctid value. The lowest will be closer to the start of the table, so duplicates will be removed from the far end of the table. When we run VACUUM, we may find that the table gets smaller because we have removed rows from the far end.

The BEGIN and COMMIT commands wrap the LOCK and DELETE commands into a single transaction, which is required. Otherwise, the lock will be released immediately after being taken.

Another reason to use a single transaction is that we can always roll back if anything goes wrong, which is a good thing when we are removing data from a live table.

There's more...

Locking the table against changes for long periods may not be possible while we remove duplicate rows. That creates some fairly hard problems with large tables. In that case, we need to do things slightly differently:

1. Identify the rows to be deleted and save them in a side table.
2. Build an index on the main table to speed up access to rows (maybe using the CONCURRENTLY keyword, as explained in the *Maintaining indexes* recipe in *Chapter 9, Regular Maintenance*).
3. Write a program that reads the rows from the side table in a loop, performing a series of smaller transactions.
4. Start a new transaction.
5. From the side table, read a set of rows that match.
6. Select those rows from the main table for updates, relying on the index to make those accesses happen quickly.

7. Delete the appropriate rows.
8. Commit, and then loop again.

The aforementioned program can't be written as a database function, as we can't have multiple transactions in a function. We need multiple transactions to ensure that we hold locks on each row for the shortest possible duration.

Preventing duplicate rows

Preventing duplicate rows is one of the most important aspects of data quality for any database. PostgreSQL offers some useful features in this area, extending beyond most relational databases.

Getting ready

Identify the set of columns that you wish to make unique. Does this apply to all rows or just a subset of rows?

Let's find out with our example table:

```
postgres=# SELECT * FROM new_cust;
 customerid
------------
          1
          2
          3
          4
(4 rows)
```

How to do it...

To prevent duplicate rows, we need to create a unique index that the database server can use to enforce the uniqueness of a particular set of columns. We can do this in the following three similar ways for basic data types:

1. Create a primary key constraint on the set of columns. We are allowed only one of these per table. The values of the data rows must not be NULL, as we force the columns to be NOT NULL if they aren't already:

   ```
   ALTER TABLE new_cust ADD PRIMARY KEY(customerid);
   ```

   This creates a new index named new_cust_pkey.

2. Create a unique constraint on the set of columns. We can use these instead of, or with, a primary key. There is no limit on the number of these per table. NULL values are allowed in the columns:

```
ALTER TABLE new_cust ADD UNIQUE(customerid);
```

This creates a new index named new_cust_customerid_key.

3. Create a unique index on the set of columns:

```
CREATE UNIQUE INDEX ON new_cust (customerid);
```

This creates a new index named new_cust_customerid_idx.

All these techniques exclude duplicates by defining constraints and structures that guarantee uniqueness, just with slightly different syntaxes. All of them create an index, but only the first two create a formal *constraint*. Each of these techniques can be used when we have a primary key or unique constraint that uses multiple columns.

The last method is important because it allows you to specify a WHERE clause on the index. This can be useful if you know that the column values are unique only in certain circumstances. The resulting index is then known as a partial index.

Suppose our data looked like this:

```
postgres=# SELECT * FROM partial_unique;
```

This gives us the following output:

```
 customerid | status | close_date
------------+--------+------------
          1 | OPEN   |
          2 | OPEN   |
          3 | OPEN   |
          3 | CLOSED | 2010-03-22
(4 rows)
```

Then, we can put a partial index on the table to enforce the uniqueness of customerid only for status = 'OPEN', like this:

```
CREATE UNIQUE INDEX ON partial_unique (customerid)
    WHERE status = 'OPEN';
```

If your uniqueness constraint needs to be enforced across more complex data types, then you may need to use a more advanced syntax. A few examples will help here.

Let's start with the simplest example: create a table of boxes and put sample data in it. This may be the first time you're seeing PostgreSQL's data type syntax, so bear with me:

```
postgres=# CREATE TABLE boxes (name text, position box);
CREATE TABLE
postgres=# INSERT INTO boxes VALUES ('First', box '((0,0), (1,1))');
INSERT 0 1
postgres=# INSERT INTO boxes VALUES ('Second', box '((2,0), (2,1))');
INSERT 0 1
postgres=# SELECT * FROM boxes;
  name  |   position
--------+-------------
 First  | (1,1),(0,0)
 Second | (2,1),(2,0)
(2 rows)
```

We can see two boxes that neither touch nor overlap, based on their *x* and *y* coordinates.

To enforce uniqueness here, we want to create a constraint that will throw out any attempt to add a position that overlaps with any existing box. The overlap operator for the box data type is defined as &&, so we use the following syntax to add the constraint:

```
ALTER TABLE boxes ADD EXCLUDE USING gist (position WITH &&);
```

This creates a new index named boxes_position_excl:

```
#\d boxes_position_excl
Index "public.boxes_position_excl"
  Column  | Type | Key? | Definition
----------+------+------+------------
 position | box  | yes  | "position"
gist, for table "public.boxes"
```

We can use the same syntax even with the basic data types. So a fourth way of performing our first example would be as follows:

```
ALTER TABLE new_cust ADD EXCLUDE (customerid WITH =);
```

This creates a new index named `new_cust_customerid_excl`, and duplicates are excluded:

```
# insert into new_cust VALUES (4);
ERROR: conflicting key value violates exclusion constraint "new_cust_
customerid_excl"
DETAIL: Key (customerid)=(4) conflicts with existing key (customerid)=(4).
```

How it works...

Uniqueness is always enforced by an index.

Each index is defined with a data type operator. When a new row is inserted or the set of column values is updated, we use the operator to search for existing values that conflict with the new data.

So, to enforce uniqueness, we need an index and a search operator defined on the data types of the columns. When we define normal UNIQUE constraints, we simply assume that we mean the equality operator (=) for the data type. The EXCLUDE syntax offers a richer syntax to allow us to express the same problem with different data types and operators.

There's more...

Unique and exclusion constraints provide strong guarantees on the data that gets written to the database, evaluating the conditions as soon as a row gets written, which consumes CPU resources. However, they can be marked as deferrable, meaning that a user can choose to postpone the check to the end of the transaction – a nice way to relax constraints without reducing data integrity.

Duplicate indexes

Note that PostgreSQL allows you to have multiple indexes with exactly the same definition. This is useful in some contexts but can also be annoying if you accidentally create multiple indexes, as each index has its own cost in terms of writes. You can also have constraints defined using each of the aforementioned different ways. Each of these ways enforces, essentially, the same constraint, so take care.

Uniqueness without indexes

It's possible to have uniqueness in a set of columns without creating an index, although without a strong guarantee to enforce it. That might be useful if all we want is to ensure uniqueness rather than allow index lookups.

To do that, you can do either of the following:

- Use a serial data type.
- Manually alter the default to be the `nextval()` function of a sequence.

Each of these will provide a unique value for use as a row's key. Remember, the uniqueness is not enforced, nor will there be a unique constraint defined. So there is still a possibility that someone might reset the sequence to an earlier value, which will eventually cause duplicate values.

Consider, also, that this method provides the unique value **as a default**, which is not used when a user specifies an explicit value. An example of this is as follows:

```
CREATE TABLE t(id serial, descr text);
INSERT INTO t(descr) VALUES ('First value');
INSERT INTO t(id,descr) VALUES (1,'Cheating!');
```

Note that the last insert bypasses the default value, creating a duplicate id. This example reinforces the value of indexes and constraints.

Finally, you might also wish to have mostly unique data, such as using the clock_timestamp() function to provide ascending times to a microsecond resolution.

A real-world example — IP address range allocation

The problem is about assigning ranges of IP addresses while at the same time ensuring that we don't allocate (or potentially allocate) the same addresses to different people or purposes. This is easy to do if we keep track of each individual IP address but much harder to do if we want to deal solely with ranges of IP addresses.

Initially, you may think of designing the database as follows:

```
CREATE TABLE iprange
(iprange_start inet
,iprange_stop inet
,owner text);
INSERT INTO iprange VALUES ('192.168.0.1','192.168.0.16', 'Simon');
INSERT INTO iprange VALUES ('192.168.0.17','192.168.0.24', 'Gianni');
INSERT INTO iprange VALUES ('192.168.0.32','192.168.0.64', 'Gabriele');
```

However, you'll realize that there is no way to create a unique constraint that enforces the model constraint of avoiding overlapping ranges. You can create an AFTER trigger that checks existing values, but it's going to be messy.

PostgreSQL offers a better solution, based on *range types*. In fact, every data type that supports a btree operator class (that is, a way of ordering any two given values) can be used to create a range type. In our case, the SQL is as follows:

```
CREATE TYPE inetrange AS RANGE (SUBTYPE = inet);
```

This command creates a new data type that can represent ranges of inet values – that is, of IP addresses. Now, we can use this new type when creating a table:

```
CREATE TABLE iprange2
(iprange inetrange
,owner text);
```

This new table can be populated as usual. We just have to group the extremes of each range into a single value, as follows:

```
INSERT INTO iprange2 VALUES ('[192.168.0.1,192.168.0.16]', 'Simon');
INSERT INTO iprange2 VALUES ('[192.168.0.17,192.168.0.24]', 'Gianni');
INSERT INTO iprange2 VALUES ('[192.168.0.32,192.168.0.64]', 'Gabriele');
```

Now, we can create a *unique exclusion constraint* on the table, using the following syntax:

```
ALTER TABLE iprange2
ADD EXCLUDE USING GIST (iprange WITH &&);
```

If we try to insert a range that overlaps with any of the existing ranges, then PostgreSQL will stop us:

```
INSERT INTO iprange2
VALUES ('[192.168.0.10,192.168.0.20]', 'Somebody else');
ERROR:  conflicting key value violates exclusion constraint "iprange2_
iprange_excl"
DETAIL:  Key (iprange)=([192.168.0.10,192.168.0.20]) conflicts with
existing key (iprange)=([192.168.0.1,192.168.0.16]).
```

A real-world example – a range of time

In many databases, there will be historical data tables with data that has a START_DATE value and an END_DATE value, or something similar. As in the previous example, we can solve this problem elegantly with a range type. Actually, this example is even shorter – we don't need to create the range type, since the most common cases are already built in – that is, integers, decimal values, dates, and timestamps with and without a time zone.

Finding a unique key for a set of data

Sometimes, it can be difficult to find a unique set of key columns that describe the data. In this recipe, we will analyze the data in a database to allow us to identify the column(s) that together form a unique key. This is useful when a key is not documented, not defined, or has been defined incorrectly.

Getting ready

Let's start with a small table, where the answer is fairly obvious:

```
postgres=# select * from ord;
```

We assume that the output is as follows:

```
 orderid | customerid |  amt
---------+------------+--------
   10677 |          2 |   5.50
    5019 |          3 | 277.44
    9748 |          3 |  77.17
(3 rows)
```

How to do it...

First of all, there's no need to do this through a brute-force approach. Checking all the permutations of columns to see which is unique might take you a long time.

Let's start by using PostgreSQL's own optimizer statistics. Run the following command on the table to get a fresh sample of statistics:

```
postgres=# analyze ord;
ANALYZE
```

This runs quickly, so we don't have to wait too long. Now, we can examine the relevant columns of the statistics:

```
postgres=# SELECT attname, n_distinct
             FROM pg_stats
            WHERE schemaname = 'public'
              AND tablename = 'ord';
  attname   | n_distinct
------------+------------
 orderid    |         -1
 customerid |  -0.666667
 amt        |         -1
(3 rows)
```

The preceding example was chosen because we have two potential answers. If the value of n_distinct is -1, then the column is thought to be unique within the sample of rows examined.

We will then need to use our judgment to decide whether one or both of these columns are unique by chance or as part of the design of the database that created them.

It's possible that there is no single column that uniquely identifies the rows. Multiple column keys are fairly common. If none of the columns were unique, then we should start looking for unique keys that are combinations of the most unique columns. The following query shows a frequency distribution for the table, where a value occurs twice in one case and another value occurs only once:

```
postgres=# SELECT num_of_values, count(*)
            FROM (SELECT customerid, count(*) AS num_of_values
                  FROM ord
                  GROUP BY customerid) s
            GROUP BY num_of_values
            ORDER BY count(*);
num_of_values | count
--------------+-------
            2 |     1
            1 |     1
(2 rows)
```

We can change the query to include multiple columns, like this:

```
SELECT num_of_values, count(*)
FROM (SELECT   customerid, orderid, amt
       ,count(*) AS num_of_values
       FROM ord
       GROUP BY customerid, orderid, amt
     ) s
GROUP BY num_of_values
ORDER BY count(*);
```

When we find a set of columns that is unique, this query will result in only one row, as shown in the following example:

```
num_of_values | count
--------------+-------
            1 |     3
```

As we get closer to finding the key, we will see that the distribution gets tighter and tighter.

So the procedure is as follows:

1. Choose one column to start with.
2. Compute the corresponding frequency distribution.
3. If the outcome is multiple rows, then add one more column and repeat from *step 2*. Otherwise, it means you have found a set of columns satisfying a uniqueness constraint.

Now, you must verify that the set of columns is minimal – for example, check whether it is possible to remove one or more columns without violating the unique constraint. This can be done using the frequency distribution as a test. To be precise, do the following:

1. Test each column by computing the frequency distribution on all the other columns.
2. If the frequency distribution has one row, then the column is not needed in the uniqueness constraint. Remove it from the set of columns and repeat from *step 1*. Otherwise, you have found a minimal set of columns, which is also called a key for that table.

How it works...

Finding a unique key is possible for a program, but in most cases, a human can do this much faster by looking at things such as column names, foreign keys, or business understanding to reduce the number of searches required by the brute-force approach.

The ANALYZE command works by taking a sample of the table data and then performing a statistical analysis of the results. The n_distinct value has two different meanings, depending on its sign: if positive, it is the estimate of the number of distinct values for the column; if negative, it is the estimate of the density of such distinct values, with the sign changed. For example, n_distinct = -0.2 means that a table of 1 million rows is expected to have 200,000 distinct values, while n_distinct = 5 means that we expect just 5 distinct values.

Generating test data

DBAs frequently need to generate test data for a variety of reasons, whether it's for setting up a test database or just for generating a test case for a SQL performance issue.

How to do it...

To create a table of test data, we need the following:

- Some rows
- Some columns

- Some order

The steps are as follows:

1. First, generate a lot of rows of data. We use something named a `set-returning` function. You can write your own, although PostgreSQL includes a couple of very useful ones.

2. You can generate a sequence of rows using a query like the following:

```
postgres=# SELECT * FROM generate_series(1,5);
 generate_series
-----------------
               1
               2
               3
               4
               5
(5 rows)
```

3. Alternatively, you can generate a list of dates, like this:

```
postgres=# SELECT date(t)
FROM generate_series(now(), now() + '1 week', '1 day') AS f(t);
    date
------------
 2023-10-25
 2023-10-26
 2023-10-27
 2023-10-28
 2023-10-29
 2023-10-30
 2023-10-31
 2023-11-01
(8 rows)
```

4. Then, we want to generate a value for each column in the `test` table. We can break that down into a series of functions, using the following examples as a guide:

    - Either of these functions can be used to generate both rows and reasonable primary key values for them.

- For a random `integer` value, this is the function:

  ```
  (random()*(2*10^9))::integer
  ```

- For a random `bigint` value, the function is as follows:

  ```
  (random()*(9*10^18))::bigint
  ```

- For random `numeric` data, the function is the following:

  ```
  (random()*100.)::numeric(5,2)
  ```

- For a random-length string, up to a maximum length, this is the function:

  ```
  repeat('1',(random()*40)::integer)
  ```

- For a random-length substring, the function is as follows:

  ```
  substr('abcdefghijklmnopqrstuvwxyz',1, (random()*25)::integer)
  ```

- Here is the function for a random string from a list of strings:

  ```
  (ARRAY['one','two','three'])[0.5+random()*3]
  ```

5. Finally, we can put both techniques together to generate our table:

   ```
   postgres=# SELECT key
           ,(random()*100.)::numeric(4,2)
           ,repeat('1',(random()*25)::integer)
   FROM generate_series(1,10) AS f(key);
   key | numeric |         repeat
   ----+---------+-------------------------
     1 |   83.05 | 1111
     2 |    5.28 | 111111111111
     3 |   41.85 | 11111111111111111111
     4 |   41.70 | 111111111111111
     5 |   53.31 | 1
     6 |   10.09 | 1111111111111111
     7 |   68.08 | 111
     8 |   19.42 | 111111111111111
     9 |   87.03 | 11111111111111111111
    10 |   70.64 | 111111111111111
   (10 rows)
   ```

6. Alternatively, we can use random ordering:

```
postgres=# SELECT key
        ,(random()*100.)::numeric(4,2)
        ,repeat('1',(random()*25)::integer)
        FROM generate_series(1,10) AS f(key)
        ORDER BY random() * 1.0;
 key | numeric |          repeat
-----+---------+-------------------------
   4 |   86.09 | 1111
  10 |   28.30 | 11111111
   2 |   64.09 | 111111
   8 |   91.59 | 111111111111111
   5 |   64.05 | 11111111
   3 |   75.22 | 1111111111111111
   6 |   39.02 | 1111
   7 |   20.43 | 1111111
   1 |   42.91 | 1111111111111111111
   9 |   88.64 | 111111111111111111111
(10 rows)
```

How it works...

Set-returning functions literally return a set of rows. That allows them to be used in either the FROM clause, as if they were a table, or the SELECT clause. The generate_series() set of functions returns either dates or integers, depending on the data types of the input parameters you use.

The :: (double colon) operator is used to cast between data types. The *random string from a list of strings* example uses PostgreSQL arrays. You can create an array using the ARRAY constructor syntax and then use an integer to reference one element in the array. In our case, we used a random subscript.

There's more...

There are also some commercial tools used to generate application-specific test data for PostgreSQL. They are available at https://www.sqlmanager.net/products/postgresql/datagenerator and https://www.datanamic.com/datagenerator/index.html.

The key features of any data generator are as follows:

- The ability to generate data in the right format for custom data types
- The ability to add data to multiple tables, while respecting foreign key constraints between tables
- The ability to add data to non-uniform distributions

The tools and tricks shown here are cool and clever, although there are some problems hiding here as well. Real data has so many strange things in it that it can be very hard to simulate. One of the most difficult things is generating data that follows realistic distributions. For example, if we had to generate data for people's heights, then we'd want to generate data to follow a normal distribution. If we were generating customers' bank balances, we'd want to use a Zipf distribution, or for the number of reported insurance claims, perhaps a Poisson distribution (or perhaps not). Replicating real quirks in data can take some time.

Finally, note that casting a float into an integer rounds it to the nearest integer, so the distribution of integers is not uniform on each extreme. For instance, the probability of (random()*10)::int being 0 is just 5%, as is its probability of being 10, while each integer between 1 and 9 occurs with a probability of 10%. This is why we put 0.5 in the last example, which is simpler than using the floor() function.

See also

You can use existing data to generate test databases using sampling. That's the subject of our next recipe, *Randomly sampling data*.

Randomly sampling data

DBAs may be asked to set up a test server and populate it with test data. Often, that server will be old hardware, possibly with smaller disk sizes. So, the subject of data sampling raises its head.

The purpose of sampling is to reduce the size of the dataset and improve the speed of later analysis. Some statisticians are so used to the idea of sampling that they may not even question whether its use is valid or if it might cause further complications.

The SQL standard way to perform sampling is by adding the TABLESAMPLE clause to the SELECT statement.

How to do it...

In this section, we will take a random sample of a given collection of data (for example, a given table). First, you should realize that there isn't a simple tool to slice off a sample of your database. It would be neat if there were, but there isn't. You'll need to read all of this to understand why:

1. We first consider using SQL to derive a sample. Random sampling is actually very simple because we can use the TABLESAMPLE clause. Consider the following example:

   ```
   postgres=# SELECT count(*) FROM mybigtable;
   count
   -------
   10000
   (1 row)
   postgres=# SELECT count(*) FROM mybigtable
                           TABLESAMPLE BERNOULLI(1);
   count
   -------
      106
   (1 row)
   postgres=# SELECT count(*) FROM mybigtable
                           TABLESAMPLE BERNOULLI(1);
   count
   -------
       99
   (1 row)
   ```

2. Here, the TABLESAMPLE clause applies to mybigtable and tells SELECT to consider only a random sample, while the BERNOULLI keyword denotes the sampling method used, and the number 1 between parentheses represents the percentage of rows that we want to consider in the sample – that is, 1%. Quite easy!

3. Now, we need to get the sampled data out of the database, which is tricky for a few reasons. Firstly, there is no option to specify a WHERE clause for pg_dump. Secondly, if you create a view that contains the WHERE clause, pg_dump dumps only the view definition, not the view itself.

4. You can use pg_dump to dump all databases, apart from a set of tables, so you can produce a sampled dump like this:

```
pg_dump --exclude-table=mybigtable > db.dmp
pg_dump --table=mybigtable --schema-only > mybigtable.schema
psql -c '\copy (SELECT * FROM mybigtable
                    TABLESAMPLE BERNOULLI (1)) to mybigtable.dat'
```

5. Then, reload onto a separate database using the following commands:

```
psql -f db.dmp
psql -f mybigtable.schema
psql -c '\copy mybigtable from mybigtable.dat'
```

Overall, our advice is to use sampling with caution. In general, it is easier to apply it to a few very large tables only, in view of both the mathematical issues surrounding the sample design and the difficulty of extracting the data.

How it works...

The extract mechanism shows off the capabilities of the psql and pg_dump PostgreSQL command-line tools, as pg_dump allows you to include or exclude objects and dump the entire table (or only its schema), whereas psql allows you to dump out the result of an arbitrary query into a file.

The BERNOULLI clause specifies the sampling method – that is, PostgreSQL takes the random sample by performing a full table scan and then selecting each row with the required probability (here, 1%).

Another built-in sampling method is SYSTEM, which reads a random sample of table pages and then includes all rows in these pages; this is generally faster, given that samples are normally quite a bit smaller than the original, but the randomness of the selection is affected by how rows are physically arranged on disk, which makes it suitable for some applications only.

Here is an example that shows what the problem is. Suppose you take a dictionary, rip out a few pages, and then select all the words in them; you will get a random sample composed of a few *clusters* of consecutive words. This is good enough if you want to estimate the average length of a word but not for analyzing the average number of words for each initial letter. The reason is that the initial letter of a word is strongly correlated with how words are arranged in pages, while the length of a word is not.

We haven't discussed how random the TABLESAMPLE clause is. This isn't the right place for such details; however, it is reasonably simple to extend PostgreSQL with extra functions or sampling methods, so if you prefer another mechanism, you can find an external random number generator and create a new sampling method for the TABLESAMPLE clause. PostgreSQL includes two extra sampling methods, tsm_system_rows and tsm_system_time, as contrib extensions; they are excellent examples to start with.

The tsm_system_rows method does not work with percentages; instead, the numeric argument is interpreted as the number of rows to be returned. Similarly, the tsm_system_time method will regard its argument as the number of milliseconds to spend retrieving the random sample.

These two methods include the word system in their name because they use block-level sampling, such as the built-in system sampling method; hence, their randomness is affected by the same *clustering* limitation as described previously.

The sampling method shown earlier is a simple random sampling technique that has an **Equal Probability of Selection** (**EPS**) design.

EPS samples are considered useful because the variance of the sample attributes is similar to the variance of the original dataset. However, bear in mind that this is useful only if you are considering variances.

Simple random sampling can make the eventual sample biased toward more frequently occurring data. For example, if you have a 1% sample of data on which some kinds of data occur only 0.001% of the time, you may end up with a dataset that doesn't have any of that outlying data.

What you might wish to do is to pre-cluster your data and take different samples from each group, ensuring that you have a sampled dataset that includes many more outlying attributes. A simple method might be to do the following:

- Include 1% of all normal data
- Include 25% of outlying data

Note that if you do this, then it is no longer an EPS sample design.

Undoubtedly, there are statisticians who will be fuming after reading this. You're welcome to use the facilities of the SQL language to create a more accurate sample. Just make sure that you know what you're doing, and check out some good statistical literature, websites, or textbooks.

Loading data from a spreadsheet

Spreadsheets are the most obvious starting place for most data stores. Studies within a range of businesses consistently show that more than 50% of smaller data stores are held in spreadsheets or small desktop databases. Loading data from these sources is a frequent and important task for many DBAs.

Getting ready

Spreadsheets combine data, presentations, and programs all into one file. That's perfect for power users wanting to work quickly. As with other relational databases, PostgreSQL is mainly concerned with the lowest level of data, so extracting just data from these spreadsheets can present some challenges.

We can easily handle spreadsheet data if that spreadsheet's layout follows a very specific form, as follows:

- Each spreadsheet column becomes one column in one table.
- Each row of the spreadsheet becomes one row in one table.
- Data is only in one worksheet of the spreadsheet.
- Optionally, the first row is a list of column descriptions/titles.

This is a very simple layout, and more often, there will be other things in the spreadsheet, such as titles, comments, constants for use in formulas, summary lines, macros, and images. If you're in this position, the best thing to do is to create a new worksheet within the spreadsheet in the pristine form described earlier and then set up cross-worksheet references to bring in the data. An example of a cross-worksheet reference would be =Sheet2.A1. You'll need a separate worksheet for each set of data, which will become one table in PostgreSQL. You can load multiple worksheets into one table, however.

Some spreadsheet users will say that all of this is unnecessary and is evidence of the problems of databases. Real spreadsheet gurus do actually advocate this type of layout – data in one worksheet and calculation and presentation in other worksheets. So, it is actually a best practice to design spreadsheets in this way; however, we must work with the world the way it is.

How to do it...

Here, we will show you an example where data in a spreadsheet is loaded into a database:

1. If your spreadsheet data is neatly laid out in a single worksheet, as shown in the following screenshot, then you can go to **File** | **Save As** and select **CSV** as the file type to be saved:

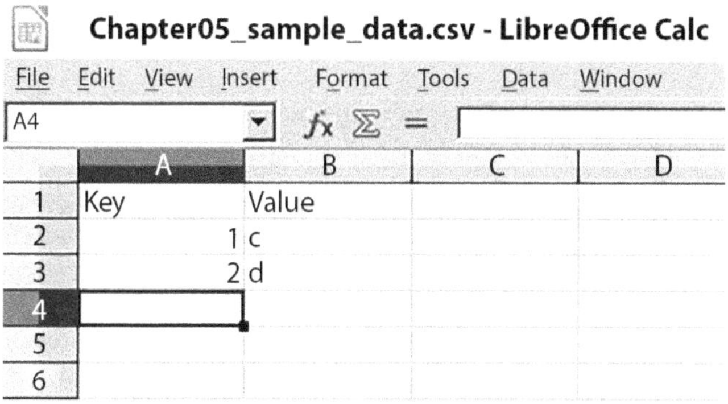

Figure 5.1: A very simple spreadsheet example

2. This will export the current worksheet to a file, as follows:

```
"Key","Value"
1,"c"
2,"d"
```

3. We can then create a table to load the data into, using psql and the following command:

```
CREATE TABLE sample
(key integer
,value text);
```

4. We can then load it into the PostgreSQL table, using the following psql command:

```
postgres=# \COPY sample FROM sample.csv CSV HEADER
postgres=# SELECT * FROM sample;
 key | value
-----+-------
   1 | c
   2 | d
```

5. Alternatively, from the command line, this would be as follows:

```
psql -c '\COPY sample FROM sample.csv CSV HEADER'
```

The filename can include a full file path if the data is in a different directory. The psql \COPY command transfers data from the client system where you run the command through to the database server, so the file is on the client. Higher privileges are not required, so this is the preferred method.

6. If you are submitting SQL through another type of connection, then you can also use the following SQL statement of the form, noting that the leading backslash is removed:

```
COPY sample FROM '/mydatafiledirectory/sample.csv' CSV HEADER;
```

The COPY statement shown in the preceding SQL statement uses an absolute path to identify data files, which is required. This method runs on the database server and can only be executed by a super user, or a user who has been granted one of the pg_read_server_files, pg_write_server_files, or pg_execute_server_program roles. So, you need to ensure that the server process is allowed to read that file, then transfer the data yourself to the server, and finally, load the file. These privileges are not commonly granted, which is why we prefer the earlier method.

The COPY (or \COPY) command does not create the table for you; that must be done beforehand. Note also that the HEADER option does nothing but ignore the first line of the input file, so the names of the columns from the .csv file don't need to match those of the Postgres table. If it hasn't occurred to you yet, this is also a problem. If you say HEADER and the file does not have a header line, then all it does is ignore the first data row. Unfortunately, there's no way for PostgreSQL to tell whether the first line of the file is truly a header or not. Be careful!

There isn't a standard tool to load data directly from the spreadsheet to the database. It's fairly simple to write a spreadsheet macro to automate the aforementioned tasks, but that's not a topic for this book.

How it works...

The \COPY command executes a COPY SQL statement, so the two methods described earlier are very similar. There's more to be said about COPY, so we'll cover that in the next recipe.

Under the covers, the \COPY command executes a COPY ... FROM STDIN command. When using this form of command, the client program must read the file and feed the data to the server. psql does this for you, but in other contexts, you can use this mechanism to avoid the need for higher privileges or additional roles, which are needed when running COPY with an absolute filename.

There's more...

There are many data extraction and loading tools available out there, some cheap and some expensive. Remember that the hardest part of loading data from any spreadsheet is separating the data from all the other things it contains. We've not yet seen a tool that can help with that! This is why the best practice for spreadsheets is to separate data into separate worksheets.

Loading data from flat files

Loading data into your database is one of the most important tasks. You need to do this accurately and quickly. Here's how.

Getting ready

For basic loading, COPY works well for many cases, including CSV files, as shown in the last recipe.

If you want advanced functionality for loading, you may wish to try pgloader, which is commonly available in all main software distributions. At the time of writing, the current stable version is 3.6.9. There are many features, but it is stable, with very few new features in recent years.

How to do it...

To load data with pgloader, we need to understand our requirements, so let's break this down into a step-by-step process, as follows:

1. Identify the data files and where they are located. Make sure that pgloader is installed in the location of the files.
2. Identify the table into which you are loading, ensure that you have the permissions to load, and check the available space. Work out the file type (examples include fixed-size fields, delimited text, and CSV) and check the encoding.
3. Specify the mapping between columns in the file and columns on the table being loaded. Make sure you know which columns in the file are not needed – pgloader allows you to include only the columns you want. Identify any columns in the table for which you don't have data. Do you need them to have a default value on the table, or does pgloader need to generate values for those columns through functions or constants?
4. Specify any transformations that need to take place. The most common issue is date formats, although it's possible that there may be other issues.
5. Write the pgloader script.

6. The pgloader script will create a log file to record whether the load has succeeded or failed, and another file to store rejected rows. You need a directory with sufficient disk space if you expect them to be large. Their size is roughly proportional to the number of failing rows.
7. Finally, consider what settings you need for performance options. This is definitely last, as fiddling with things earlier can lead to confusion when you're still making the load work correctly.
8. You must use a script to execute pgloader. This is not a restriction; actually, it is more like a best practice, because it makes it much easier to iterate toward something that works. Loads never work the first time, except in the movies!

Let's look at a typical example from the quick-start documentation of pgloader, the csv.load file.

Define the required operations in a command and save it in a file, such as csv.load:

```
LOAD CSV
    FROM '/tmp/file.csv' (x, y, a, b, c, d)
    INTO postgresql://postgres@localhost:5432/postgres?csv (a, b, d, c)
      WITH truncate,
           skip header = 1,
           fields optionally enclosed by '"',
           fields escaped by double-quote,
           fields terminated by ','
       SET client_encoding to 'latin1',
           work_mem to '12MB',
           standard_conforming_strings to 'on'
    BEFORE LOAD DO
     $$ drop table if exists csv; $$,
     $$ create table csv (
          a bigint,
          b bigint,
          c char(2),
          d text
        );
     $$;
```

This command allows us to load the following CSV file content. Save this in a file, such as `file.csv`, under the `/tmp` directory:

```
Header, with a © sign
"2.6.190.56","2.6.190.63","33996344","33996351","GB","United Kingdom"
"3.0.0.0","4.17.135.31","50331648","68257567","US","United States"
"4.17.135.32","4.17.135.63","68257568","68257599","CA","Canada"
"4.17.135.64","4.17.142.255","68257600","68259583","US","United States"
"4.17.143.0","4.17.143.15","68259584","68259599","CA","Canada"
"4.17.143.16","4.18.32.71","68259600","68296775","US","United States"
```

We can use the following `load` script:

```
pgloader csv.load
```

Here's what gets loaded in the PostgreSQL database:

```
postgres=# select * from csv;
    a     |    b     | c  |       d
----------+----------+----+----------------
 33996344 | 33996351 | GB | United Kingdom
 50331648 | 68257567 | US | United States
 68257568 | 68257599 | CA | Canada
 68257600 | 68259583 | US | United States
 68259584 | 68259599 | CA | Canada
 68259600 | 68296775 | US | United States
(6 rows)
```

How it works...

pgloader copes gracefully with errors. The COPY command loads all rows in a single transaction, so only a single error is enough to abort the load. pgloader breaks down an input file into reasonably sized chunks and loads them piece by piece. If some rows in a chunk cause errors, then pgloader will split it iteratively until it loads all the good rows and skips all the bad rows, which are then saved in a separate rejects file for later inspection. This behavior is very convenient if you have large data files with a small percentage of bad rows – for instance, you can edit the rejects, fix them, and finally, load them with another pgloader run.

Versions from the 2.x iteration of pgloader were written in Python and connected to PostgreSQL through the standard Python client interface. Version 3.x is written in Common Lisp. Yes, pgloader is less efficient than loading data files using a COPY command, but running a COPY command has many more restrictions: the file has to be in the right place on the server, has to be in the right format, and must be unlikely to throw errors on loading. pgloader has additional overhead, but it also has the ability to load data using multiple parallel threads, so it can be faster to use as well. The ability of pgloader to reformat data via user-defined functions is often essential; a straight COPY command may not be enough.

pgloader also allows loading from fixed-width files, which COPY does not.

If you need to reload the table completely from scratch, then specify the WITH TRUNCATE clause in the pgloader script.

There are also options to specify SQL to be executed before and after loading data. For instance, you can have a script that creates the empty tables before, you can add constraints after, or you can have both.

There's more...

After loading, if we have load errors, then there will be bloat in the PostgreSQL tables. You should think about whether you need to add a VACUUM command after the data load, although this will possibly make the load take much longer.

We need to be careful to avoid loading data twice. The only easy way of doing so is to make sure that there is at least one unique index defined on every table that you load. The load should then fail very quickly.

String handling can often be difficult because of the presence of formatting or non-printable characters. The default setting for PostgreSQL is to have a parameter named standard_conforming_strings set to off, which means that backslashes will be assumed to be escape characters. Put another way, by default, the \n string means a line feed, which can cause data to appear truncated. You'll need to turn standard_conforming_strings to on, or you'll need to specify an escape character in the load-parameter file.

If you are reloading data that has been unloaded from PostgreSQL, then you may want to use the pg_restore utility instead. The pg_restore utility has the option to reload data in parallel, -j number_of_threads, although this is only possible if the dump was produced using the directory format of pg_dump. Refer to the recipes in *Chapter 11*, *Backup and Recovery*, for more details. This can be useful to reload dumps, although it lacks almost all of the other pgloader features discussed here.

If you need to use rows from a read-only text file that does not have errors, then you may consider using the file_fdw contrib module. The short story is that it lets you create a *virtual* table that will parse the text file every time it is scanned. This is different from filling a table once and for all, either with COPY or pgloader; therefore, it covers a different use case. For example, think about an external data source that is maintained by a third party and needs to be shared across different databases.

Another option would be EDB*Loader, which also contains a wide range of load options: https://www.enterprisedb.com/docs/epas/latest/database_administration/02_edb_loader/.

Making bulk data changes using server-side procedures with transactions

In some cases, you'll need to make bulk changes to your data. In many cases, you will need to scroll through the data, making changes according to a complex set of rules. You have a few choices in that case:

- Write a single SQL statement that can do everything.
- Open a cursor, read the rows out, and then make changes with a client-side program.
- Write a procedure that uses a cursor to read the rows, and make changes using server-side SQL.

Writing a single SQL statement that does everything is sometimes possible, but if you need to do more than just use UPDATE, then it becomes difficult very quickly. The main difficulty is that the SQL statement isn't restartable, so if you need to interrupt it, you will lose all of your work.

Reading all the rows back to a client-side program can be very slow – if you need to write this kind of program, it is better to do it all on the database server.

Getting ready

Create an example table, and fill it with nearly 1,000 rows of test data:

```
CREATE TABLE employee (
empid     BIGINT NOT NULL PRIMARY KEY
,job_code  TEXT NOT NULL
,salary    NUMERIC NOT NULL
);
INSERT INTO employee VALUES (1, 'A1', 50000.00);
INSERT INTO employee VALUES (2, 'B1', 40000.00);
INSERT INTO employee SELECT generate_series(10,1000), 'A2', 10000.00;
```

How to do it...

We're going to write a procedure in PL/pgSQL. A procedure is similar to a function, except that it doesn't return any value or object. We use a procedure because it allows you to run multiple server-side transactions. By using procedures in this way, we can break a problem down into a set of smaller transactions that cause less of a problem with database bloat and long-running transactions.

As an example, let's consider a case where we need to update all employees with the A2 job grade, giving each person a 2% pay rise:

```
CREATE PROCEDURE annual_pay_rise (percent numeric)
LANGUAGE plpgsql AS $$
DECLARE
c CURSOR FOR
SELECT * FROM employee
    WHERE job_code = 'A2';
BEGIN
FOR r IN c LOOP
UPDATE employee
SET salary = salary * (1 + (percent/100.0))
WHERE empid = r.empid;
IF mod (r.empid, 100) = 0 THEN
COMMIT;
END IF;
END LOOP;
END;
$$;
```

Execute the preceding procedure like this:

```
CALL annual_pay_rise(2);
```

We want to issue regular commits as we go. The preceding procedure is coded so that it issues commits roughly every 100 rows. There's nothing magical about that number; we just want to break it down into smaller pieces, whether it is the number of rows scanned or rows updated.

There's more...

You can use both COMMIT and ROLLBACK in a procedure. Each new transaction will see the changes from prior transactions and any other concurrent commits that have occurred.

What happens if your procedure is interrupted? Since we are using multiple transactions to complete the task, we won't expect the whole task to be atomic. If the execution is interrupted, we need to rerun the parts that didn't execute successfully. What happens if we accidentally rerun parts that have already been executed? We will give some people a double pay rise, but not everyone.

To cope, let's invent a simple job restart mechanism. This uses a persistent table to track changes as they are made, accessed by a simple API:

```
CREATE TABLE job_status
(id bigserial not null primary key,status text not null,restartdata
bigint);
CREATE OR REPLACE FUNCTION job_start_new ()
RETURNS bigint
LANGUAGE plpgsql
AS $$
DECLARE
  p_id BIGINT;
BEGIN
  INSERT INTO job_status (status, restartdata)
    VALUES ('START', 0)
   RETURNING id INTO p_id;
  RETURN p_id;
END; $$;
CREATE OR REPLACE FUNCTION job_get_status (jobid bigint)
RETURNS bigint
LANGUAGE plpgsql
AS $$
DECLARE
rdata BIGINT;
BEGIN
  SELECT restartdata INTO rdata
    FROM job_status
    WHERE status != 'COMPLETE' AND id = jobid;
  IF NOT FOUND THEN
    RAISE EXCEPTION 'job id does not exist';
  END IF;
  RETURN rdata;
END; $$;
```

```
CREATE OR REPLACE PROCEDURE
job_update (jobid bigint, rdata bigint)
LANGUAGE plpgsql
AS $$
BEGIN
  UPDATE job_status
    SET status = 'IN PROGRESS'
        ,restartdata = rdata
    WHERE id = jobid;
END; $$;
CREATE OR REPLACE PROCEDURE job_complete (jobid bigint)
LANGUAGE plpgsql
AS $$
BEGIN
  UPDATE job_status SET status = 'COMPLETE'
    WHERE id = jobid;
END; $$;
```

First of all, we start a new job:

```
SELECT job_start_new();
```

Then, we execute our procedure, passing the job number to it. Let's say this returns 8474:

```
CALL annual_pay_rise(8474);
```

If the procedure is interrupted, we will restart from the correct place, without needing to specify any changes:

```
CALL annual_pay_rise(8474);
```

The existing procedure needs to be modified to use the new restart API, as shown in the following code block. Note, also, that the cursor has to be modified to use an ORDER BY clause, making the procedure sensibly repeatable:

```
CREATE OR REPLACE PROCEDURE annual_pay_rise (job bigint)
LANGUAGE plpgsql AS $$
DECLARE
        job_empid bigint;
        c NO SCROLL CURSOR FOR
            SELECT * FROM employee
```

```
                WHERE job_code='A2'
                AND empid > job_empid
                ORDER BY empid;
BEGIN
      SELECT job_get_status(job) INTO job_empid;
      FOR r IN c LOOP
            UPDATE employee
            SET salary = salary * 1.02
            WHERE empid = r.empid;
            IF mod (r.empid, 100) = 0 THEN
                  CALL job_update(job, r.empid);
                  COMMIT;
            END IF;
      END LOOP;
      CALL job_complete(job);
END; $$;
```

For extra practice, follow the execution using the debugger in `pgAdmin`.

The `CALL` statement can also be used to call functions that return `void`, but other than that, functions and procedures are separate concepts. Procedures also allow you to execute transactions in PL/Python and PL/Perl.

When using this form of command, the client program must read the file and feed the data to the server. `psql` does this for you, but in other contexts, you can use this mechanism to avoid the need for higher privileges or additional roles, which are needed when running `COPY` with an absolute filename.

Dealing with large tables with table partitioning

When we say that PostgreSQL supports table partitioning, we mean the division of a table into distinct independent tables. You want to do this because it makes large tables easier to manage, but also for performance reasons.

Postgres has had partitioning since version 8.1, and that was known as *inheritance-based* partitioning, which was complex and came with some serious limitations. Since version 10, *declarative* partitioning has become available, which lets you specify a partitioning method, the partitioning key, and the partition boundaries simply by creating tables through DDL.

How to do it...

Let's create an example partitioned table. We want to partition our large table of transactions by their timestamp:

```
CREATE TABLE transactions (
tstamp TIMESTAMP WITH TIME ZONE PRIMARY KEY
,amount NUMERIC
)
PARTITION BY RANGE (tstamp);
```

Great, that's the base table created; however, it has no partitions yet. We can create some so that we can keep transactions separated by the year in which they took place:

```
CREATE TABLE transactions_2021
PARTITION OF transactions
FOR VALUES FROM ('2021-01-01') TO ('2022-01-01');
CREATE TABLE transactions_2022
PARTITION OF transactions
FOR VALUES FROM ('2022-01-01') TO ('2023-01-01');
CREATE TABLE transactions_2023
PARTITION OF transactions
FOR VALUES FROM ('2023-01-01') TO ('2024-01-01');
```

Now, we have a fully partitioned table that can accept transactions from the years 2021, 2022, and 2023. Obviously, we will need to create a new partition before the year 2024 comes around. But what about if we have older transactions that we want to move into this table? If we don't care about creating individual partitions for them, we can lump them into the DEFAULT partition, like so:

```
CREATE TABLE transactions_older
PARTITION OF transactions DEFAULT;
```

Rows will now be routed to the correct table, but we'll be able to transparently select them from the parent table:

```
=> INSERT INTO transactions values ('2023-10-25 20:00:31 UTC', 12.95),
('2020-01-01 00:00:01 UTC', 55.55);
INSERT 0 2
=> TABLE transactions;
         tstamp         | amount
------------------------+--------
```

```
 2023-10-25 21:00:31+01 |  12.95
 2020-01-01 00:00:01+00 |  55.55
(2 rows)
=> TABLE transactions_2023;
        tstamp          | amount
------------------------+--------
 2023-10-25 21:00:31+01 |  12.95
(1 row)

=> TABLE transactions_older;
        tstamp          | amount
------------------------+--------
 2020-01-01 00:00:01+00 |  55.55
(1 row)
```

How it works...

By having a large amount of data organized by ranges, we are now free to do maintenance such as removing old transactions by simply dropping the oldest partition, while avoiding costly DELETE statements that need to scan a very large table. Or, we can archive them by moving their rows to the transactions_older table, if these rows don't hold particular significance for us anymore and are not useful for further retrieval or processing.

Performance can also be enhanced by PostgreSQL's *partition pruning* feature, which ensures that only relevant tables will be examined for queries that reference the partitioning key. So if we're looking for transactions, for example, between February and March 2022, the query planner will only look at the transactions_2022 table.

There's more...

This was RANGE partitioning, but other methods are available, such as LIST, where we specify a list of values to be included in each partition, and HASH, which is used to create a number of partitions of roughly equal size.

Keep in mind that partitions may be partitioned themselves for finer granularity, which is also known as *sub-partitioning*.

The partitioning key must be carefully chosen because it affects the maintainability of the table as well as the performance of queries against it. Please see the *Finding good candidates for partition keys* recipe in this chapter for more details.

You can find out more about the many possibilities that partitioning opens up in the PostgreSQL documentation at https://www.postgresql.org/docs/16/ddl-partitioning.html#DDL-PARTITIONING-DECLARATIVE and https://www.postgresql.org/docs/16/sql-createtable.html#SQL-CREATETABLE-PARMS-PARTITION-BY.

Finding good candidates for partition keys

What is a good partition key? The answer is probably close to one that enables you to have the right number of partitions for ease of use and maintainability, but also one that enables partition pruning for your queries for optimal performance.

Getting ready

In order to choose a good partition key for your tables, you need to understand the nature of your data. This means you need to analyze your queries to determine the main keys that you're using for data retrieval from those tables. Two desirable characteristics of partition keys are:

- They need to have a high enough cardinality, or range of values, for the number of partitions desired.
- They need to be columns that don't change often, in order to avoid having to move rows among partitions.

How to do it...

First off, determine your access patterns for the table in question. See which keys you are selecting data by (the ones that are in your WHERE predicate).

In order to determine the cardinality, you can use the techniques to determine uniqueness, mentioned in the *Finding a unique key for a set of data* recipe in this chapter. Then, you need to exercise your judgment to find a suitable key that makes sense. For example, the transaction amount will have high cardinality, but stop to think how useful it would be, depending on your application, to partition a table of transactions by amount? Probably not much. Obviously, it's not very useful to partition by a key that can only take a handful of values; e.g., True and False will only give you two partitions.

It's good to be aware of the frequency distribution of the key as well. For example, you may have values that are very hot all gathered together in the same partition, which can cause performance issues.

Generally speaking, if you have a table that grows over time, and a temporal value is recorded, it's a pretty safe bet that you'll want to partition over the time dimension for an interval that is appropriate – so you don't have too many small partitions or too few huge ones.

If you choose a column that is supposed to change over the lifetime of a row, such as the last time an inventory item was ordered, that will not be a very good key to partition by because then you'll have to shuffle rows around constantly between partitions.

There's more...

You can use multiple columns as the partitioning key for more granularity, as it allows you to split the data more finely into a larger number of partitions.

Consolidating data with MERGE

The `MERGE` command is a SQL standard way to insert new data into a table, even when new data partially overlaps with the existing data.

We will illustrate `MERGE` using the example of a shop manager who records the amount of each article in a database table. The manager will use the table to know which articles are running out and to order the correct amount of articles at the right time.

Getting ready

Create the initial table with some plausible data, representing the current inventory of the shop:

```
CREATE TABLE articles
( id int PRIMARY KEY
, description text NOT NULL
, quantity int NOT NULL
);

INSERT INTO articles (id, quantity, description) VALUES
  (1, 15, 'Aubergines')
, (2,  6, 'Bananas')
, (3, 34, 'Carrots')
, (4, 22, 'Dates (kg)')
;
```

Now, we create a temporary table representing the list of supplies that have been ordered:

```
CREATE TEMP TABLE new_articles (LIKE articles);

INSERT INTO new_articles (id, quantity, description) VALUES
  (2, 24, 'Bananas')
, (4, 50, 'Dates')
, (5, 12, 'Eclairs')
;
```

How to do it...

The `articles` table contains the following data:

```
SELECT * FROM articles ORDER BY id;

 id | description | quantity
----+-------------+----------
  1 | Aubergines  |       15
  2 | Bananas     |        6
  3 | Carrots     |       34
  4 | Dates (kg)  |       22
(4 rows)
```

While the `new_articles` table contains the following rows:

```
SELECT * FROM new_articles ORDER BY id;

 id | description | quantity
----+-------------+----------
  2 | Bananas     |       24
  4 | Dates       |       50
  5 | Eclairs     |       12
(3 rows)
```

In order to merge new articles into articles, you can issue the following query:

```
MERGE INTO articles a
USING new_articles n
ON a.id = n.id
WHEN MATCHED THEN
```

```
      UPDATE SET quantity = a.quantity + n.quantity
  WHEN NOT MATCHED THEN
      INSERT VALUES (n.id, n.description, n.quantity);
```

This will change the `articles` table as follows:

```
SELECT * FROM articles ORDER BY id;

 id | description | quantity
----+-------------+----------
  1 | Aubergines  |       15
  2 | Bananas     |       30
  3 | Carrots     |       34
  4 | Dates (kg)  |       72
  5 | Eclairs     |       12
(5 rows)
```

There's more...

A similar outcome could be achieved with the INSERT ... ON CONFLICT syntax, which existed already. However, MERGE is standard SQL, so it is preferable to a PostgreSQL-only syntax. Moreover, those two commands are not entirely equivalent in terms of what they allow. The other SQL statements that modify the rows matching a given subquery are DELETE FROM ... USING and UPDATE ... WITH. They are also not equivalent to MERGE, as they don't allow inserts.

Deciding when to use JSON data types

This will be an unusual recipe, being more of a list of best practices and tips, based on experience.

Getting ready

If you are thinking of using a JSON data type, then you already know what JSON is, and you probably have in mind several examples of JSON values.

However, if for some reason you want to run this recipe without knowing what JSON is, you can get a very good idea by visiting its official website (https://www.json.org/), which has been translated into many languages.

There is nothing else you need to do on PostgreSQL because the JSON data types and operators are already built in by default.

How to do it...

When deciding on a database model, you have two opposite approaches.

In the relational approach, the database schema is normalized, by ensuring that each piece of information has its own separate column. The column metadata defines some information on what restrictions apply:

- Whether that piece of information is required (i.e., NOT NULL)
- What data type is allowed in the column
- Any other restrictions (including uniqueness and referential constraints)

Using JSON data types is equivalent to encoding complex data values into a single value, which is then inserted as one field into a row of a database table. By design, JSON data types are not subject to the three kinds of restrictions listed above. This approach is referred to as *schema freedom* or *schema dynamism*.

In theory, it is possible to encode an entire table into a single JSON value, and PostgreSQL allows single values to be as large as 1 GB in compressed form. However, this is unlikely to be useful. More common choices include storing each row in a separate JSON value, or even doing that only for a subset of the row (i.e., some columns).

We denote them as "choices" because the data stored in one way can be converted to the other way, without losing information, so the choices are always all available. However, depending on the actual data that you store, some choices might be preferrable to others, so you should be aware of the pro/con balance in some common scenarios.

For instance, if your data is sparse and has a large number of attributes, then if you store the data in a relational way, most of the rows will have most of the columns empty. It will not be easy for a user to inspect and understand a table, due to the high number of columns; and perhaps the storage is also inefficient. So you can replace the sparse columns with a single JSON value that encodes them all, while leaving the non-sparse columns as they are. This way, the table becomes readable, while retaining the restrictions where they are most useful (i.e., on columns that are mostly not null).

On the other hand, it is slower and more complicated to apply constraints to the single components of a JSON column. Previously, we mentioned a hypothetical user who finds a table easier to inspect and understand because the sparse columns have been moved to a single JSON column. That same user will find some constraints harder to inspect and understand if they involve the JSON column.

Example: moving sparse columns to JSON

We continue the example from the previous recipe, adding some columns and placing data sparsely in them:

```
ALTER TABLE articles
ADD COLUMN feet float,
ADD COLUMN inches float,
ADD COLUMN meters float,
ADD COLUMN pints float,
ADD COLUMN liters float,
ADD COLUMN gallons float,
ADD COLUMN kg float,
ADD COLUMN ounces float,
ADD COLUMN pounds float,
ALTER COLUMN quantity DROP NOT NULL;
```

So, the idea is that the quantity becomes optional because it now is just one of the possible units.

We are not saying that this is the best way to design a database; it's just a quick way to create an example of sparse columns.

We need to change existing data, by removing the unit from the name of row 4, and by moving the value in quantity to the kg column:

```
UPDATE articles
SET description = 'Dates'
, kg = quantity
, quantity = null
WHERE id = 4;
```

We then insert a few new articles, demonstrating the use of the new units:

```
INSERT INTO articles (id, liters, description)
VALUES (6, 15, 'Fresh orange juice');

INSERT INTO articles (id, meters, description)
VALUES (7, 4.15, 'Green celery sticks');

INSERT INTO articles (id, pounds, description)
VALUES (8, 19, 'Hazelnuts');
```

With all the new columns, our table has become quite sparse:

```
SELECT * FROM articles ORDER BY id;
 id |    description    | quantity | feet | inches | meters | pints | liters | gallons | kg | ounces | pounds
----+-------------------+----------+------+--------+--------+-------+--------+---------+----+--------+--------
  1 | Aubergines        |       15 |      |        |        |       |        |         |    |        |
  3 | Carrots           |       34 |      |        |        |       |        |         |    |        |
  2 | Bananas           |       30 |      |        |        |       |        |         |    |        |
  5 | Eclairs           |       12 |      |        |        |       |        |         |    |        |
  4 | Dates             |          |      |        |        |       |        |      72 |    |        |
  6 | Fresh orange juice|          |      |        |        |       |     15 |         |    |        |
  7 | Green celery stick|          |      |        |   4.15 |       |        |         |    |        |
  8 | Hazelnuts         |          |      |        |        |       |        |         |    |     19 |
(8 rows)
```

Now, we can move the sparse columns to a single JSON column. Note that we will use the **jsonb** data type. It is an optimized binary representation of JSON that allows indexing, searching, and operations to manipulate JSON objects. Rather than modifying `articles`, we will create a new `articles2` table so that we can retain the original and compare it to the new one:

```
CREATE TABLE articles2
( id int PRIMARY KEY
, description text NOT NULL
, j jsonb
);
```

We populate `articles2` with three SQL statements:

```
INSERT INTO articles2
SELECT id, description, to_jsonb(articles)
FROM articles;
```

```
UPDATE articles2 SET j['id'] = null, j['description'] = null;

UPDATE articles2 SET j = jsonb_strip_nulls(j);
```

The first statement copies the data from articles to articles2. There isn't a quick and simple way to tell PostgreSQL to create a jsonb value with all the columns, except id and description, so we create a jsonb value with all the columns and then remove id and description from the jsonb, which is done by setting these keys to NULL and then removing all the keys that are set to NULL, in accordance with the typical semantics of JSON data.

This is the result:

```
SELECT * FROM articles2 ORDER BY id;
 id |    description      |         j
----+---------------------+-------------------
  1 | Aubergines          | {"quantity": 15}
  2 | Bananas             | {"quantity": 30}
  3 | Carrots             | {"quantity": 34}
  4 | Dates               | {"kg": 72}
  5 | Eclairs             | {"quantity": 12}
  6 | Fresh orange juice  | {"liters": 15}
  7 | Green celery sticks | {"meters": 4.15}
  8 | Hazelnuts           | {"pounds": 19}
(8 rows)
```

In order to appreciate the simplification, you can compare it with the output of:

```
SELECT * FROM articles ORDER BY id;
```

Example: expose JSON data using a view

If some of your columns are collated in a single JSON column, you can still use a view to expose them as if they were stored in a relational way, as in this example:

```
CREATE VIEW articles3 AS
SELECT id, description
, CAST(j->>'quantity' AS int  ) AS quantity
, CAST(j->>'feet'     AS float) AS feet
, CAST(j->>'inches'   AS float) AS inches
, CAST(j->>'meters'   AS float) AS meters
```

```
, CAST(j->>'pints'     AS float) AS pints
, CAST(j->>'liters'    AS float) AS liters
, CAST(j->>'gallons'   AS float) AS gallons
, CAST(j->>'kg'        AS float) AS kg
, CAST(j->>'ounces'    AS float) AS ounces
, CAST(j->>'pounds'    AS float) AS pounds
FROM articles2;
```

We extract values as text, using the double-tip arrows, and then cast them to the data type that we expect.

There's more...

You can use column statistics to predict which columns are sparse, so you know which columns might benefit from being moved into a JSON column.

First, we need to run `ANALYZE` to collect the statistics on the `articles` table, because it didn't get enough row changes to trigger automatic `ANALYZE`:

```
ANALYZE articles;
```

At this point, we can see the fraction of `NULL` values in each column using this simple query:

```
SELECT attname, null_frac
FROM pg_stats
WHERE tablename = 'articles'
ORDER BY null_frac;
   attname   | null_frac
-------------+-----------
 id          |         0
 description |         0
 quantity    |       0.5
 pounds      |     0.875
 meters      |     0.875
 liters      |     0.875
 kg          |     0.875
 ounces      |         1
 gallons     |         1
 inches      |         1
 feet        |         1
 pints       |         1
(12 rows)
```

You can see how the statistics correctly record the fact that id and description do not have NULL values, while the other columns have at least 50% NULLs, with most columns being entirely NULL except, at most, one value.

JSON values with several fields can be indexed using GIN indexes, similar to arrays. This allows queries using operators that check the existence of given keys, such as ?, ?|, and ?&, plus the containment operator and the jsonpath operators, which can perform more elaborate checks. For more details, see:

https://www.postgresql.org/docs/16/datatype-json.html#JSON-INDEXING

Of course, it is also possible to use traditional B-tree indexes to optimize access to the values of specific keys. Therefore, if you want to index a single field within a JSON object, we recommend using a B-tree index. Just write an expression that extracts that value, like the ones we used in the definition of the view. More details are available here:

https://www.postgresql.org/docs/16/indexes-expressional.html

For completeness, we also mention the EAV model, which has the benefit of storing sparse data in an entirely relational way but, at the same time, is more complex to query and aggregate.

The EAV model is, today, considered too complicated to be used, and using JSON data types as explained above is definitely easier and more performant. On the other hand, the second option was not available until a decade or so ago, so in the absence of the best alternative, the second best ends up being chosen.

Learn more on Discord

To join the Discord community for this book – where you can share feedback, ask questions to the author, and learn about new releases – follow the QR code below:

https://discord.gg/pQkghgmgdG

6
Security

This chapter will present a few common recipes to secure your database server. Taken together, these will cover the main areas around security in PostgreSQL that you should be concerned with. The last recipe will cover some cloud-specific topics.

This chapter includes the following recipes:

- An overview of PostgreSQL security
- The PostgreSQL superuser
- Revoking user access to a table
- Granting user access to a table
- Granting user access to specific columns
- Granting user access to specific rows
- Creating a new user
- Temporarily preventing a user from connecting
- Removing a user without dropping their data
- Checking whether all users have a secure password
- Giving limited superuser powers to specific users
- Auditing database access
- Always knowing which user is logged in
- Integrating with the **Lightweight Directory Access Protocol (LDAP)**
- Connecting using encryption (SSL / GSSAPI)
- Using SSL certificates to authenticate

- Mapping external usernames to database roles
- Using column-level encryption
- Setting up cloud security using predefined roles

An overview of PostgreSQL security

Security is a huge area of related methods and technologies, so we will take a practical approach, covering the most common issues related to database security.

First, we set up access rules in the database server. PostgreSQL allows you to control access based on the host that is trying to connect, using the pg_hba.conf file. You can specify SSL/GSSAPI connections if needed or skip that if the network is secure. Passwords are encrypted using **SCRAM-SHA-256**, but many other authentication methods are available.

Next, set up the role and privileges to access your data. Modern databases should be configured using the **Principle Of Least Privilege (POLP)**. Data access is managed by a privilege system, where users are granted different privileges for different tables or other database objects, such as schemas or functions. Thus, some records or tables can only be seen by certain users, and even those tables that are visible to everyone can have restrictions in terms of who can insert new data or change existing data.

It is good practice not to grant privileges directly to users but, instead, to use an intermediate role to collect a set of privileges. This is easier to audit and is more extensible. Then, instead of granting all the same privileges to the actual user, the entire role is granted to users needing these privileges. For example, a clerk role may have the right to both insert data and update existing data in the user_account table but may have the right to only insert data in the transaction_history table.

Fine-grained control over access can be managed using the **Row-Level Security (RLS)** feature, which allows a defined policy on selected tables.

Another aspect of database security concerns the management of this access to the database: making sure that only the right people can access the database, that one user can't see what other users are doing (unless they are an administrator or auditor), and deciding whether users can or cannot pass on the roles granted to them.

You should consider auditing the actions of administrators using pgaudit, although there is also audit functionality within **EDB** Postgres Advanced Server.

Some aspects of security are also covered in *Chapter 7, Database Administration,* and *Chapter 8, Monitoring and Diagnosis,* of this book.

Typical user roles

The minimal production database setup contains at least two types of users—namely, administrators and end users—where administrators can do many things and end users can only do very little, usually just modifying data in only a few tables and reading from a few more.

It is not a good idea to let ordinary users create or change database object definitions, meaning that they should not have the CREATE privilege on any schema, including PUBLIC.

There can be more roles for different types of end users, such as analysts, who can only select from a single table or view, or some maintenance script users who see no data at all and just have the ability to execute a few functions.

Alternatively, there can also be a manager role, which can grant and revoke roles for other users but is not supposed to do anything else.

The PostgreSQL superuser

A PostgreSQL superuser is a user that bypasses all permission checks, except the right to log in. Superuser is a dangerous privilege and should not be used carelessly. Many cloud databases do not allow this level of privilege to be granted at all. It is normal to place strict controls on users of this type. If you are using PostgreSQL in a cloud service, then please read the *Setting up cloud security using predefined roles* recipe instead.

In this recipe, you will learn how to grant a user the right to become a superuser.

How to do it...

Follow the next steps to add or remove superuser privileges for any user:

- A user becomes a superuser when they are created with the SUPERUSER attribute set:
    ```
    CREATE USER username SUPERUSER;
    ```
- A user can be deprived of their superuser status by removing the SUPERUSER attribute using this command:
    ```
    ALTER USER username NOSUPERUSER;
    ```
- A user can be restored to superuser status later using the following command:
    ```
    ALTER USER username SUPERUSER;
    ```
- When neither SUPERUSER nor NOSUPERUSER is given in the CREATE USER command, then the default is to create a user who is not a superuser (regular database user).

How it works...

The rights to some operations in PostgreSQL are not available by default and need to be granted specifically to users. They must be performed by a special user who has this special attribute set. The preceding commands set and reset this attribute for the user.

There's more...

The PostgreSQL system comes set up with at least one superuser. Most commonly, this superuser is named postgres, but occasionally, it adopts the same name as the system user who owns the database directory and with whose rights the PostgreSQL server runs.

Other superuser-like attributes

In addition to SUPERUSER, there are two lesser attributes—CREATEDB and CREATEUSER—that give the user only some of the power reserved for superusers, namely, creating new databases and users. See the *Giving limited superuser powers to specific users* recipe for more information on this.

See also

For more insights into user management and security, please refer to the *Always knowing which user is logged in* recipe in this chapter.

Revoking user access to tables

This recipe answers the question, *How do I make sure that user X cannot access table Y?*

Getting ready

The current user must either be a superuser, the owner of the table, or a user with a GRANT option for the table.

Also, bear in mind that you can't revoke the rights of a superuser.

How to do it...

To revoke all rights on the table1 table from the user2 user, you must run the following **SQL** command:

```
REVOKE ALL ON table1 FROM user2;
```

However, if user2 has been granted another role that gives them some rights on table1—say, role3—this command is not enough; you must also choose one of the following options:

- Fix the user—that is, revoke role3 from user2

- Fix the role—that is, revoke privileges on table1 from role3

Both choices are imperfect because of their side effects. The former will revoke all of the privileges associated with role3, not just the privileges concerning table1; the latter will revoke the privileges on table1 from all other users that have been granted role3, not just from user2.

It is normally better to avoid damaging other legitimate users, so we'll opt for the first solution. We'll now look at a working example.

Using psql, display a list of roles that have been granted at least one privilege on table1 by issuing \z table1. For instance, you can obtain the following output (an extra column about column privileges has been removed from the right-hand side because it was not relevant here):

```
postgres=#\z table1
                      Access privileges
 Schema |  Name  | Type  |     Access privileges      | ...
--------+--------+-------+----------------------------+ ...
 public | table1 | table | postgres=arwdDxt/postgres+ | ...
        |        |       | role3=r/postgres          +| ...
        |        |       | role5=a/postgres           | ...
(1 row)
```

Then, we check whether user2 is a member of any of those roles by typing \du user2:

```
          List of roles
 Role name | Attributes |  Member of
-----------+------------+---------------
 user2     |            | {role3, role4}
```

In the previous step, we noticed that role3 had been granted the SELECT privilege (r for read) by the postgres user, so we must revoke it as follows:

```
REVOKE role3 FROM user2;
```

We must also inspect role4. Even if it doesn't have privileges on table1, in theory, it could be a member of one of the three roles that have privileges on that table. We issue \du role4 and get the following output:

```
             List of roles
 Role name |  Attributes  | Member of
-----------+--------------+-----------
 role4     | Cannot login | {role5}
```

Our suspicion was well founded: user2 can get the INSERT privilege (*a* for *append*) on table1, first via role4 and then via role5. So, we must break this two-step chain as follows:

```
REVOKE role4 FROM user2;
```

This example may seem too unlikely to be true. We unexpectedly gained access to the table via a chain of two different role memberships, which was made possible by the fact that a non-login role, such as role4, was made a member of another non-login role—that is, role5. In most real-world cases, superusers will know whether such chains exist at all, so there will be no surprises; however, the goal of this recipe is to make sure that the user cannot access the table, meaning we cannot exclude less-likely options. See also the later recipe, *Auditing database access*.

How it works...

The \z command, as well as its synonym, \dp, displays all privileges granted on tables, views, and sequences. If the Access privileges column is empty, it means we use **default privileges**—that is, all privileges are given to the owner (and the superusers, as always).

The \du command shows you the attributes and roles that have been granted to roles.

Both commands accept an optional name or pattern to restrict the display.

There's more...

Here, we'll cover some good practices on user and role management.

Database creation scripts

For production systems, it is usually a good idea to always include GRANT and REVOKE statements in the database creation script so that you can be sure that only the right set of users has access to the table. If this is done manually, it is easy to forget. Also, in this way, you can be sure that the same roles are used in development and testing environments so that there are no surprises at deployment time.

Here is an extract from the database creation script:

```
CREATE TABLE table1(
...
);
GRANT SELECT ON table1 TO webreaders;
GRANT SELECT, INSERT, UPDATE, DELETE ON table1 TO editors;
GRANT ALL ON table1 TO admins;
```

Default search path

It is always good practice to use a fully qualified name when revoking or granting rights; otherwise, you may be inadvertently working with the wrong table.

To see the effective search path for the current database, run the following code:

```
pguser=# show search_path ;
   search_path
-----------------
 "$user",public
(1 row)
```

To see which table will be affected if you omit the schema name, run the following code in psql:

```
pguser=# \d x
       Table "public.x"
 Column | Type | Modifiers
--------+------+-----------
```

The `public.x` table name in the response contains the full name, including the schema.

From version 15 onward, only the database owner can create objects in the public schema by default.

Securing views

It is a common technique to use a view to disclose only some parts of a secret table; however, a clever attacker can use access to the view to display the rest of the table using log messages. For instance, consider the following example:

```
CREATE VIEW for_the_public AS
   SELECT * FROM reserved_data WHERE importance < 10;
GRANT SELECT ON for_the_public TO PUBLIC;
```

A malicious user could define the following function:

```
CREATE FUNCTION f(text)
RETURNS boolean
COST 0.00000001
LANGUAGE plpgsql AS $$
BEGIN
  RAISE INFO '$1: %', $1;
```

```
    RETURN true;
END;
$$;
```

They could use it to filter rows from the view:

```
SELECT * FROM for_the_public x WHERE f(x :: text);
```

The PostgreSQL optimizer will then internally rearrange the query, expanding the definition of the view and then combining the two filter conditions into a single WHERE clause. The trick here is that the function has been told to be very cheap using the COST keyword, so the optimizer will choose to evaluate that condition first. In other words, the function will access all of the rows in the table, as you will realize when you see the corresponding INFO lines on the console if you run the code yourself.

This security leak can be prevented using the security_barrier attribute:

```
ALTER VIEW for_the_public SET (security_barrier = on);
```

This means that the conditions that define the view will always be computed first, irrespective of cost considerations.

The performance impact of this fix can be mitigated by the LEAKPROOF attribute for functions. In short, a function that cannot leak information other than its output value can be marked as LEAKPROOF by a superuser, ensuring that the planner will know it's secure enough to compute the function before the other view conditions.

Granting user access to a table

A user needs to have access to a table in order to perform any actions on it.

Getting ready

Make sure that you have the appropriate roles defined and that privileges are revoked from the PUBLIC role:

```
CREATE GROUP webreaders;
CREATE USER tim;
CREATE USER bob;
REVOKE ALL ON SCHEMA someschema FROM PUBLIC;
```

How to do it...

We had to grant access to the schema in order to allow access to the table. This suggests that access to a given schema can be used as a fast and extreme way of preventing any access to any object in that schema. Otherwise, if you want to allow some access, you must use specific GRANT and REVOKE statements, as needed:

```
GRANT USAGE ON SCHEMA someschema TO webreaders;
```

It is often desirable to give a group of users' similar permissions to a group of database objects. To do this, you will first assign all the permissions to a proxy role (also known as a **permission group**) and then assign the group to selected users, as follows:

```
GRANT SELECT ON someschema.pages TO webreaders;
GRANT INSERT ON someschema.viewlog TO webreaders;
GRANT webreaders TO tim, bob;
```

Now, both tim and bob have the SELECT privilege on the pages table and INSERT on the viewlog table. You can also add privileges to the group role after assigning it to users. Consider the following command:

```
GRANT INSERT, UPDATE, DELETE ON someschema.comments TO webreaders;
```

After running this command, both bob and tim have all of the aforementioned privileges on the comments table.

This assumes that both the bob and tim roles were created with the INHERIT default setting. Otherwise, they do not automatically inherit the rights of roles but need to explicitly set their role to the granted user, making use of the privileges granted to that role.

We can grant privileges to all objects of a certain kind in a specific schema, as follows:

```
GRANT SELECT ON ALL TABLES IN SCHEMA someschema TO bob;
```

You still need to grant the privileges on the schema itself in a separate GRANT statement.

How it works...

The preceding sequence of commands first grants access to a schema for a group role, then gives appropriate viewing (SELECT) and modifying (INSERT) rights on certain tables to the role, and finally, grants membership in that role to two database users.

There's more...

There is no requirement in PostgreSQL to have some privileges in order to have others. This means that you may well have write-only tables where you are allowed to insert, but you can't select. This can be used to implement a mail-queue-like functionality, where several users post messages to one user but can't see what other users have posted.

Alternatively, you could set up a situation where you can write a record but you can't change or delete it. This is useful for auditing log-type tables, where all changes are recorded but cannot be tampered with.

Granting user access to specific columns

A user can be given access to only some table columns.

Getting ready

We will continue the example from the previous recipe, so we assume that there is already a schema called `someschema` and a role called `somerole`, with `USAGE` privileges on it. We create a new table on which we will grant column-level privileges:

```
CREATE TABLE someschema.sometable2(col1 int, col2 text);
```

How to do it...

We want to grant `somerole` the ability to view existing data and insert new data; we also want to provide the ability to amend existing data, limited to the `col2` column only. We use the following self-evident statements:

```
GRANT SELECT, INSERT ON someschema.sometable2
TO somerole;
GRANT UPDATE (col2) ON someschema.sometable2
TO somerole;
```

We can then test whether this has worked successfully as follows:

1. Let's assume the identity of the `somerole` role and test these privileges with the following commands:

    ```
    SET ROLE TO somerole;
    INSERT INTO someschema.sometable2 VALUES (1, 'One');
    SELECT * FROM someschema.sometable2 WHERE col1 = 1;
    ```

2. As expected, we are able to insert a new row and view its contents. Let's now check our ability to update individual columns. We start with the second column, which we have authorized:

```
UPDATE someschema.sometable2 SET col2 = 'The number one';
```

This command returns the familiar output:

```
UPDATE 1
```

3. This means that we were able to successfully update that column in one row. Now, we try to update the first column:

```
UPDATE someschema.sometable2 SET col1 = 2;
```

This time, we get the following error message:

```
ERROR:  permission denied for relation sometable2
```

This confirms that, as planned, we only authorized updates to the second column.

How it works...

The GRANT command has been extended to allow the specification of a list of columns, meaning that the privilege is granted on that list of columns rather than on the whole table.

There's more...

Consider a table, t, with c1, c2, and c3 columns; there are two different ways of authorizing the user (u) to perform the following query:

```
SELECT * FROM t;
```

The first is by granting a table-level privilege, as follows:

```
GRANT SELECT ON TABLE t TO u;
```

The alternative way is by granting column-level privileges, as follows:

```
GRANT SELECT (c1,c2,c3) ON TABLE t TO u;
```

Despite these two methods having overlapping effects, table-level privileges are distinct from column-level privileges, which is correct since the meaning of each is different. Granting privileges on a table means giving them *to all columns present and future*, while column-level privileges require the explicit indication of columns and, therefore, don't extend automatically to new columns.

The way privileges work in PostgreSQL means that a given role will be allowed to perform a given action if it matches one of its privileges. This creates some ambiguity in overlapping areas. For example, consider the following command sequence:

```
GRANT SELECT ON someschema.sometable2 TO somerole;
REVOKE SELECT (col1) ON someschema.sometable2 FROM
  somerole;
```

The outcome, somehow surprisingly, will be that `somerole` is allowed to view all of the columns of that table, using the table-level privilege granted by the first command. The second command was ineffective because it tried to revoke a column-level privilege (`SELECT` on `col1`) that was never granted in the first place.

Granting user access to specific rows

PostgreSQL supports granting privileges on a subset of rows in a table using RLS.

Getting ready

Just as we did for the previous recipe, we assume that there is already a schema called `someschema` and a role called `somerole` with `USAGE` privileges on it. We create a new table to experiment with row-level privileges:

```
CREATE TABLE someschema.sometable3(col1 int, col2 text);
```

RLS must also be enabled on that table:

```
ALTER TABLE someschema.sometable3 ENABLE ROW LEVEL SECURITY;
```

How to do it...

First, we grant `somerole` the privilege to view the contents of the table, as we did in the previous recipe:

```
GRANT SELECT ON someschema.sometable3 TO somerole;
```

Let's assume that the contents of the table are as shown by the following command:

```
SELECT * FROM someschema.sometable3;
 col1 |   col2
------+-----------
    1 | One
   -1 | Minus one
(2 rows)
```

In order to grant the ability to access some rows only, we create a **policy** specifying what is allowed and on which rows. For instance, this way, we can enforce the condition that someone is only allowed to select rows with positive values of col1:

```
CREATE POLICY example1 ON someschema.sometable3
FOR SELECT
TO somerole
USING (col1 > 0);
```

The effect of this command is that the rows that do not satisfy the policy are silently skipped, as shown when somerole issues the following command:

```
SELECT * FROM someschema.sometable3;
 col1 |  col2
------+----------
    1 | One
(1 row)
```

What if we want to introduce a policy on the INSERT clause? The preceding policy shows how the USING clause specifies which rows are affected. There is also a WITH CHECK clause that can be used to specify which inserts are accepted. More generally, the USING clause applies to pre-existing rows, while WITH CHECK applies to rows that are generated by the statement being analyzed. So, the former works with SELECT, UPDATE, and DELETE, while the latter works with INSERT and UPDATE.

Coming back to our example, we may want to allow inserts only where col1 is positive:

```
CREATE POLICY example2 ON someschema.sometable3
FOR INSERT
TO somerole
WITH CHECK (col1 > 0);
```

We must also remember to allow INSERT commands on the table, as we did before with SELECT:

```
GRANT INSERT ON someschema.sometable3 TO somerole;
SELECT * FROM someschema.sometable3;
 col1 |  col2
------+----------
    1 | One
(1 row)
```

Now, we are able to insert a new row and see it afterward:

```
INSERT INTO someschema.sometable3 VALUES (2, 'Two');
SELECT * FROM someschema.sometable3;
 col1 |  col2
------+----------
    1 | One
    2 | Two
(2 rows)
```

How it works...

RLS policies are created and dropped on a given table using the `CREATE POLICY` syntax. The RLS policy itself must also be enabled explicitly on the given table because it is disabled by default.

In the previous example, we needed to grant privileges on the table or on the columns, in addition to creating an RLS policy. This is because RLS is not one more privilege to be added to the other; rather, it works as an additional check. In this sense, it is convenient that it is disabled by default, as we have to create policies only on the tables where our access logic depends on the row contents.

There's more...

RLS can lead to very complex configurations for a variety of reasons, as in the following instances:

- An `UPDATE` policy can specify both the rows on which we act and which changes can be accepted.
- `UPDATE` and `DELETE` policies, in some cases, require visibility as granted by an appropriate `SELECT` policy.
- `UPDATE` policies are also applied to `INSERT ... ON CONFLICT DO UPDATE`.

We recommend reading the finer details at the following URL: https://www.postgresql.org/docs/current/ddl-rowsecurity.html.

Creating a new user

In this recipe, we will show you two ways of creating a new database user—one with a dedicated command-line utility and another using SQL commands.

Getting ready

To create new users, you must either be a superuser or have the `CREATEROLE` privilege.

How to do it...

From the command line, you can run the `createuser` command:

```
pguser@hvost:~$ createuser bob
```

If you add the `--interactive` command-line option, you activate interactive mode, which means you will be asked some questions as follows:

```
pguser@hvost:~$ createuser --interactive alice
Shall the new role be a superuser? (y/n) n
Shall the new role be allowed to create databases? (y/n) y
Shall the new role be allowed to create more new roles? (y/n) n
```

Without `--interactive`, the preceding questions get *no* as the default answer; you can change that with the `-s` (`--superuser`), `-d` (`--createdb`), and `-r` (`--createrole`) command-line options.

In interactive mode, questions are asked only if they make sense. One example is when the user is a superuser; no other questions are asked because a superuser is not subject to privilege checks. Another example is when one of the preceding options is used to specify a non-default setting, the corresponding question will not be asked.

How it works...

The `createuser` program is just a shallow wrapper around the executing SQL against the database cluster. It connects to the `postgres` database and then executes SQL commands for user creation. To create the same users through SQL, you can issue the following commands:

```
CREATE USER bob;
CREATE USER alice CREATEDB;
```

There's more...

You can check the attributes of a given user in `psql`, as follows:

```
pguser=# \du alice
```

This gives you the following output:

```
             List of roles
 Role name | Attributes | Member of
-----------+------------+-----------
 alice     | Create DB  | {}
```

The `CREATE USER` and `CREATE GROUP` commands are actually variations of `CREATE ROLE`. The `CREATE USER username;` statement is equivalent to `CREATE ROLE username LOGIN;`, and the `CREATE GROUP groupname;` statement is equivalent to `CREATE ROLE groupname NOLOGIN;`.

Temporarily preventing a user from connecting

Sometimes, you need to temporarily revoke a user's connection rights without deleting the user or changing their password. This recipe presents ways of doing this.

Getting ready

To modify other users, you must either be a superuser or have the `CREATEROLE` privilege (in the latter case, only non-superuser roles can be altered).

How to do it...

Follow these steps to temporarily prevent and reissue the login capability of a user:

1. To temporarily prevent the user from logging in, run this command:

    ```
    pguser=# alter user bob nologin;
    ALTER ROLE
    ```

2. To let the user connect again, run the following command:

    ```
    pguser=# alter user bob login;
    ALTER ROLE
    ```

How it works...

This sets a flag in the system catalog, telling PostgreSQL not to let the user log in. It does not kick out already connected users.

There's more...

Here are some additional remarks.

Limiting the number of concurrent connections by a user

The same result can be achieved by setting the connection limit for that user to 0:

```
pguser=# alter user bob connection limit 0;
ALTER ROLE
```

To allow 10 concurrent connections for the bob user, run this command:

```
pguser=# alter user bob connection limit 10;
ALTER ROLE
```

To allow an unlimited number of connections for this user, run the following command:

```
pguser=# alter user bob connection limit -1;
ALTER ROLE
```

Allowing unlimited connections to PostgreSQL concurrently could allow a **Denial-of-Service (DoS)** attack by exhausting connection resources; also, a system could fail or degrade by an overload of legitimate users. To reduce these risks, you may wish to limit the number of concurrent sessions per user.

Revoking a user's database access

A user's connection to a database can be restricted by revoking the database privilege from that role. However, before revoking the CONNECT privilege from an individual user, ensure that the CONNECT privilege is revoked from the PUBLIC role first. By default, the PUBLIC role possesses the CONNECT privilege for all databases. Removing this privilege from PUBLIC will consequently revoke it from all users, potentially preventing them from connecting unless they have the CONNECT privilege explicitly granted to them. It's highly advised to revoke all privileges from PUBLIC after creating a database.

How it works...

1. Check if a database grants the CONNECT privilege to PUBLIC:

   ```
   SELECT has_database_privilege('public', pguser, 'CONNECT');
   ```

2. If the above command returns true, revoke the CONNECT privilege from PUBLIC:

   ```
   REVOKE ALL ON DATABASE pguser FROM PUBLIC;
   ```

3. Now, grant the CONNECT privilege to specific roles that you wish to permit database connections:

   ```
   GRANT CONNECT ON DATABASE pguser TO alice;
   ```

4. Revoke the CONNECT privilege from users you temporarily wish to block from the database:

   ```
   REVOKE CONNECT ON DATABASE pguser FROM bob;
   ```

Forcing NOLOGIN users to disconnect

In order to make sure that all users whose login privileges have been revoked are disconnected right away, run the following SQL statement as a superuser:

```
SELECT pg_terminate_backend(pid)
  FROM pg_stat_activity a
    JOIN pg_roles r ON a.usename = r.rolname AND NOT rolcanlogin;
```

This disconnects all users who are no longer allowed to connect by terminating the backends opened by these users.

Removing a user without dropping their data

When trying to drop a user who owns some tables or other database objects, you get the following error, preventing the user from being dropped:

```
testdb=# drop user bob;
ERROR:  role "bob" cannot be dropped because some objects depend on it
DETAIL:  owner of table bobstable
owner of sequence bobstable_id_seq
```

This recipe presents two solutions to this problem.

Getting ready

To modify users, you must either be a superuser or have the CREATEROLE privilege.

How to do it...

The easiest solution to this problem is to refrain from dropping the user and use the trick from the *Temporarily preventing a user from connecting* recipe, which prevents the user from connecting:

```
pguser=# alter user bob nologin;
ALTER ROLE
```

This has the added benefit of the original owner of the table being available later, if needed, for auditing or debugging purposes (*Why is this table here? Who created it?*).

Then, you can assign the rights of the deleted user to a new user, using the following code:

```
pguser=# GRANT bob TO bobs_replacement;
GRANT
```

How it works...

As noted previously, a user is implemented as a role with the login attribute set. This recipe works by removing that attribute from the user, which is then kept just as a role.

If you really need to get rid of a user, you have to assign all ownership to another user. To do so, run the following query, which is a PostgreSQL extension to standard SQL:

```
REASSIGN OWNED BY bob TO bobs_replacement;
```

It does exactly what it says: it assigns ownership of all database objects currently owned by the bob role to the bobs_replacement role.

However, you need to have privileges on both the old and new roles to do that, and you need to do it in all databases where bob owns any objects, as the REASSIGN OWNED command works only on the current database. After this, you can delete the original user, bob.

Checking whether all users have a secure password

By default, as of PostgreSQL 16, passwords are encrypted using the SCRAM-SHA-256 login method for users, which was added in PostgreSQL 10. Any servers upgrading from earlier versions should upgrade from **MD5** to SCRAM-SHA-256 password encryption, since the MD5 authentication method is considered insecure for many applications.

For client applications connecting from trusted private networks, either real or a **Virtual Private Network** (**VPN**), you may use host-based access, provided that you know that the machine on which the application runs is not used by some non-trusted individuals. For remote access over public networks, it may be a better idea to use SSL client certificates. See the later recipe, *Using SSL certificates to authenticate*, for more on this.

How to do it...

To see which users don't yet have SCRAM-encrypted passwords, use this query:

```
test2=# select usename,passwd from pg_shadow where passwd
not like 'SCRAM%' or passwd is null;
 usename  |    passwd
----------+---------------
 tim      | weakpassword
 asterisk | md5chicken
(2 rows)
```

How it works...

The password_encryption parameter here, decides how the ALTER USER statement will encrypt the password. This should be set globally in the postgresql.conf file or by using ALTER SYSTEM. As of PostgreSQL 16, the default value is scram-sha-256.

Having the passwords encrypted in the database is just half of the equation. The bigger problem is making sure that users actually use passwords that are hard to guess. Passwords such as password, secret, or test are out of the question, and most common words are not good passwords either.

As of PostgreSQL16, passwords can be of arbitrary length. However, on PgBouncer, there is a limit of 996 characters, so that is a reasonable limit. Note that usernames can be—at most—63 characters.

If you don't trust your users to select strong passwords, you can write a wrapper application that checks the password strength and makes them use that when changing passwords. A contrib module lets you do this for a limited set of cases (the password is sent from client to server in plain text). Visit http://www.postgresql.org/docs/current/static/passwordcheck.html for more information on this.

Giving limited superuser powers to specific users

The superuser role has some privileges that can also be granted to non-superuser roles separately.

To give the bob role the ability to create new databases, run this:

```
ALTER ROLE BOB WITH CREATEDB;
```

To give the bob role the ability to create new users, run the following command:

```
ALTER ROLE BOB WITH CREATEROLE;
```

Note that the PostgreSQL documentation warns against doing the preceding action:

> *"Be careful with the CREATEROLE privilege. There is no concept of inheritance for the privileges of a CREATEROLE-role. That means that even if a role does not have a certain privilege but is allowed to create other roles, it can easily create another role with different privileges than its own (except for creating roles with superuser privileges). For example, if the role "user" has the CREATEROLE privilege but not the CREATEDB privilege, nonetheless, it can create a new role with the CREATEDB privilege. Therefore, regard roles that have the CREATEROLE privilege as almost-superuser-roles."*

For more information about the CREATEROLE privilege see https://www.postgresql.org/docs/current/sql-createrole.html.

It is also possible to give ordinary users more fine-grained and controlled access to an action reserved for superusers, using security definer functions. The same trick can also be used to pass partial privileges between different users.

Getting ready

First, you must have access to the database as a superuser in order to delegate powers. Here, we assume we are using a default superuser named postgres.

We will demonstrate two ways to make some superuser-only functionality available to a selected ordinary user.

How to do it...

An ordinary user cannot tell PostgreSQL to copy table data from a file. Only a superuser or user with pg_read_server_files can do that, as follows:

```
pguser@hvost:~$ psql -U postgres
test2
...
test2=# create table lines(line text);
CREATE TABLE
test2=# copy lines from '/home/bob/names.txt';
COPY 37
test2=# GRANT ALL ON TABLE lines TO bob;
test2=# SET ROLE to bob;
SET
test2=> copy lines from '/home/bob/names.txt';
ERROR:  must be superuser or have privileges of the pg_read_server_files role to COPY from a file
HINT:  Anyone can COPY to stdout or from stdin. psql's \copy command also works for anyone.
```

To let bob copy directly from the file, the superuser can grant pg_read_server_files privileges to bob as follows:

```
test2=# GRANT pg_read_server_files TO bob;
```

When granting the above permission, you may also want to verify that bob imports files only from his home directory.

Assigning backup privileges to a user

For security and auditing reasons, it's beneficial to have a dedicated role or user specifically for database backups. PostgreSQL 16 offers predefined roles for both logical and physical backups, enabling database users to perform backup tasks without being a superuser.

How it works...

Create a backup user using the following command:

```
CREATE USER backup_user;
```

Grant the pg_read_all_data role to backup_user for logical backups:

```
GRANT pg_read_all_data TO backup_user;
```

Add the REPLICATION attribute to backup_user for physical backups:

```
ALTER USER backup_user WITH REPLICATION;
```

The pg_read_all_data role allows backup_user to access all data and table definitions in a database. This role also ensures that the backup_user can read any new tables and data. The REPLICATION attribute enables backup_user to take physical backups using the pg_basebackup command.

There's more...

If you wish to let a user verify or reload all PostgreSQL settings, you can assign the pg_read_all_settings and pg_signal_backend roles.

If you want to allow a regular user to initiate a CHECKPOINT in the PostgreSQL database, the pg_checkpoint role should be assigned.

To monitor and, if necessary, terminate connected users in PostgreSQL, privileges associated with the pg_monitor and pg_signal_backend roles can be granted to the desired user. This enables commands like pg_terminate_backend.

For a comprehensive understanding of predefined roles in PostgreSQL, please refer to the official documentation or the provided links:

https://www.postgresql.org/docs/current/predefined-roles.html

Auditing database access

Auditing database access is a much bigger topic than you might expect because it can cover a whole range of requirements.

Getting ready

First, decide which of these you want and look at the appropriate subsection:

- Which privileges can be executed? (*Auditing access*)
- Which SQL statements were executed? (*Auditing SQL*)
- Which tables were accessed? (*Auditing table access*)
- Which data rows were changed? (*Auditing data changes*)
- Which data rows were viewed? (Not described here—usually too much data)

Auditing just SQL produces the lowest volume of audit log information, especially if you choose to log only **Data Definition Language** (**DDL**). Higher levels accumulate more information very rapidly, so you may quickly decide not to do this in practice. Read each section to understand the benefits and trade-offs.

Auditing access

Reviewing which users have access to which information is important. There are a few ways of doing this:

- Write scripts that access the database catalog tables. **Access Control List** (**ACL**) information is not held in one place, so you have lots of places to look at:

```
cookbook=# select relname, attname
from pg_attribute join pg_class c on attrelid = c.oid
where attname like '%acl%' and relkind = 'r';
         relname         |  attname
-------------------------+----------
 pg_proc                 | proacl
 pg_type                 | typacl
 pg_attribute            | attacl
 pg_class                | relacl
 pg_language             | lanacl
 pg_largeobject_metadata | lomacl
 pg_namespace            | nspacl
```

```
pg_database              | datacl
pg_tablespace            | spcacl
pg_foreign_data_wrapper  | fdwacl
pg_foreign_server        | srvacl
pg_default_acl           | defaclrole
pg_default_acl           | defaclnamespace
pg_default_acl           | defaclobjtype
pg_default_acl           | defaclacl
(15 rows)
```

- Write scripts that test access conforms to a specific definition. This can be achieved by writing tests using the database information functions provided by PostgreSQL—for example, has_table_privilege(), has_column_privilege(), and so on.

Auditing SQL statements

There are a few ways to capture SQL statements:

- Using the PostgreSQL log_statement parameter—a fairly crude approach
- Using the pgaudit extension's pgaudit.log parameter
- Using EDB Postgres' audit facility

The log_statement parameter can be set to one of the following options:

- ALL: Logs all SQL statements executed at the top level
- MOD: Logs all SQL statements for INSERT, UPDATE, DELETE, and TRUNCATE
- ddl: Logs all SQL statements for DDL commands
- NONE: No statements logged

For example, to log all DDL commands, edit your postgresql.conf file to set the following:

```
log_statement = 'ddl'
```

log_statement SQL statements are explicitly given in top-level commands. It is still possible to perform SQL without it being logged by this setting if you use any of the **Python Languages** (**PLs**), either through DO statements or by calling a function that includes SQL statements.

Was the change committed? It is possible to have some statements recorded in the log file, but these are not visible in the database structure. Most DDL commands in PostgreSQL can be rolled back, so what is in the log is just a list of commands executed by PostgreSQL—not what was actually committed. The log file is not transactional, and it keeps commands that were rolled back.

It is possible to display the **Transaction Identifier** (TID) on each log line by including %x in the log_line_prefix setting, although that has some difficulties in terms of usage.

Who made the changes? To know which database user made the DDL changes, you have to make sure that this information is logged as well. In order to do so, you may have to change the log_line_prefix parameter to include the %u format string.

A recommended minimal log_line_prefix format string to audit DDL is %t %u %d, which tells PostgreSQL to log the timestamp, database user, and database name at the start of every log line.

The pgaudit extension provides two levels of audit logging: session and object levels. The session level was designed to solve some of the problems of log_statement. The pgaudit extension will log all access, even if it is not executed as a top-level statement, and it will log all dynamic SQL. pgaudit.log can be set to include zero or more of the following settings:

- READ: SELECT and COPY
- WRITE: INSERT, UPDATE, DELETE, TRUNCATE, and COPY
- FUNCTION: Function calls and DO blocks
- ROLE: GRANT, REVOKE, CREATE/ALTER/DROP ROLE
- DDL: All DDL not already included in the ROLE category
- MISC: Miscellaneous—DISCARD, FETCH, CHECKPOINT, VACUUM, and so on

For example, to log all DDL commands, edit your postgresql.conf file to set the following:

```
pgaudit.log = 'role, ddl'
```

You should set these parameters to reduce the overhead of logging:

```
pgaudit.log_catalog = off
pgaudit.log_relation = off
pgaudit.log_statement_once = on
```

The pgaudit extension was originally written by Simon Riggs and Abhijit Menon-Sen of 2nd *Quadrant* as part of the **Advanced Analytics for Extremely Large European Databases** (**AXLE**) project for the **European Union** (**EU**). The next version was designed by Simon Riggs and David Steele to provide object-level logging. The original version was deprecated and is no longer available. The new version is fully supported and has been adopted by the **United States Department of Defense** (**US DoD**) as the tool of choice for PostgreSQL audit logging.

pgaudit is available in binary form via postgresql.org repositories.

Auditing table access

pgaudit can log access to each table. So, if an SQL table touches three tables, then it can generate three log records, one for each table. This is important because, otherwise, you might have to try to parse the SQL to find out which tables it touched, which would be difficult without access to the schema and the search_path settings.

To make it easier to access the audit log per table, adjust these settings:

```
pgaudit.log_relation = on
pgaudit.log_statement_once = off
```

If you want even finer-grained auditing, pgaudit allows you to control which tables are audited. The user cannot tell which tables are logged and which are not, so it is possible for investigators to quietly enhance the level of logging once they are alerted to a suspect or a potential attack.

First, set the role that will be used by the auditor:

```
pgaudit.role = 'investigator'
```

Then, you can define logging through the privilege system, as in the following command:

```
GRANT INSERT, UPDATE, DELETE on <vulnerable_table> TO investigator;
```

Remove it again when no longer required.

Privileges may be set at the individual column level to protect **Personally Identifiable Information (PII)**.

Managing the audit log

Both log_statement and pgaudit output audit log records to the server log. This is the most flexible approach, since the log can be routed in various ways to ensure it is safe and separate from normal log entries.

If you allow the log entries to go to the normal server log, you can find all occurrences of the CREATE, ALTER, and DROP commands in the log:

```
postgres@hvost:~$ egrep -i "create|alter|drop" \
/var/log/postgresql/postgresql-16-main.log
```

If log rotation is in effect, you may need to use grep on older logs as well.

If the available logs are too new and you haven't saved the older logs in some other place, you are out of luck.

The default settings in the postgresql.conf file for log rotation looks like this:

```
log_filename = 'postgresql-%Y-%m-%d_%H%M%S.log'
log_rotation_age = 1d
log_rotation_size = 10MB
```

Log rotation can also be implemented with third-party utilities. For instance, the default behavior on Debian and Ubuntu distributions is to use the logrotate utility to compress or delete old log files, according to the rules specified in the /etc/logrotate.d/postgresql-common file.

To make sure you have the full history of DDL commands, you may want to set up a cron job that saves the DDL statements extracted from the main PostgreSQL log to a separate DDL audit log. You would still want to verify that the logs are not rotating too fast for this to catch all DDL statements.

If you use syslog, you can then route audit messages using various **Operating System (OS)** utilities.

Alternatively, you can use the pgaudit_analyze extension to load data back into a special audit log database. Various other options exist.

Auditing data changes

This section of the recipe provides different ways of collecting changes to data contained in tables for auditing purposes.

First, you must make the following decisions:

- Do you need to audit all changes or only some?
- What information about the changes do you need to collect? Only the fact that the data has changed?
- When recording the new value of a field or tuple, do you also need to record the old value?
- Is it enough to record which user made the change, or do you also need to record the **Internet Protocol (IP)** address and other connection information?
- How secure (tamper-proof) must the auditing information be? For example, does it need to be kept separately, away from the database being audited?

Changes can be collected using triggers that collect new (and, if needed, old) values from tuples and save them to auditing table(s). Triggers can be added to whichever tables need to be tracked.

The audit_trigger extension provides a handy universal audit trigger, so you do not need to write your own. It logs both old and new values of rows in any table, serialized as hstore data type values. The latest version and its documentation are both available at https://github.com/2ndQuadrant/audit-trigger.

The extension creates a schema called audit into which all of the other components of the audit trigger code are placed, after which we can enable auditing on specific tables.

As an example, we create standard pgbench tables by running the pgbench utility:

```
pgbench -i
```

Next, we connect to PostgreSQL as a superuser and issue the following SQL to enable auditing on the pgbench_account table:

```
SELECT audit.audit_table('pgbench_accounts');
```

Now, we will perform some writing activities to see how it is audited. The easiest choice is to run the pgbench utility again, this time to perform some transactions as follows:

```
pgbench -t 1000
```

We expect the audit trigger to have logged the actions on pgbench_accounts, as we have enabled auditing on it. In order to verify this, we connect again with psql and issue the following SQL:

```
cookbook=# SELECT count(*) FROM audit.logged_actions;
 count
-------
  1000
(1 row)
```

This confirms that we have indeed logged 1,000 actions. Let's inspect the information that is logged by reading one row of the logged_actions table. First, we enable expanded mode, as the query produces a large number of columns:

```
cookbook=# \x on
```

Then, we issue the following command:

```
cookbook=# SELECT * FROM audit.logged_actions LIMIT 1;
-[ RECORD 1 ]-----+------------------------------------------
event_id          | 1
schema_name       | public
table_name        | pgbench_accounts
relid             | 246511
session_user_name | gianni
action_tstamp_tx  | 2017-01-18 19:48:05.626299+01
action_tstamp_stm | 2017-01-18 19:48:05.626446+01
```

```
action_tstamp_clk  | 2017-01-18 19:48:05.628488+01
transaction_id     | 182578
application_name   | pgbench
client_addr        |
client_port        |
client_query       | UPDATE pgbench_accounts SET abalance = abalance + -758
WHERE aid = 86061;
action             | U
row_data           | "aid"=>"86061", "bid"=>"1", "filler"=>"   ",
"abalance"=>"0"
changed_fields     | "abalance"=>"-758"
statement_only     | f
```

Always knowing which user is logged in

In the preceding recipes, we just logged the value of the user variable in the current PostgreSQL session to log the current user role.

This does not always mean that this particular user was the user who was actually authenticated at the start of the session. For example, a superuser can execute the SET ROLE TO ... command to set its current role to any other user or role in the system. As you might expect, non-superusers can only assume roles that they own.

It is possible to differentiate between the logged-in role and the assumed role, using the current_user and session_user session variables:

```
postgres=# select current_user, session_user;
 current_user | session_user
--------------+--------------
 postgres     | postgres
postgres=# set role to bob;
SET
postgres=> select current_user, session_user;
 current_user | session_user
--------------+--------------
 bob          | postgres
```

Sometimes, it is desirable to let each user log in with their own username and just assume the role needed on a case-by-case basis.

Getting ready

Prepare the required group roles for different tasks and access levels by granting the necessary privileges and options.

How to do it...

Follow these steps:

1. Create user roles with no privileges and with the NOINHERIT option:

   ```
   postgres=# create user alice noinherit;
   CREATE ROLE
   postgres=# create user bob noinherit;
   CREATE ROLE
   ```

2. Then, create roles for each group of privileges that you need to assign:

   ```
   postgres=# create group sales;
   CREATE ROLE
   postgres=# create group marketing;
   CREATE ROLE
   postgres=# grant postgres to marketing;
   GRANT ROLE
   ```

3. Now, grant each user the roles they may need:

   ```
   postgres=# grant sales to alice;
   GRANT ROLE
   postgres=# grant marketing to alice;
   GRANT ROLE
   postgres=# grant sales to bob;
   GRANT ROLE
   ```

After you do this, the alice and bob users have no rights after login, but they can assume the sales role by executing SET ROLE TO sales, and alice can additionally assume the superuser role.

How it works...

If a role or user is created with the NOINHERIT option, this user will not automatically get the rights that have been granted to the other roles that have been granted to them. To claim these rights from a specific role, they have to set their role to one of those other roles.

In some sense, this works a bit like the su (set user) command in Unix and Linux systems—that is, you (may) have the right to become that user, but you do not automatically have the rights of the aforementioned user.

This setup can be used to get better audit information, as it lets you know who the actual user was. If you just allow each user to log in as the role needed for a task, there is no good way to know later which of the users was really logged in as clerk1 when a **USD** $100,000 transfer was made.

There's more...

The SET ROLE command works both ways—that is, you can both gain and lose privileges. A superuser can set their role to any user defined in the system. To get back to your original login role, just use RESET ROLE.

Not inheriting user attributes

Not all rights come to users via GRANT commands. Some important rights are given via user attributes (SUPERUSER, CREATEDB, and CREATEROLE), and these are never inherited.

If your user has been granted a superuser role and you want to use the superuser powers of this granted role, you have to use SET ROLE TO mysuperuserrole before anything that requires the superuser attribute to be set.

In other words, the user attributes always behave as if the user had been a NOINHERIT user.

Integrating with LDAP

This recipe shows you how to set up your PostgreSQL system so that it uses the LDAP for authentication.

Getting ready

Ensure that the usernames in the database and your LDAP server match, as this method works for user authentication checks of users who are already defined in the database.

How to do it...

In the pg_hba.conf PostgreSQL authentication file, we define some address ranges to use LDAP as an authentication method, and we configure the LDAP server for this address range:

```
host    all        all         10.10.0.1/16          ldap \
  ldapserver=ldap.our.net ldapprefix="cn=" ldapsuffix=",
    dc=our,dc=net"
```

How it works...

This setup makes the PostgreSQL server check passwords from the configured LDAP server.

User rights are not queried from the LDAP server but have to be defined inside the database, using the ALTER USER, GRANT, and REVOKE commands.

There's more...

We have shown you how PostgreSQL can use an LDAP server for password authentication. It is also possible to use some more information from the LDAP server, as shown in the next two examples.

Setting up the client to use LDAP

If you are using the pg_service.conf file to define your database access parameters, you may define some to be queried from the LDAP server by including a line similar to the following in your pg_service.conf file:

```
ldap://ldap.mycompany.com/dc=mycompany,dc=com?uniqueMember?one?(cn=mydb)
```

Replacement for the User Name Map feature

Although we cannot use the User Name Map feature with LDAP, we can achieve a similar effect on the LDAP side. Use ldapsearchattribute and the search and bind mode to retrieve the PostgreSQL role name from the LDAP server.

See also

- For server setup, including the search and bind mode, visit http://www.postgresql.org/docs/current/static/auth-methods.html#AUTH-LDAP.
- For client setup, visit http://www.postgresql.org/docs/current/static/libpq-ldap.html.

Connecting using encryption (SSL / GSSAPI)

Here, we will demonstrate how to enable PostgreSQL to use SSL for the protection of database connections, by encrypting all of the data passed over that connection. Using SSL makes it much harder to sniff database traffic, including usernames, passwords, and other sensitive data. Otherwise, everything that is passed unencrypted between a client and the database can be observed by someone listening to a network somewhere between them. An alternative to using SSL is running the connection over a VPN.

Using SSL makes the data transfer on the encrypted connection a little slower, so you may not want to use it if you are sure that your network is safe. The performance impact can be quite large if you are creating lots of short connections, as setting up an SSL connection is quite **Central-Processing-Unit (CPU)**-heavy. In this case, you may want to run a local connection-pooling solution, such as PgBouncer, to which the client connects without encryption, and then configure PgBouncer for server connections using SSL. Older versions of PgBouncer did not support SSL; the solution was to channel server connections through `stunnel`, as described in the PgBouncer **FAQs** at `https://pgbouncer.github.io/faq.html`.

Getting ready

Get, or generate, an SSL server key and certificate pair for the server, and store these in the `data` directory of the current database instance as `server.key` and `server.crt` files.

On some platforms, this is unnecessary; the key and certificate pair may already be generated by the packager. For example, in Ubuntu, PostgreSQL is set up to support SSL connections by default.

How to do it...

Set `ssl = on` in `postgresql.conf` and restart the database if not already set.

How it works...

If `ssl = on` is set, then PostgreSQL listens to both plain and SSL connections on the same port (5432, by default) and determines the type of connection from the first byte of a new connection. Then, it proceeds to set up an SSL connection if an incoming request asks for it.

PostgreSQL 13+ now defaults to use `ssl_min_protocol_version = TLSv1.2`, although valid values are `TLSv1.3`, `TLSv1.2`, `TLSv1.1`, and `TLSv1`. Now, SSL2 and SSL3 are always disabled.

There's more...

You can leave the choice of whether or not to use SSL up to the client, or you can force SSL usage from the server side.

To let the client choose, use a line of the following form in the `pg_hba.conf` file:

```
host  database  user  IP-address/IP-mask  auth-method
```

If you want to allow only SSL clients, use the `hostssl` keyword instead of the host. If connecting using GSSAPI, you would use `hostgssenc` rather than `hostssl`, as shown previously. Details of connecting with GSSAPI are not otherwise covered in this recipe.

The contents of pg_hba.conf can be seen using the pg_hba_file_rules view, so you can run queries to check that you have configured it correctly and it is actually working!

Entries in pg_hba.conf can now span multiple lines by specifying a backslash at the end of the line.

The following fragment of pg_hba.conf enables both non-SSL and SSL connections from the 192.168.1.0/24 local subnet, but it requires SSL from everybody accessing the database from other networks:

```
host      all     all     192.168.1.0/24     scram-sha-256
hostssl   all     all     0.0.0.0/0          scram-sha-256
```

Getting the SSL key and certificate

For web servers, you must usually get your SSL certificate from a recognized **Certificate Authority (CA)**, as most browsers complain if the certificate is not issued by a known CA. They warn the user of the most common security risks, and they require confirmation before connecting to a server with a certificate issued by an unknown CA.

For your database server, it is usually sufficient to generate a certificate yourself using OpenSSL. The following commands generate a self-signed certificate for your server:

```
openssl genrsa 2048 > server.key
openssl req -new -x509 -key server.key -out server.crt
```

Read more on X.509 keys and certificates by visiting OpenSSL's *HOWTO* pages at https://github.com/openssl/openssl/tree/master/doc/HOWTO.

Setting up a client to use SSL

The behavior of the client application regarding SSL is controlled by a PGSSLMODE environment variable. This can have the following values, as defined in the official PostgreSQL documentation:

SSL mode	Eavesdrop protection	Man-in-the-Middle (MITM) protection	Statement
`disabled`	No	No	I don't care about security, and I don't want to pay the overhead of encryption.
`allow`	Maybe	No	I don't care about security, but I will pay the overhead of encryption if the server insists on it.
`prefer`	Maybe	No	I don't care about encryption, but I will pay the overhead of encryption if the server supports it.
`require`	Yes	No	I want my data to be encrypted, and I accept the overhead. I trust that the network will ensure that I always connect to the server I want.
`verify-ca`	Yes	Depends on the CA policy	I want my data encrypted, and I accept the overhead. I want to be sure that I connect to a server that I trust.
`verify-full`	Yes	Yes	I want my data encrypted, and I accept the overhead. I want to be sure that I connect to a server I trust and that the server is the one I specify.

Table 6.1: Explanation of ssl_mode

An MITM attack refers to when someone poses as your server, perhaps by manipulating **Domain Name System (DNS)** records or IP routing tables, but actually just observes and forwards traffic.

For this to be possible with an SSL connection, this person needs to have obtained a certificate that your client considers valid.

Checking server authenticity

The last two SSL modes allow you to be reasonably sure that you are actually talking to your server, by checking the SSL certificate presented by the server.

In order to enable this useful security feature, the following files must be available on the client side. On Unix systems, they are located in the client home directory, in a subdirectory named `~/.postgresql`. On Windows, they are in `%APPDATA%\postgresql\`:

File	Contents	Effect
`root.crt`	Certificates of one or more trusted CAs	PostgreSQL verifies that the server certificate is signed by a trusted CA.
`root.crl`	Certificates revoked by CAs	The server certificate must not be on this list.

Table 6.2: Certificate files

Only the root.crt file is required for the client to authenticate the server certificate. It can contain multiple root certificates, against which the server certificate is compared.

Using SSL certificates to authenticate

This recipe shows you how to set up your PostgreSQL system so that it *requires* clients to present a valid X.509 certificate before allowing them to connect.

This can be used as an additional security layer, using double authentication, where the client must both have a valid certificate to set up the SSL connection and know the database user's password. It can also be used as the sole authentication method, where the PostgreSQL server will first verify the client connection, using the certificate presented by the client, and then retrieve the username from the same certificate.

Getting ready

Get, or generate, a root certificate and a client certificate to be used by the connecting client.

How to do it...

For testing purposes or to set up a single trusted user, you can use a self-signed certificate:

```
openssl genrsa  2048  >  client.key
openssl req  -new -x509 -key  server.key  -out client.crt
```

In the server, set up a line in the pg_hba.conf file with the hostssl method, and the clientcert option set to 1:

```
hostssl  all  all    0.0.0.0/0     scram-sha-256 clientcert=1
```

Put the client root certificate in the root.crt file in the server data directory ($PGDATA/root.crt). This file may contain multiple trusted root certificates.

If you are using a central CA, you probably also have a certificate revocation list, which should be put in a root.crl file and regularly updated.

In the client, put the client's private key and certificate in ~/.postgresql/postgresql.key and ~/.postgresql/postgresql.crt. Make sure that the private key file is not world-readable or group-readable by running the following command:

```
chmod 0600 ~/.postgresql/postgresql.key
```

In a Windows client, the corresponding files are %APPDATA%\postgresql\postgresql.key and %APPDATA%\postgresql\postgresql.crt. No permission check is done, as the location is considered secure.

If the client certificate is not signed by the root CA but by an intermediate CA, then all of the intermediate CA certificates up to the root certificate must be placed in the postgresql.crt file as well.

How it works...

If the clientcert=1 option is set for a hostssl row in pg_hba.conf, then PostgreSQL accepts only connection requests accompanied by a valid certificate.

The validity of the certificate is checked against certificates present in the root.crt file in the server data directory.

If there is a root.crl file, then the presented certificate is searched for in this file and, if found, is rejected.

After the client certificate is validated and the SSL connection is established, the server proceeds to validate the actual connecting user, using whichever authentication method is specified in the corresponding hostssl line.

In the following example, clients from a special address can connect as any user when using an SSL certificate, and they must specify a SCRAM-SHA-256 password for non-SSL connections. Clients from all other addresses must present a certificate and use the SCRAM-SHA-256 password authentication:

```
host     all all  10.10.10.10/32 scram-sha-256
hostssl  all all  10.10.10.10/32 trust clientcert=1
hostssl  all all  all            scram-sha-256 clientcert=1
```

There's more...

In this section, we provide some additional content, describing an important optimization for an SSL-only database server, plus two extensions of the basic SSL configuration.

Avoiding duplicate SSL connection attempts

In the *Setting up a client to use SSL* section of the previous *Connecting using encryption (SSL/GSSAPI)* recipe, we saw how the client's SSL behavior is affected by environment variables. Depending on how the SSLMODE environment variable is set on the client (either via compile-time settings, the PGSSLMODE environment variable, or the sslmode connection parameter), the client may attempt to connect without SSL first and then attempt an SSL connection only after the server rejects the non-SSL connection. This duplicates a connection attempt every time a client accesses an SSL-only server.

To make sure that the client tries to establish an SSL connection on the first attempt, SSLMODE must be set to prefer or higher.

Using multiple client certificates

You may sometimes need different certificates to connect to different PostgreSQL servers.

The location of the certificate and key files in postgresql.crt and postgresql.key in the table from the *Checking server authenticity* section (*Table 6.2*) is just the default, and it can be overridden by specifying alternative file paths, using the sslcert and sslkey connection parameters or the PGSSLCERT and PGSSLKEY environment variables.

Using the client certificate to select a database user

It is possible to use the client certificate for two purposes at once: proving that the connecting client is a valid one and selecting a database user to be used for the connection.

To do this, set the authentication method to cert in the hostssl line:

```
hostssl    all     all     0.0.0.0/0               cert
```

As you can see, the clientcert=1 option used with hostssl to require client certificates is no longer required, as it is implied by the cert method itself.

When using the cert authentication method, a valid client certificate is required, and the cn (short for **Common Name**) attribute of the certificate will be compared to the requested database username. The login will be allowed only if they match.

It is possible to use a User Name Map to map common names in the certificates to database usernames, by specifying the map option:

```
hostssl    all     all     0.0.0.0/0       cert  map=x509cnmap
```

Here, x509cnmap is the name that we have arbitrarily chosen for our mapping. More details on User Name Maps are provided in the *Mapping external usernames to database roles* recipe.

See also

To understand more about SSL in general, and the OpenSSL library used by PostgreSQL in particular, visit http://www.openssl.org or get a good book about SSL.

To get started with the generation of simple SSL keys and certificates, see https://github.com/openssl/openssl/blob/master/doc/HOWTO/certificates.txt.

Mapping external usernames to database roles

In some cases, the authentication username is different from the PostgreSQL username. For instance, this can happen when using an external system for authentication, such as certificate authentication (as described in the previous recipe) or any other external or **Single Sign-On (SSO)** system authentication method from http://www.postgresql.org/docs/current/static/auth-methods.html (GSSAPI, **Security Support Provider Interface** (**SSPI**), Kerberos, Radius, or **Privileged Access Management** (**PAM**)). You may just need to enable an externally authenticated user to connect as multiple database users. In such cases, you can specify rules to map the external username to the appropriate database role.

Getting ready

Prepare a list of usernames from the external authentication system and decide which database users they are allowed to connect to—that is, which external users map to which database users.

How to do it...

Create a pg_ident.conf file in the usual place (PGDATA), with lines in the following format:

```
map-name system-username database-username
```

This should be read as "system-username is allowed to connect as database-username," rather than "every time system-username connects, they will be forced to use database-username."

Here, map-name is the value of the map option from the corresponding line in pg_hba.conf, system-username is the username that the external system authenticated the connection as, and database-username is the database user this system user is allowed to connect as. The same system user may be allowed to connect as multiple database users, so this is not a 1:1 mapping but, rather, a list of allowed database users for each system user.

If system-username starts with a slash (/), then the rest of it is treated as a **Regular Expression** (**regex**) rather than a directly matching string, and it is possible to use the \1 string in database-username to refer to the part captured by the parentheses in the regex. For example, consider the following lines:

```
salesmap    /^(.*)@sales\.comp\.com$      \1
salesmap    /^(.*)@sales\.comp\.com$      sales
salesmap    manager@sales.comp.com        auditor
```

These will allow any user authenticated with a `@sales.comp.com` email address to connect both as a database user equal to the name before the `@` sign-in their email address and as the `sales` user. They will additionally allow `anager@sales.comp.com` to connect as the auditor user. Then, edit the `pg_hba.conf` line to specify the `map=salesmap` option.

How it works…

After authenticating the connection using an external authentication system, PostgreSQL will usually proceed to check that the externally authenticated username matches the database username that the user wishes to connect to, and will reject the connection if these two do not match.

If there is a `map=` parameter specified for the current line in `pg_hba.conf`, then the system will scan the map line by line and will let the client proceed to connect if a match is found.

There's more…

By default, the map file is called `pg_ident.conf` (because it was first used for the `ident` authentication method).

Nowadays, it is possible to change the name of this file via the `ident_file` configuration parameter in `postgresql.conf`. It can also be located outside the `PGDATA` directory by setting `ident_file` to a full path.

A relative path can also be used, but since it is relative to where the `postgres` process starts, this is usually not a good idea.

Using column-level encryption

A user can encrypt data in a database so that it is not visible to the hosting provider. In general, this means that the data cannot then be used for searching or indexing unless you use homomorphic encryption.

The strictest form of encryption would be client-side encryption so that all the database knows about is a blob of data, which would then normally be stored in a `bytea` database column but could be others.

Data can also be encrypted server-side before it is returned to the user, using the `pgcrypto` contrib package, provided as an extension with PostgreSQL.

Getting ready

Make sure you (and/or your database server) are in a country where encryption is legal—in some countries, it is either banned completely or a license is required.

In order to create and manage **Pretty Good Privacy** (**PGP**) keys, you also need the well-known GnuPG command-line utility, which is available on practically all distributions.

pgcrypto is part of the contrib collection. Starting from version 10, on Debian and Ubuntu, it is part of the main postgresql-NN server package.

Install it on the database in which you want to use it, following the *Adding an external module to PostgreSQL* recipe from *Chapter 3, Server Configuration*.

You also need to have PGP keys set up:

```
pguser@laptop:~$ gpg --gen-key
```

Answer some questions here (the defaults are OK unless you are an expert), select the key type as DSA and Elgamal, and enter an empty password.

Now, export the keys:

```
pguser@laptop:~$ gpg -a --export "PostgreSQL User (test key for PG Cookbook) <pguser@somewhere.net>" > public.key
pguser@laptop:~$ gpg -a --export-secret-keys "PostgreSQL User (test key for PG Cookbook) <pguser@somewhere.net>" > secret.key
```

Make sure only you and the postgres database user have access to the secret key:

```
pguser@laptop:~$ sudo chgrp postgres secret.key
pguser@laptop:~$ chmod 440 secret.key
pguser@laptop:~$ ls -l *.key
-rw-r--r-- 1 pguser pguser   1718 2016-03-26 13:53 public.key
-r--r----- 1 pguser postgres 1818 2016-03-26 13:54 secret.key
```

Last but not least, make a copy of the public and the secret key; if you lose them, you'll lose the ability to encrypt/decrypt.

How to do it...

To ensure that secret keys are never visible in database logs, you should write a wrapper function to retrieve the keys from the file. You need to create a SECURITY DEFINER function as a user who has the pg_read_server_files privilege or role assigned to them. For convenience, below is an example that illustrates how to write a function to read a key from a secret file:

```
CREATE OR REPLACE FUNCTION get_my_public_key() RETURNS TEXT
SECURITY DEFINER
LANGUAGE SQL
```

```
AS
$function_body$
    SELECT pg_read_file ('/home/pguser/public.key');
$function_body$;


REVOKE ALL ON FUNCTION get_my_public_key() FROM PUBLIC;

CREATE OR REPLACE FUNCTION get_my_secret_key() RETURNS TEXT
SECURITY DEFINER
LANGUAGE SQL
AS
$function_body$
    SELECT pg_read_file ('/home/pguser/secret.key');
$function_body$;
REVOKE ALL ON FUNCTION get_my_secret_key() FROM PUBLIC;
```

This can also be fully implemented in PL/pgSQL, using the built-in pg_read_file (filename) PostgreSQL system function. To use this function, you must place the files in the data directory, as required by that function for added security, so that the database user cannot access the rest of the filesystem directly. However, using that file needs pg_read_server_files privilege unless granted via a role or accessed using security definer functions.

If you don't want other database users to be able to see the keys, you also need to write wrapper functions for encryption and decryption, and then give end users access to these wrapper functions.

The encryption function could look like this:

```
create or replace function encrypt_using_my_public_key(
    cleartext text,
    ciphertext out bytea
)
AS $$
DECLARE
    pubkey_bin bytea;
BEGIN
    -- text version of public key needs to be passed through function
dearmor() to get to raw key
    pubkey_bin := dearmor(get_my_public_key());
    ciphertext := pgp_pub_encrypt(cleartext, pubkey_bin);
```

```
END;
$$ language plpgsql security definer;
revoke all on function encrypt_using_my_public_key(text) from public;
grant execute on function encrypt_using_my_public_key(text) to bob;
```

The decryption function could look like this:

```
create or replace function decrypt_using_my_secret_key(
    ciphertext bytea,
    cleartext out text
)
AS $$
DECLARE
    secret_key_bin bytea;
BEGIN
    -- text version of secret key needs to be passed through function
dearmor() to get to raw binary key
    secret_key_bin := dearmor(get_my_secret_key());
    cleartext := pgp_pub_decrypt(ciphertext, secret_key_bin);
END;
$$ language plpgsql security definer;
revoke all on function decrypt_using_my_secret_key(bytea) from public;
grant execute on function decrypt_using_my_secret_key(bytea) to bob;
```

Finally, we test the encryption:

```
test2=# select encrypt_using_my_public_key('X marks the spot!');
```

This function returns a bytea (that is, raw binary) result that looks something like this:

```
encrypt_using_my_public_key |
\301\301N\003\223o\215\2125\203\252;\020\007\376-z\233\211H...
```

To see that it actually works, you must run both commands:

```
test2=# select decrypt_using_my_secret_key(encrypt_using_my_public_key('X
marks the spot!'));
decrypt_using_my_secret_key
-------------------------------
X marks the spot!
(1 row)
```

Yes—we got back our initial string!

How it works...

What we have done here is this:

- Hidden the keys from non-superuser database users
- Provided wrappers for authorized users to use encryption and decryption functionalities

To ensure that your sensitive data is not stolen while in transit between the client and the database server, make sure you connect to PostgreSQL, either using an SSL-encrypted connection or from `localhost`.

You also have to trust your server administrators and all of the other users with superuser privileges to be sure that your encrypted data is safe. And, of course, you must trust the safety of the entire environment; PostgreSQL can decrypt the data, so any other user or software that has access to the same files can do the same.

If you are using **EDB Postgres Advanced Server (EPAS)**, you can create this function using EDB-SPL, and you can also utilize EDB*Wrap. EDB*Wrap obfuscates code and protects the secret key code from superusers or any other privileged users. For more information, please refer to the following link:

`https://www.enterprisedb.com/docs/epas/latest/epas_security_guide/03_edb_wrap/`

There's more...

A higher level of security is possible with more complex procedures and architecture, as shown in the next sections. We also mention a limited `pgcrypto` version that does not use OpenSSL.

For really sensitive data

For some data, you wouldn't want to risk keeping the decryption password on the same machine as the encrypted data.

In those cases, you can use **public-key cryptography**, also known as **asymmetric cryptography**, and carry out only the encryption part on the database server. This also means that you only have the encryption key on the database host and not the key needed for decryption. Alternatively, you can deploy a separate, extra-secure encryption server in your server infrastructure that provides just the encrypting and decrypting functionality as a remote call.

This solution is secure because, in asymmetric cryptography, the private (that is, decryption) key cannot be derived from the corresponding public (that is, encryption) key, hence the names public and private, which denote the appropriate dissemination policies.

If you wish to prove the identity of the author of a file, the correct method is to use a digital signature, which is an entirely different application of cryptography. Note that this is not currently supported by pgcrypto, so you must implement your own methods as C functions, or in a procedural language capable of using cryptographic libraries.

For really, really, really sensitive data

For even more sensitive data, you may never want the data to leave the client's computer unencrypted; therefore, you need to encrypt the data before sending it to the database. In that case, PostgreSQL receives already encrypted data and never sees the unencrypted version. This also means that the only useful indexes you can have are for use in WHERE encrypted_column = encrypted_data and to ensure uniqueness.

Even these forms can be used only if the encryption algorithm always produces the same ciphertext (output) for the same plaintext (input), which is true only for weaker encryption algorithms. For example, it would be easy to determine the age or sex of a person if the same value were always encrypted into the same ciphertext. To avoid this vulnerability, strong encryption algorithms are able to produce a different ciphertext for the same value.

The versions of pgcrypto are usually compiled to use the *OpenSSL library* (http://www.openssl.org). If, for some reason, you don't have OpenSSL or just don't want to use it, it is possible to compile pgcrypto without it, with a smaller number of supported encryption algorithms and a slightly reduced performance.

See also

- The page on pgcrypto in the PostgreSQL online documentation, available at http://www.postgresql.org/docs/current/static/pgcrypto.html
- The OpenSSL web page, accessed at http://www.openssl.org/
- The *GNU Privacy Handbook* at http://www.gnupg.org/gph/en/manual.html

Setting up cloud security using predefined roles

Many **Databases as a Service** (**DBaaS**)/database clouds restrict the use of superusers, with good reason. Administrators in a database cloud need to use an intermediate level of authority.

For example, in the EDB BigAnimal cloud service, a user called edb_admin holds most privileges, including CREATEROLE and CREATEDB. BigAnimal runs within your own account on cloud platforms, so the service provides data isolation, which in turn makes it easier and safer to administer than other clouds.

In prior releases of PostgreSQL, many functions were superuser-only, but these functions and views are now just superuser by default.

Rather than have administrators work out for themselves how to set up admin privileges, PostgreSQL now provides predefined roles, previously known as default roles, which can be thought of as useful groupings of privileges to grant to different types of administrators.

Getting ready

Set up a cloud account (for example, using BigAnimal) that supports PostgreSQL 16: https://www.biganimal.com.

How to do it…

The edb_admin user can be used to create two new "group" roles:

- ops_dba—This will monitor and control PostgreSQL, but without being able to see any of the data in the database, so it may not be counted as a data processor under the **General Data Protection Regulation (GDPR)**.
- app_dba—This will supervise the data in PostgreSQL, allowing the authority to correct data quality issues and add new and remove old data, which would be a data processor under GDPR.

These two roles can be configured like this:

```
CREATE ROLE ops_dba LOGIN;
CREATE ROLE app_dba LOGIN;
GRANT pg_monitor, pg_signal_backend TO ops_dba;
GRANT pg_read_all_data, pg_write_all_data TO app_dba;
```

We can then assign individual users to each group role:

```
GRANT app_dba TO donald;
```

NOTE

Don't put your company name in the usernames. This gets especially confusing if you need to grant access to someone from a service company contracted to assist in managing the database.

How it works...

PostgreSQL has the following predefined roles:

Role	Notes
pg_read_all_settings	Read all configuration settings (pg_settings), even if usually superuser only.
pg_read_all_stats	Read all pg_stat_* views and use various statistics-related extensions, even those normally visible only to superusers. Also include pg_statistic_ext_data().
pg_stat_scan_tables	Execute monitoring functions that may take ACCESS SHARE locks on tables, potentially for a long time. Shouldn't normally be required.
pg_monitor	Read/execute various monitoring views and functions. Includes pg_read_all_settings, pg_read_all_stats, and pg_stat_scan_tables.
pg_signal_backend	Allows execution of both pg_cancel_backend() and pg_terminate_backend().
pg_read_all_data	PG14+. SELECT on all tables, views, sequences, and USAGE on all schemas. Does not bypass RLS.
pg_write_all_data	PG14+. INSERT/UPDATE/DELETE on all tables, views, sequences, and USAGE on all schemas. Does not bypass RLS.

Table 6.3: Main predefined roles

There are also certain roles that should not often be granted to users. The pg_read_server_files, pg_write_server_files, and pg_execute_server_program roles are intended to allow administrators to have trusted but non-superuser roles that can access files and run programs, on the same database server on which the user database runs. As these roles are able to access any file on the server filesystem, they bypass all database-level permission checks when accessing files directly, and they could be used to gain superuser-level access; therefore, great care should be taken when granting these roles to users.

Here are additional roles that users can use to perform specific operations.

Role	Notes
pg_checkpoint	PG 15+. Allows you to perform CHECKPOINT.
pg_use_reserved_connections	PG15+. Allows you to use reserved connection slots using the reserved_connection parameter.
pg_create_subscription	PG16+. Allows users to create a subscription.

Table 6.4: Additional roles

Some other aspects of security in BigAnimal come preconfigured:

- **Data encryption**: All data in BigAnimal is encrypted in motion and at rest. Network traffic is encrypted using **Transport Layer Security** (**TLS**) v1.2 or greater, where applicable. Data at rest is encrypted using the **Advanced Encryption Standard** (**AES**) with 256-bit keys. Data encryption keys are envelope-encrypted, and the wrapped data encryption keys are securely stored in a **Key Management System** (**KMS**).
- **Portal audit logging**: Activities in the cloud user portal, such as those related to user roles, organization updates, and cluster creation and deletion, are tracked and viewed in the activity log. **Command-Line Interface** (**CLI**) actions are also logged.
- **Database logging and auditing**: The functionality to track and analyze database activities is enabled automatically. For PostgreSQL, the **PostgreSQL Audit Extension** (pgaudit) is enabled for you when deploying a Postgres cluster. For EDB Postgres Advanced Server, the **EDB Audit Extension** (edbaudit) is enabled for you. All DDL is logged.

There's more...

Some PostgreSQL privileges can only be granted directly by superusers, which may need special actions in a cloud-based service. These exceptions are shown here for completeness.

For replication management functions (described here: https://www.postgresql.org/docs/devel/functions-admin.html#FUNCTIONS-REPLICATION), slot functions are available to users with the REPLICATION privilege—for example, streaming_replica.

Replication origins functions are available by default to superusers and could be granted to other users.

Execute privileges could be granted on this to allow you to plan for **Point-in-Time Recovery (PITR)**:

- pg_create_restore_point()

Execute privileges could be granted on these functions, although backup is already provided by the cloud service:

- pg_start_backup()
- pg_stop_backup()

Execute privileges could be granted by superusers on these functions, but this would likely interfere with the orchestration of high availability features, so it would be dangerous:

- pg_switch_wal()
- pg_promote()
- pg_wal_replay_pause()
- pg_wal_replay_resume()

Generic file access functions are insecure and should not typically be granted: https://www.postgresql.org/docs/devel/functions-admin.html#FUNCTIONS-ADMIN-GENFILE.

Learn more on Discord

To join the Discord community for this book – where you can share feedback, ask questions to the author, and learn about new releases – follow the QR code below:

https://discord.gg/pQkghgmgdG

7

Database Administration

In *Chapter 5*, *Tables and Data*, we looked at the contents of tables and their various complexities. Now, we'll turn our attention to larger administration tasks that we need to perform from time to time, such as creating things, moving things around, storing things neatly, and removing them when they're no longer required.

The most sensible way to perform major administrative tasks is to write a script to do what you think is required. This allows you to run the script on a system test server, and then run it again on the production server once you're happy with it. Devising and typing commands against production database servers, especially when under pressure, isn't wise. Worse, using an admin tool can lead to serious issues if that tool doesn't show you the SQL you're about to execute. If you haven't dropped your first live table yet, don't worry; there is still time. Perhaps you might want to read *Chapter 11*, *Backup and Recovery*, first, eh? Back it up using scripts.

Scripts are great because you can automate common tasks, and there's no need to sit there with a mouse, working your way through hundreds of changes. If you're drawn to the discussion about the command line versus GUI, then my thoughts and reasons are completely orthogonal to that. I want to encourage you to avoid errors and save time by executing small administration programs or scripts repetitively and automatically. If it were safe or easy to record a macro using mouse movements in a script, then that would be an option, but it's not. The only viable way to write a repeatable script is by writing SQL commands in a text file.

The choice of the scripting tool to be used is a debatable topic. Here we will consider psql because it's a great scripting tool and if you've got PostgreSQL, then you've already got it, without needing to install additional software. We will also discuss GUI tools and explain how and when they are relevant.

Let's move on to the recipes! First, we'll start by looking at some scripting techniques that are valuable in PostgreSQL.

In this chapter, we will cover the following recipes:

- Writing a script that either succeeds entirely or fails entirely
- Writing a psql script that exits on the first error
- Using psql variables
- Placing query output into psql variables
- Writing a conditional psql script
- Investigating a psql error
- Setting the psql prompt with useful information
- Using pgAdmin for DBA tasks
- Scheduling jobs for regular background execution
- Performing actions on many tables
- Adding/removing columns on a table
- Changing the data type of a column
- Changing the definition of an enum data type
- Adding a constraint concurrently
- Adding/removing schemas
- Moving objects between schemas
- Adding/removing tablespaces
- Moving objects between tablespaces
- Accessing objects in other PostgreSQL databases
- Accessing objects in other foreign databases
- Making views updatable
- Using materialized views
- Using GENERATED data columns
- Using data compression

Writing a script that either succeeds entirely or fails entirely

Database administration often involves applying a coordinated set of changes to the database. One of PostgreSQL's greatest strengths is its transaction system, wherein almost all actions can be executed inside a transaction. This allows us to build a script with many actions that will either *all* succeed or *all* fail. This means that if any of these actions fail, then all the other actions in the script are rolled back and never become visible to any other user, which can be critically important in a production system. This property is referred to as **atomicity** in the sense that the script is intended as a single unit that cannot be split. This is the meaning of the *A* in the **ACID** properties of database transactions.

Transactions apply to **Data Definition Language** (**DDL**), which refers to the set of SQL commands that are used to define, modify, and delete database objects. The term DDL goes back many years, but it persists because that subset is a useful short name for the commands that most administrators need to execute: CREATE, ALTER, DROP, and so on.

NOTE

Although most commands in PostgreSQL are transactional, there are a few that cannot be so. One example is sequence allocation. It cannot be transactional because when a new sequence number is allocated, the effect of having consumed that number must become visible immediately, without waiting for that transaction to be committed. Otherwise, the same number will be given to another transaction. Other examples include CREATE INDEX CONCURRENTLY and CREATE DATABASE.

How to do it...

The basic way to ensure that all the commands are successful or that none are is to wrap our script into a transaction, as follows:

```
BEGIN;
command 1;
command 2;
command 3;
COMMIT;
```

Writing a transaction control command involves editing the script, which you may not want to do or even have access to. There are, however, other ways to do this.

Using psql, you can do this by simply using the -1 or --single-transaction command-line options, as follows:

```
bash $ psql -1 -f myscript.sql
bash $ psql --single-transaction -f myscript.sql
```

The effect of this switch is equivalent to adding a BEGIN; statement and a COMMIT; statement respectively at the beginning and the end of the contents of the file (myscript.sql in this case). So if you use this switch then these statements shouldn't be used inside the script.

The -1 option is short, but I recommend using --single-transaction as it's much clearer regarding which option is being selected.

How it works...

The entire script will fail if, at any point, one of the commands gives an error message, or another failure of higher severity. Almost all of the SQL that's used to define objects (DDL) provides a way to avoid throwing errors. More precisely, commands that begin with the DROP keyword have an IF EXISTS option. This allows you to execute the DROP commands, regardless of whether or not the object already exists.

Thus, by the end of the command, that object will not exist:

```
DROP VIEW IF EXISTS cust_view;
```

Similarly, most commands that begin with the CREATE keyword have the optional OR REPLACE suffix. This allows the CREATE statement to overwrite the definition if one already exists, or add the new object if it doesn't exist yet, like this:

```
CREATE OR REPLACE VIEW cust_view AS SELECT * FROM cust;
```

In cases where both the DROP IF EXISTS and CREATE OR REPLACE options are present, you may think that CREATE OR REPLACE is usually sufficient. However, if you change the output definition of a function or a view, then using OR REPLACE is not sufficient. In that case, you must use DROP and recreate it, as shown in the following example:

```
postgres=# CREATE OR REPLACE VIEW cust_view AS
SELECT col as title1 FROM cust;
CREATE VIEW
postgres=# CREATE OR REPLACE VIEW cust_view
AS SELECT col as title2 FROM cust;
ERROR:  cannot change name of view column "title1" to "title2"
```

Also, note that CREATE INDEX does not have an OR REPLACE option. If you run it twice, you'll get two indexes on your table, unless you specifically name the index. There is a DROP INDEX IF EXISTS option, but it may take a long time to drop and recreate an index. An index exists just for optimization, and it does not change the actual result of any query, so this different behavior is very convenient. This is also reflected in the fact that the SQL standard doesn't mention indexes at all, even though they exist in practically all database systems, because they do not affect the logical layer.

PostgreSQL does not support nested transaction control commands, which can lead to unexpected behavior. For instance, consider the following code, which has been written in a **nested transaction** style:

```
postgres=# BEGIN;
BEGIN
postgres=# CREATE TABLE a(x int);
CREATE TABLE
postgres=# BEGIN;
WARNING:  there is already a transaction in progress
BEGIN
postgres=# CREATE TABLE b(x int);
CREATE TABLE
postgres=# COMMIT;
COMMIT
postgres=# ROLLBACK;
NOTICE:  there is no transaction in progress
ROLLBACK
```

The hypothetical author of such code probably meant to create table a first, and then create table b. Then, they changed their mind and rolled back both the *inner* transaction and the *outer* transaction. However, what PostgreSQL does is discard the second BEGIN statement so that the COMMIT statement is matched with the first BEGIN statement, and what looks like an inner transaction is part of the top-level transaction. Hence, right after the COMMIT statement, we are outside a transaction block, so the next statement is assigned a separate transaction. When ROLLBACK is issued as the next statement, PostgreSQL notices that the transaction is empty.

The danger in this particular example is that the user inadvertently committed a transaction, thus waiving the right to roll it back; however, note that a careful user would have noticed this warning and paused to think before going ahead.

From this example, you have learned a valuable lesson: if you have used transaction control commands in your script, then wrapping them again in a higher-level script or command can cause problems of the worst kind, such as committing stuff that you wanted to roll back. This is important enough to deserve a boxed warning.

NOTE

PostgreSQL accepts nested transactional control commands but does not act on them. After the first commit, the commands will be assumed to be transactions in their own right and will persist, should the script fail. Be careful!

There's more...

These commands cannot be included in a script that uses transactions in the way we just described because they execute multiple database transactions and cannot be used in a transaction block:

- CREATE DATABASE/DROP DATABASE
- CREATE TABLESPACE/DROP TABLESPACE
- CREATE INDEX CONCURRENTLY
- VACUUM
- REINDEX DATABASE/REINDEX SYSTEM
- CLUSTER

None of these actions need to be run manually regularly within complex programs, so this shouldn't be a problem for you.

Also, note that these commands do not substantially alter the *logical* content of a database; that is, they don't create new user tables or alter any rows, so there's less need to use them inside complex transactions.

While PostgreSQL does not support nested transaction commands, it supports the notion of SAVEPOINT, which can be used to achieve the same behavior. Suppose we wanted to implement the following pseudocode:

```
(begin transaction T1)
   (statement 1)
   (begin transaction T2)
     (statement 2)
   (commit transaction T2)
```

```
    (statement 3)
(commit transaction t1)
```

The effect we seek has the following properties:

- If statements 1 and 3 succeed, and `statement 2` fails, then statements 1 and 3 will be committed.
- If all three statements succeed, then they will all be committed.
- Otherwise, no statement will be committed.

These properties also hold with the following PostgreSQL commands:

```
BEGIN;
    (statement 1)
  SAVEPOINT T2;
    (statement 2)
  RELEASE SAVEPOINT T2; /* we assume that statement 2 does not fail */
    (statement 3)
COMMIT;
```

This form, as noted in the preceding code, applies only if `statement 2` does not fail. If it fails, we must replace `RELEASE SAVEPOINT` with `ROLLBACK TO SAVEPOINT`, or we will get an error. This is a slight difference between top-level transaction commands; a `COMMIT` statement is silently converted into a `ROLLBACK` when the transaction is in a failed state.

Writing a psql script that exits on the first error

The default mode for the `psql` script tool is to continue processing when it finds an error. This sounds silly, but it exists for historical compatibility only. There are some easy and permanent ways to avoid this, so let's look at them.

Getting ready

Let's start with a simple script, with a command we know will fail:

```
$ $EDITOR test.sql
mistake1;
mistake2;
mistake3;
```

Execute the following script using psql to see what the results look like:

```
$ psql -f test.sql
psql:test.sql:1: ERROR:  syntax error at or near "mistake1"
LINE 1: mistake1;
        ^
psql:test.sql:2: ERROR:  syntax error at or near "mistake2"
LINE 1: mistake2;
        ^
psql:test.sql:3: ERROR:  syntax error at or near "mistake3"
LINE 1: mistake3;
        ^
```

How to do it...

Let's perform the following steps:

1. To exit the script on the first error, we can use the following command:

   ```
   $ psql -f test.sql -v ON_ERROR_STOP=on
   psql:test.sql:1: ERROR:  syntax error at or near "mistake1"
   LINE 1: mistake1;
           ^
   ```

2. Alternatively, we can edit the test.sql file with the initial line that's shown here:

   ```
   $ vim test.sql
   \set ON_ERROR_STOP on
   mistake1;
   mistake2;
   mistake3;
   ```

3. Note that the following command will *not* work because we have missed the crucial ON value:

   ```
   $ psql -f test.sql -v ON_ERROR_STOP
   ```

How it works...

The ON_ERROR_STOP variable is a psql special variable that controls the behavior of psql as it executes in script mode. When this variable is set, a SQL error will generate an OS return code 3, whereas other OS-related errors will return code 1.

There's more...

When you run `psql`, a startup file will be executed, sometimes called a profile file. You can place your `psql` commands in that startup file to customize your environment. Adding `ON_ERROR_STOP` to your profile will ensure that this setting is applied to all `psql` sessions:

```
$ $EDITOR ~/.psqlrc
\set ON_ERROR_STOP
```

You can forcibly override this and request `psql` to execute without a startup file using `-X`. This is probably the safest thing to do for the batch execution of scripts so that they always work in the same way, irrespective of the local settings.

`ON_ERROR_STOP` is one of some special variables that affect the way `psql` behaves. The full list is available at the following URL: https://www.postgresql.org/docs/current/static/app-psql.html#APP-PSQL-VARIABLES.

Using psql variables

In the previous recipe, you learned how to use the `ON_ERROR_STOP` variable. Here, we will show you how to work with any variable, including user-defined ones.

Getting ready

As an example, we will create a script that takes a table name as a parameter. We will keep it simple because we just want to show how variables work.

For instance, we might want to add a text column to a table and then set it to a given value. So, we must write the following lines in a file called `vartest.sql`:

```
ALTER TABLE mytable ADD COLUMN mycol text;
UPDATE mytable SET mycol = 'myval';
```

The script can be run as follows:

```
psql -f vartest.sql
```

How to do it...

We change `vartest.sql` as follows:

```
\set tabname mytable
\set colname mycol
\set colval 'myval'
```

```
ALTER TABLE :tabname ADD COLUMN :colname text;
UPDATE :tabname SET :colname = :'colval';
```

How it works...

What do these changes mean? We have defined three variables, setting them to the table name, column name, and column value. Then, we replaced the mentions of those specific values with the name of the variable preceded by a colon, which in `psql` means *replace with the value of this variable*. In the case of `colval`, we have also surrounded the variable name with single quotes, meaning *treat the value as a string*.

If we want `vartest.sql` to add a different column, we just have to make one change to the top of the script, where all the variables are conveniently set. Then, the new column name will be used.

There's more...

This was just one way to define variables. Another is to indicate them in the command line when running the script:

```
psql -v tabname=mytab2 -f vartest.sql
```

Variables can also be set interactively. The following line will prompt the user, and then set the variable to whatever is typed before hitting *Enter*:

```
\prompt 'Insert the table name: ' tabname
```

In the next recipe, we will learn how to set variables using a SQL query.

Placing query output into psql variables

It is also possible to store some values that have been produced by a query into variables – for instance, to reuse them later in other queries.

In this recipe, we will demonstrate this approach with a concrete example.

Getting ready

In the *Controlling automatic database maintenance* recipe of *Chapter 9, Regular Maintenance*, we will describe VACUUM, showing that it runs regularly on each table based on the number of rows that might need vacuuming (**dead rows**). The VACUUM command will run if that number exceeds a given threshold, which by default is just above 20% of the row count.

In this recipe, we will create a script that picks the table with the largest number of dead rows and runs VACUUM on it, assuming you have some tables already in existence.

How to do it...

The script is as follows:

```
SELECT schemaname
, relname
, n_dead_tup
, n_live_tup
FROM pg_stat_user_tables
ORDER BY n_dead_tup DESC
LIMIT 1
\gset
\qecho Running VACUUM on table :"relname" in schema :"schemaname"
\qecho Rows before: :n_dead_tup dead, :n_live_tup live
VACUUM ANALYZE :schemaname.:relname;
\qecho Waiting 1 second...
SELECT pg_sleep(1);
SELECT n_dead_tup AS n_dead_tup_now
,      n_live_tup AS n_live_tup_now
FROM pg_stat_user_tables
WHERE schemaname = :'schemaname'
AND relname = :'relname'
\gset
\qecho Rows after: :n_dead_tup_now dead, :n_live_tup_now live
```

How it works...

You may have noticed that the first query does not end with a semicolon, as usual. This is because we end it with \gset instead, which means to *run the query and assign each returned value to a variable that has the same name as the output column*.

This command expects the query to return exactly one row, as you might expect it to, and if not, it does not set any variable.

The script waits 1 second before reading the updated number of dead and live rows. The reason for the wait is that such statistics are updated after the end of the transaction that makes the changes, which sends a signal to the statistics collector, which then does the update. There's no guarantee that the stats will be updated in 1 second, though in most cases they will be.

There's more...

See the next recipe on how to improve the script with iterations so that it vacuums more than one table.

Writing a conditional psql script

psql supports the conditional \if, \elif, \else, and \endif meta-commands. In this recipe, we will demonstrate some of them.

Getting ready

We want to improve the vartest.sql script so that it runs VACUUM if there are dead rows in that table.

How to do it...

We can add conditional commands to vartest.sql, resulting in the following script:

```
\set needs_vacuum false
SELECT schemaname
, relname
, n_dead_tup
, n_live_tup
, n_dead_tup > 0 AS needs_vacuum
FROM pg_stat_user_tables
ORDER BY n_dead_tup DESC
LIMIT 1
\gset
\if :needs_vacuum
\qecho Running VACUUM on table :"relname" in schema :"schemaname"
\qecho Rows before: :n_dead_tup dead, :n_live_tup live
VACUUM ANALYZE :schemaname.:relname;
\qecho Waiting 1 second...
SELECT pg_sleep(1);
SELECT n_dead_tup AS n_dead_tup_now
,      n_live_tup AS n_live_tup_now
FROM pg_stat_user_tables
WHERE schemaname = :'schemaname' AND relname = :'relname'
\gset
```

```
\qecho Rows after: :n_dead_tup_now dead, :n_live_tup_now live
\else
\qecho Skipping VACUUM on table :"relname" in schema :"schemaname"
\endif
```

How it works...

We have added an extra column, needs_vacuum, to the first query, resulting in one more variable that we can use to make the VACUUM part conditional.

This is just an example where we assume that the table needs vacuuming if there is at least one dead row. In reality, the actual threshold is more variable and depends on cost/benefit, as explained in *Chapter 9, Regular Maintenance*.

There's more...

Conditional statements are usually part of flow-control statements, which also include iterations.

While iterating is not directly supported by psql, a similar effect can be achieved in other ways.

For instance, a script called file.sql (for instance) can be iterated by adding some lines at the end, as shown in the following fragment:

```
SELECT /* add a termination condition as appropriate */ AS do_loop
\gset
\if do_loop
\ir file.sql
\endif
```

Instead of iterating, you can follow the approach described later in this chapter in the *Performing actions on many tables* recipe.

Investigating a psql error

Error messages can sometimes be cryptic, and you may be left wondering, *why did this error happen at all?*

For this purpose, psql recognizes two variables:

- VERBOSITY, which can be set to terse, default, or verbose
- CONTEXT, which can be set to never, errors, or always

These variables control how much detail is displayed to the user in case of an error.

Here is an example to show the difference:

```
postgres=# \set VERBOSITY terse
postgres=# \set CONTEXT never
postgres=# select * from missingtable;
ERROR:  relation "missingtable" does not exist at character 15
```

This is quite a simple error, so we don't need the extra details, but it is nevertheless useful for illustrating the extra detail you get when raising verbosity and enabling context information:

```
postgres=# \set VERBOSITY verbose
postgres=# \set CONTEXT errors
postgres=# select * from missingtable;
ERROR:  42P01: relation "missingtable" does not exist
LINE 1: select * from missingtable;
                      ^
LOCATION:  parserOpenTable, parse_relation.c:1159
```

Now, you get the SQL error code 42P01, which you can look up in the PostgreSQL manual. You will even find a reference to the file and the line in the PostgreSQL source code where this error has been raised so that you can investigate it (the beauty of open source!).

However, there is a problem with having to enable verbosity in advance: you need to do so before running the command. If all the errors were reproducible, this would not be a huge inconvenience. But in certain cases, you may hit a transient error, such as a **serialization failure**. That kind of error is difficult to detect, and it could sometimes happen that you struggle to reproduce the error, let alone analyze it.

The \errverbose meta-command in psql was introduced to avoid these problems.

Serialization failures occur only under special circumstances that are quite complex to describe here. We can give a simple (yet incomplete) explanation by saying that PostgreSQL supports different transaction isolation modes, and that some of those modes are more convenient for the application; unfortunately, though, they are also vulnerable to certain anomalies that the database is able to detect. Should one such anomaly occur, the database causes an error with the intention of forcing the application to retry the exact same transaction. In fact, a good application should be programmed to retry or not retry a transaction that fails based on what error code is returned.

Getting ready

There isn't much to do, as the point of the \errverbose meta-command is to capture information about the error without requiring any prior activity.

How to do it...

Follow these steps to understand the usage of the \errverbose meta-command:

1. Suppose you hit an error, as shown in the following query, and verbose reporting was not enabled:

    ```
    postgres=# create table wrongname();
    ERROR:  relation "wrongname" already exists
    ```

2. The extra detail that is not displayed is remembered by psql, so you can view it as follows:

    ```
    postgres=# \errverbose
    ERROR:  42P07: relation "wrongname" already exists
    LOCATION:  heap_create_with_catalog, heap.c:1067
    ```

There's more...

The error and source codes for this recipe can be found at the following links:

- The list of PostgreSQL error codes is available at the following URL: https://www.postgresql.org/docs/current/static/errcodes-appendix.html.
- The PostgreSQL source code can be downloaded from or inspected at the following URL: https://git.postgresql.org/.

Setting the psql prompt with useful information

When you're connecting to multiple systems, it can be useful to configure your psql prompt so that it tells you what you are connected to.

To do this, we will edit the psql profile file so that we can execute commands when we first start psql. In the profile file, we will set values for two special variables, called PROMPT1 and PROMPT2, that control the command-line prompt.

Getting ready

Identify and edit the ~/.psqlrc file that will be executed when you start psql.

How to do it...

My `psql` prompt looks like this:

```
gciolli@laptop680:~$ psql cookbook
 _____   ____    _____
|  _____) __  \ |  __   \
|  |__    |  |  \ \|  |_)  )
|   __)   |  |   |  |    __(
|  |_____ |  |__/ /|  |_)  )
|_____)_____/  |_____/

EnterpriseDB     https://www.enterprisedb.com/

Timing is on.
psql (16.0 (Debian 16.0-1.pgdg120+1))
Type "help" for help.

[16/main] gciolli@cookbook  =#
```

Figure 7.1: A customized psql prompt

As you can see, it has a banner that highlights my employer's company name, an idea I read from Simon Riggs in a previous edition of this book, which he set for when we do demos. You can skip that part, or you can create some word art, being careful with backslashes since they are escape characters:

```
\echo ' _____   ____    _____ '
\echo '|  _____) __  \\ |  __   \\'
\echo '|  |__    |  |  \\ \\\\|  |_)  )'
\echo '|   __)   |  |   |  |    __('
\echo '|  |_____ |  |__/ /|  |_)  )'
\echo '|_____)_____/  |_____/'
\echo ''
\echo 'EnterpriseDB     https://www.enterprisedb.com/'
\echo ''
select current_setting('cluster_name') as nodename,
       case current_setting('cluster_name') when '' then 'true' else
'false' end as nodename_unset
\gset
\if :nodename_unset
  \set nodename unknown
```

```
\endif
\set PROMPT1 '[%:nodename:] %n@%/ %x %R%# '
\set PROMPT2 '[%:nodename:] %n@%/ %x %R%# '
\timing
```

How it works...

The last part of the file runs a SQL query to retrieve the value of the cluster_name parameter. This is usually set to something sensible, but if not, it will return the word true in the nodename variable. I then use a \if condition to check whether nodename is set correctly. If not, it uses the unknown string.

The prompts are set from multiple variables and fields:

```
Nodename      as set above
%n            current session username
%/            current databasename
%x            transaction status - mostly blank, * if transaction block, ! if
aborted, ? if disconnected
%R            multi-line status - mostly =, shows if in a continuation/quote/
double-quote/comment
%#            set to # if user is a superuser, else set to >
```

Lastly, I turn on timing automatically for all future SQL commands.

Using pgAdmin for DBA tasks

In this recipe, we will show you how to use pgAdmin for some administration tasks in your database. pgAdmin is one of the two graphical interfaces that we introduced in the *Using graphical administration tools* recipe in *Chapter 1, First Steps*.

Getting ready

You should have already installed pgAdmin as part of the *Using graphical administration tools* recipe of *Chapter 1, First Steps*, which includes website pointers. If you haven't done so, please read it now.

Remember to install pgAdmin 4, which is the last generation of the software; the previous one, pgAdmin 3, is no longer supported and hasn't been for a few years, so it will give various errors on PostgreSQL 10 and above.

How to do it...

The first task of a DBA is to get access to the database and get a first glance at its contents. In that respect, we have already learned how to create a connection, access the dashboard, and display some database statistics. We also mentioned the **Grant Wizard** and the graphical **Explain** tool:

1. The list of schemas in a given database can be obtained by opening a database and selecting **Schemas**:

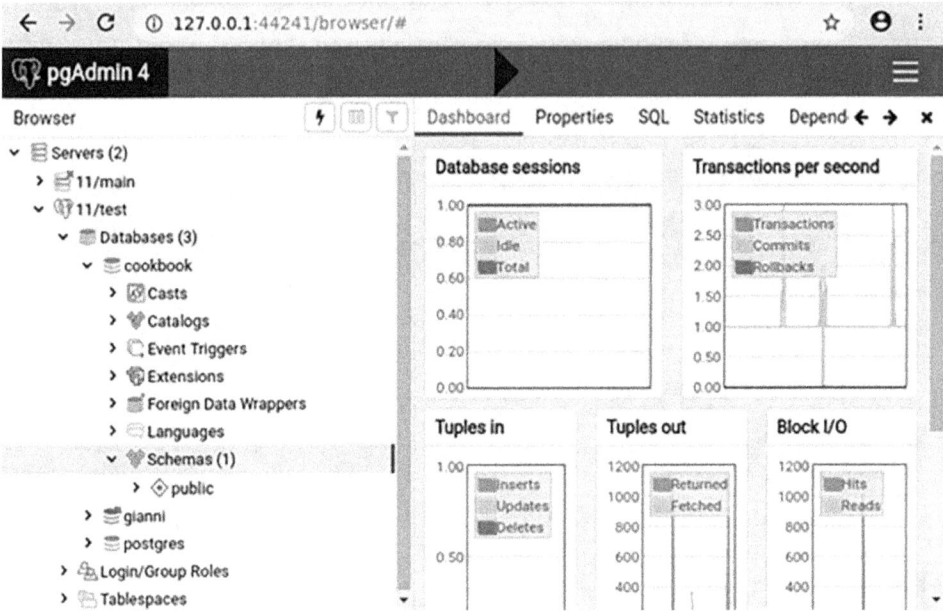

Figure 7.2: The pgAdmin 4 dashboard

2. If you right-click on an individual schema, you will see several possible actions that you can perform. For instance, you can take a backup of that schema only:

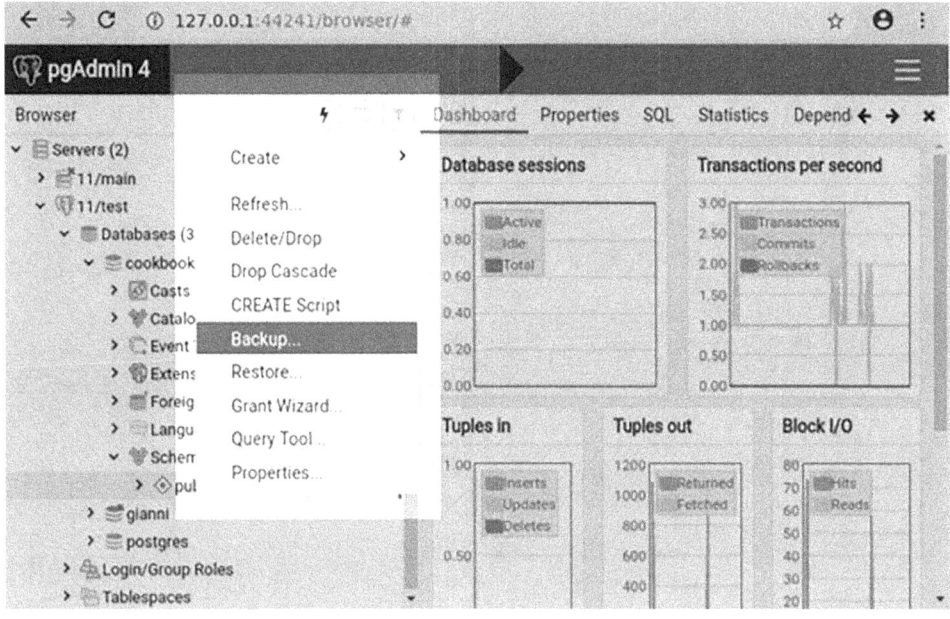

Figure 7.3: pgAdmin 4 context-sensitive menus

3. Clicking the left mouse button will drill down inside the schema and show you several object types. You will probably want to start from **Tables**:

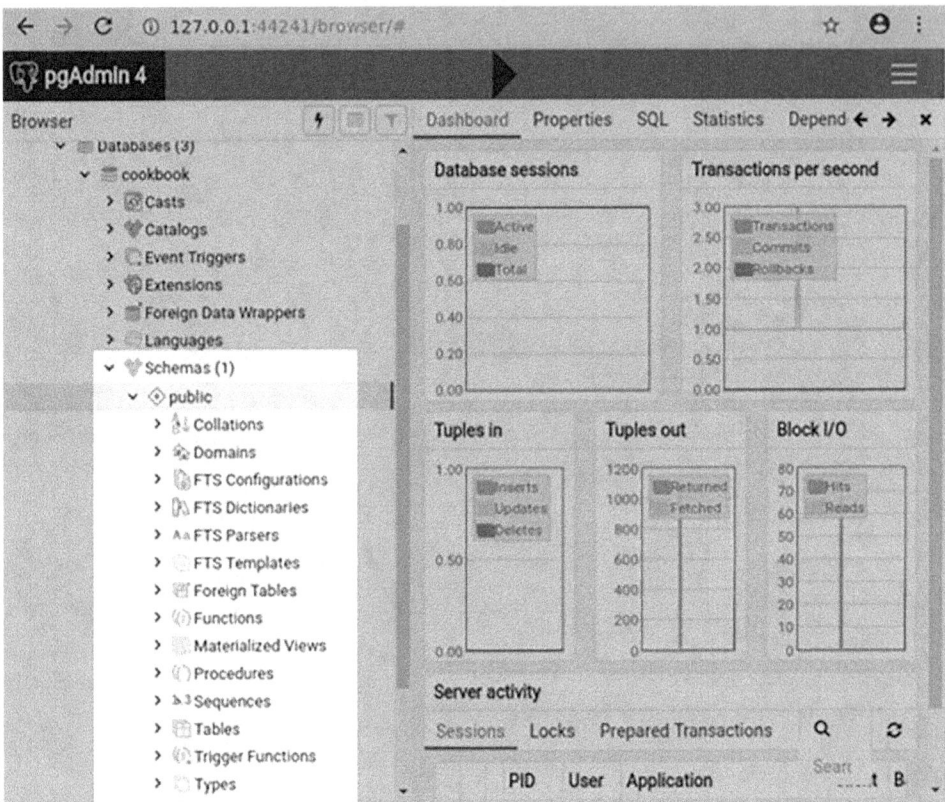

Figure 7.4: pgAdmin 4 tree view of schema contents

4. A PostgreSQL table supports a wide range of operations. For instance, you can count the number of rows:

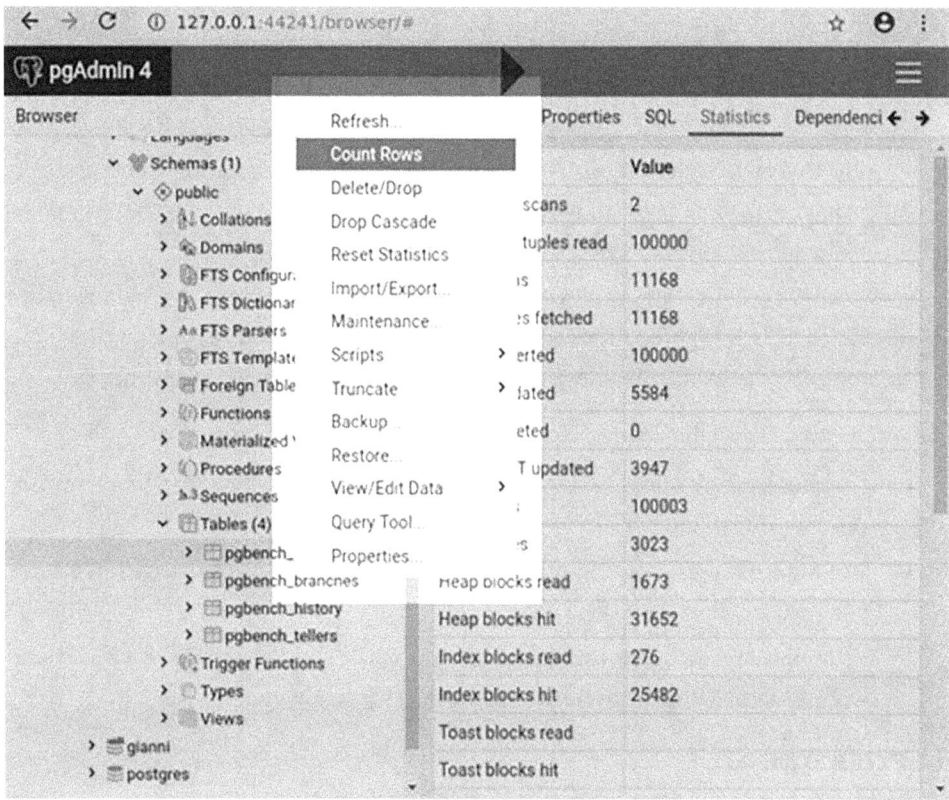

Figure 7.5: pgAdmin 4 table-context menu

Note that this is just an example of a pgAdmin feature; we are not suggesting that counting table rows is the best way to gather information on your database. See the *How many rows are there in a table?* recipe of *Chapter 2, Exploring the Database*, for a discussion on this topic.

How it works...

PostgreSQL is a complex database system, with many features and even more actions, so we can't discuss them all; we will just mention three table actions of interest here:

- The **Maintenance...** option opens a dialog box that includes actions such as VACUUM and ANALYZE, which will be discussed in various recipes in *Chapter 9, Regular Maintenance*.
- The **Import/Export...** option leads to a dialog box where you can export and import data using the COPY command, which includes CSV format, as demonstrated in *Chapter 5, Tables and Data*.
- With **View/Edit Data**, you can edit the contents of the table as you would do in a spreadsheet. This is slightly different than the CSV import/export feature because you edit the data directly inside the database without having to export it to another tool.

Finally, we would also like to mention these other three options as well:

- Each server (for example, connection) offers the option to **Backup Globals**, meaning roles (users/groups) and tablespaces.
- The **Maintenance...** option inside **Indexes**, which itself is a sub-entry of **Tables**, allows you to REINDEX or CLUSTER a given index.
- You can create SQL scripts to perform some of the specific actions, such as if you want to execute a procedure or write an INSERT query on a given table.

There's more...

As you can see, the general idea of pgAdmin is that right-clicking on an object or a group of objects opens a menu presenting several actions for that particular object or group.

Browsing the available actions is a very good way to become more familiar with what PostgreSQL can do, although not all the actions that are available in PostgreSQL can be researched through pgAdmin's interface.

Scheduling jobs for regular background execution

Normal user tasks cause the user to wait while the task executes. Frequently, there is a requirement to run tasks or "jobs" in the background without the user present, which is referred to as a job scheduler component. You can use cron, but some users look for an in-database solution.

pgAgent is our recommended job scheduler for Postgres, which isn't supplied as part of the pgAdmin package, but as a separate component. pgAgent can be operated from the pgAdmin GUI or using a simple command-line API. pgAgent keeps a history of job executions so that you can see what is happening and what is not happening.

Getting ready

If you want to manage a new database from an existing pgAgent installation, then you don't need to prepare anything. If you want to set up a new pgAgent database, execute the following command:

```
CREATE EXTENSION pgagent;
```

pgAgent is an external program, not a binary plugin, so you do not need to modify the shared_preload_libraries parameter – allowing it to work easily with all cloud databases.

Further information is available at https://www.pgadmin.org/docs/pgadmin4/latest/pgagent_install.html.

How to do it...

Each job has a name, can be configured to have one or more job steps, and can be configured to have multiple schedules that specify when it will run – but most jobs just have one step and one schedule. If more than one job step exists, they are executed serially in alphanumeric order.

Jobs are scheduled using UTC.

Each job that's executed keeps a log that can be inspected to see what has run. Jobs can be enabled/disabled and schedules can have defined start/end dates to allow you to switch from one schedule to another at a planned point in time.

You can do this using the GUI, as described in the pgAdmin docs: https://www.pgadmin.org/docs/pgadmin4/latest/pgagent_jobs.html.

But since I encourage scripting, you can add a simple job like this:

```
SELECT pgagent.add_job('reindex weekly', '30 1 * * 7',
                      'REINDEX DATABASE postgres');
```

Here, we have used code from https://github.com/simonriggs/pgagent_add_job/.

This will create a job that runs at 01:30 A.M. every Sunday and re-indexes the local database.

The parameters here are as follows:

- Jobname
- Jobschedule
- SQL

Jobschedule uses the same syntax as the `cron(1)` command in Linux:

- Minutes (0-59)
- Hours (0-23)
- Day of month (1-31)
- Month of year (1-12, 1=January)
- Day of week (1-7, 1=Monday)

You can test a job in pgAdmin by right-clicking and then selecting **Run now**.

Once the jobs have been executed, you will see the result in the `pgagent.pga_joblog` and `pgagent.pga_jobsteplog` tables.

How it works...

pgAgent is an external program that connects to the database server that stores its metadata inside the database.

pgAgent polls the database each minute to see what jobs need to be started. pgAgent will run multiple jobs in parallel when needed, each with a different thread. If a job is still running when its next scheduled time arrives, the next job will wait for the first to finish and then start immediately afterward.

pgAgent can be used to manage multiple databases or just the local database, as you choose. pgAgent can be configured for high availability using two agents accessing the same database server(s). Locking prevents the same job from being executed by multiple hosts.

There's more...

SQL jobs that have been executed will use the connection string supplied with that job, which requires you to provision how passwords or certificates are set up. Batch jobs use the OS user for the pgAgent program.

Security will always be an important consideration, so we strongly recommend limiting how many users can add/remove jobs. This will probably be a small list of maintenance activities that are agreed upon in advance for each application, rather than a long list of jobs with many users adding/removing jobs.

You can separate the roles responsible for adding/removing jobs and for checking they have run correctly. This can be accomplished with two roles, as shown in the following code block:

```
CREATE ROLE pgggent_admin;
GRANT ALL ON pgagent to pgagent_admin;
CREATE ROLE pgagent_operator;
GRANT SELECT ON
    pgagent.pga_joblog,
    pgagent.pga_jobsteplog
TO pgagent_operator;
```

If you want to prevent pgAgent from using duplicate job names, you may wish to add the following code:

```
CREATE UNIQUE INDEX ON pgagent.pga_job (jobname);
```

Performing actions on many tables

As a database administrator, you will often need to apply multiple commands as part of the same overall task. This task could be one of the following:

- Performing many different actions on multiple tables
- Performing the same action on multiple tables
- Performing the same action on multiple tables in parallel
- Performing different actions, one on each table, in parallel

The first is a general case where you need to make a set of coordinated changes. The solution is to *write a script*, as we've already discussed. We can also call this **static scripting** because you write the script manually and then execute it.

The second type of task can be achieved very simply with dynamic scripts, where we write a script that writes another script. This technique is the main topic of this recipe.

Performing actions in parallel sounds cool, and it would be useful if it were easy. In some ways, it is, but trying to run multiple tasks concurrently and trap and understand all the errors is much harder. And if you're thinking it won't matter if you don't check for errors, think again. If you run tasks in parallel, then you cannot run them inside the same transaction, so you need error handling in case one part fails.

Don't worry! Running in parallel is usually not as bad as it may seem after reading the previous paragraph, and we'll explain it after looking at a few basic examples.

Getting ready

Let's create a basic schema to run some examples on:

```
postgres=# create schema test;
CREATE SCHEMA
postgres=# create table test.a (col1 INTEGER);
CREATE TABLE
postgres=# create table test.b (col1 INTEGER);
CREATE TABLE
postgres=# create table test.c (col1 INTEGER);
CREATE TABLE
```

How to do it...

Our task is to run a SQL statement using this form, with X as the table name, against each of our three test tables:

```
ALTER TABLE X
ADD COLUMN last_update_timestamp TIMESTAMP WITH TIME ZONE DEFAULT current_timestamp;
```

The steps are as follows:

1. Our starting point is a script that lists the tables that we want to perform tasks against – something like the following:

   ```
   postgres=# SELECT n.nspname, c.relname
       FROM pg_class c
       JOIN pg_namespace n
       ON c.relnamespace = n.oid
       WHERE n.nspname = 'test'
       AND c.relkind = 'r';
   ```

2. This displays the list of tables that we will act upon (so that you can check it):

   ```
   relname
   ---------
   a
   b
   c
   (3 rows)
   ```

3. We can then use the preceding SQL to generate the text for a SQL script, substituting the schema name and table name in the SQL text:

```
postgres=# SELECT format('ALTER TABLE %I.%I ADD COLUMN last_update_
timestamp TIMESTAMP WITH TIME ZONE DEFAULT current_timestamp;'
, n.nspname, c.relname )
FROM pg_class c
JOIN pg_namespace n
ON c.relnamespace = n.oid
WHERE n.nspname = 'test'
AND c.relkind = 'r';
```

4. Finally, we can run the script and watch the results (success!):

```
postgres=# \gexec
ALTER TABLE
ALTER TABLE
ALTER TABLE
```

How it works…

Overall, this is just an example of dynamic scripting, and it has been used by DBAs for many decades, even before PostgreSQL was born.

The \gexec command means to *execute the results of the query*, so be very careful that you test your query before you run it in production.

The format function takes a template string as its first argument and replaces all occurrences of %I with the values supplied as additional arguments (in our case, the values of n.nspname and r.relname).

%I treats the value as a SQL identifier, adding double quotes as appropriate. This is extremely important if some joker or attacker creates a table like this:

```
postgres=# create table test.";  DROP TABLE customer;" (col1 INTEGER);
```

If the script used just %s rather than %I, then the script will generate this SQL, which will result in you dropping the customer table if it exists. So, for security purposes, you should use %I:

```
ALTER TABLE test.a ADD COLUMN last_update_timestamp TIMESTAMP WITH TIME
ZONE DEFAULT current_timestamp;
ALTER TABLE test.; drop table customer; ADD COLUMN last_update_timestamp
TIMESTAMP WITH TIME ZONE DEFAULT current_timestamp;
```

```
ALTER TABLE test.b ADD COLUMN last_update_timestamp TIMESTAMP WITH TIME
ZONE DEFAULT current_timestamp;
ALTER TABLE test.c ADD COLUMN last_update_timestamp TIMESTAMP WITH TIME
ZONE DEFAULT current_timestamp;
```

Dynamic scripting can also be called a **quick and dirty** approach. The previous scripts didn't filter out views and other objects in the test schema, so you'll need to add that yourself, or not, as required.

There is another way of doing this as well:

```
DO $$
DECLARE t record;
BEGIN
    FOR t IN SELECT c.*, n.nspname
        FROM pg_class c JOIN pg_namespace n
        ON c.relnamespace = n.oid
        WHERE n.nspname = 'test'
        AND c.relkind = 'r'   /* ; not needed */
    LOOP
        EXECUTE format(
           'ALTER TABLE %I.%I
            ADD COLUMN last_update_timestamp
         TIMESTAMP WITH TIME ZONE'
         , t.nspname, t.relname);
    END LOOP;
END $$;
```

I don't prefer using this method because it executes the SQL directly and doesn't allow you to review it before, or keep the script afterward.

The preceding syntax with DO is called an **anonymous code block** because it's like a function without a name.

There's more...

Earlier, I said I'd explain how to run multiple tasks in parallel. Some practical approaches to this are possible, with a bit of discussion.

Making tasks run in parallel can be thought of as subdividing the main task so that we run x2, x4, x8, and other subscripts, rather than one large script.

First, you should note that error-checking gets worse when you spawn more parallel tasks, whereas performance improves the most for the first few subdivisions. Also, we're often constrained by CPU, RAM, or I/O resources for intensive tasks. This means that splitting the main task into two to four parallel subtasks isn't practical without some kind of tool to help us manage them.

There are two approaches here, depending on the two types of tasks:

- A task consists of many smaller tasks, all roughly of the same size.
- A task consists of many smaller tasks, and the execution times vary according to the size and complexity of the database object.

If we have lots of smaller tasks, then we can simply run our scripts multiple times using a simple round-robin split so that each subscript runs a part of all the subtasks. Here is how to do it: each row in pg_class has a hidden column called oid, whose value is a 32-bit number that's allocated from an internal counter on table creation. Therefore, about half of the tables will have even values of oid, and we can achieve an even split by adding the following clauses:

- **Script 1**: Add WHERE c.oid % 2 = 0
- **Script 2**: Add WHERE c.oid % 2 = 1

Here, we added a column to many tables. In the previous example, we were adding the column with no specified default; so, the new column will have a NULL value, and as a result, it will run very quickly with ALTER TABLE, even on large tables. If we change the ALTER TABLE statement to specify a default, then we should choose a non-volatile expression for the default value; otherwise, PostgreSQL will need to rewrite the entire table. So, the runtime will vary according to the table's size (approximately, and also according to the number and type of indexes).

Now that our subtasks vary at runtime according to their size, we need to be more careful when splitting the subtasks so that we end up with multiple scripts that will run for about the same time.

If we already know that we have just a few big tables, it's easy to split them manually into scripts.

If the database contains many large tables, then we can sort SQL statements by table size and then distribute them using round-robin distribution into multiple subscripts that will have approximately the same runtime. The following is an example of this technique, which assumes you have multiple large tables in a schema called test.

First, create a table with all the SQL you would like to run:

```
CREATE TABLE run_sql AS
SELECT format('ALTER TABLE %I.%I ADD COLUMN
last_update_timestamp TIMESTAMP WITH TIME ZONE
```

```
         DEFAULT now();' , n.nspname, c.relname) as sql,
         row_number() OVER (ORDER BY pg_relation_size(c.oid))
         FROM pg_class c
         JOIN pg_namespace n
            ON c.relnamespace = n.oid
         WHERE n.nspname = 'test'
           AND c.relkind = 'r';
```

Then, create a file called `exec-script.sql` and place the following code in it:

```
SELECT sql FROM run_sql
WHERE row_number % 2 = :i
ORDER BY row_number DESC
\gexec
```

Then, we run the script twice, as follows:

```
$ psql -v i=0 -f make-script.sql &
$ psql -v i=1 -f make-script.sql &
```

Note how we used the `psql` parameters – via the `-v` command-line option – to select different rows using the same script.

Also, note how we used the `row_number()` window function to sort the data by size. Then, we split the data into pieces using the following line:

```
WHERE row_number % N = i;
```

Here, N is the total number of scripts we're producing, and i ranges between 0 and N minus 1 (we are using modulo arithmetic to distribute the subtasks).

Adding/removing columns on a table

As designs change, we may want to add or remove columns from our data tables. These are common operations in development, though they need more careful planning on a running production database server as they take full locks and may run for long periods.

How to do it...

You can add a new column to a table using the following command:

```
ALTER TABLE mytable
ADD COLUMN last_update_timestamp TIMESTAMP WITHOUT TIME ZONE;
```

You can drop the same column using the following command:

```
ALTER TABLE mytable
DROP COLUMN last_update_timestamp;
```

You can combine multiple operations when using ALTER TABLE, which then applies the changes in a sequence. This allows you to perform a useful trick, which is to add a column unconditionally using IF EXISTS, which is useful because ADD COLUMN does not allow IF NOT EXISTS:

```
ALTER TABLE mytable
DROP COLUMN IF EXISTS last_update_timestamp,ADD COLUMN last_update_
timestamp TIMESTAMP WITHOUT TIME ZONE;
```

Note that this will have almost the same effect as the following command:

```
UPDATE mytable SET last_update_timestamp = NULL;
```

However, ALTER TABLE runs much faster. This is very cool if you want to perform an update, but it's not much fun if you want to keep the data in the existing column.

How it works...

The ALTER TABLE statement, which is used to add or drop a column, takes a full table lock (at the AccessExclusiveLock lock level) so that it can prevent all other actions on the table. So, we want it to be as fast as possible.

The DROP COLUMN command doesn't remove the column from each row of the table; it just marks the column as dropped. This makes DROP COLUMN a very fast operation.

The ADD COLUMN command is also very fast if we are adding a column with a non-volatile default value, such as a NULL value or a constant. A non-volatile expression always returns the same value when it's computed multiple times within the same SQL statement; this means that PostgreSQL can compute the default value once and write it into the table metadata. Conversely, if the default is a volatile expression, then it is not guaranteed to evaluate the same result for each of the existing rows; therefore, PostgreSQL needs to rewrite every row of the table, which can be quite slow.

If we rewrite the table, then the dropped columns are removed. If not, they may stay there for some time. Subsequent INSERT and UPDATE operations will ignore the dropped column(s). Updates will reduce the size of the stored rows if they were not null already. So, in theory, you just have to wait, and the database will eventually reclaim the space. In practice, this only works if all the rows in the table are updated within a given period. Many tables contain historical data, so space may not be reclaimed at all without additional actions.

To reclaim space from dropped columns, the PostgreSQL manual recommends changing the data type of a column to the same type, which forces everything to be rewritten. I don't recommend this because it will completely lock the table for a long period, at least on larger databases. If you're looking for alternatives, then VACUUM will not rewrite the table, though a VACUUM FULL or a CLUSTER statement will. Be careful in those cases as well, because they also hold a full table lock.

There's more...

Indexes that depend on a dropped column are automatically dropped as well. This is what you would expect if all the columns in the index are dropped, but it can be surprising if some columns in the index are not dropped. All other objects that depend on the column(s), such as foreign keys from other tables, will cause the ALTER TABLE statement to be rejected. You can override this and drop everything in sight using the CASCADE option, as follows:

```
ALTER TABLE x
DROP COLUMN last_update_timestamp
CASCADE;
```

Adding a column with a non-null default value can be done with ALTER TABLE ... ADD COLUMN ... DEFAULT ..., as we have just shown, but this holds an AccessExclusive lock for the duration of the command, which can take a long time if DEFAULT is a volatile expression, because in that case 100% of the rows must be rewritten in order to add that value.

The script that we introduced in the *Using psql variables* recipe in this chapter is an example of how to do the same without holding an AccessExclusive lock for a long time. This lighter solution has only one other tiny difference: it doesn't use a single transaction, which would be pointless since it would hold the lock until the end.

If any row is inserted by another session between ALTER TABLE and UPDATE and that row has a NULL value for the new column, then that value will be updated together with all the rows that existed before ALTER TABLE, which is OK in most cases, though not in all, depending on the data model of the application.

For a completely safe, optimal solution to this problem, you need to open two distinct sessions, and take explicit locks to close any window of opportunity for those anomalies. More specifically, our plan is to take two locks in sequence: the first lock will be exclusive as it needs to perform DDL, but the second one will only be a shared lock, which only prevents other DML. We will do that in a way that does not allow any DML in between.

Here is an example:

1. Open a psql session, and note its PID:

   ```
   cookbook=# SELECT pg_backend_pid();
    pg_backend_pid
   ----------------
           3997901
   (1 row)
   ```

2. Do the same again, because we need two separate sessions:

   ```
   cookbook=# SELECT pg_backend_pid();
    pg_backend_pid
   ----------------
           3997906
   (1 row)
   ```

3. In the first session (the one whose PID is 3997901), start a transaction, and take an AccessExclusive lock on the table, which PostgreSQL will grant (after the end of concurrent activities, if there are any):

   ```
   cookbook=# BEGIN;
   BEGIN
   cookbook=# LOCK mytable IN ACCESS EXCLUSIVE MODE;
   LOCK TABLE
   ```

4. Now we switch to the second session and try to acquire a Share lock on the same table. We know it won't be granted, because it conflicts with the AccessExclusive lock that the first session already holds, so the session will hang without returning:

   ```
   cookbook=# BEGIN;
   BEGIN
   cookbook=# LOCK mytable IN SHARE MODE;
   ```

5. We switch back to the first session, which is not blocked, and use the following SQL query to display the ordered wait queue for the locks that are blocked by session 1 (that is, the session having a PID equal to 3997901):

   ```
   cookbook=# SELECT pid, state, query, waitstart
   FROM pg_stat_activity a
   JOIN pg_locks l USING(pid)
   WHERE pg_blocking_pids(pid) @> array[3997901]
   ```

```
            AND l.relation = regclass 'mytable'
ORDER BY waitstart \gx

-[ RECORD 1 ]-----------------------------------
pid       | 3997906
state     | active
query     | LOCK mytable IN SHARE MODE;
waitstart | 2023-11-13 12:42:23.339828+01
```

This output confirms that the second session is the only one currently blocked by the first session because a lock on the mytable relation. This is good enough for us, because it means that anybody else trying to write to this table will now join the queue after the second session, i.e. it will be safe to use the second session to populate new columns created in the first session.

For instance, if a third session was used to issue the following SQL:

```
cookbook=# INSERT INTO mytable VALUES (-1,'xyz');
```

Then the preceding query will now see two entries:

```
cookbook=# SELECT pid, state, query, waitstart
FROM pg_stat_activity a
JOIN pg_locks l USING(pid)
WHERE pg_blocking_pids(pid) @> array[3997901]
    AND l.relation = regclass 'mytable'
ORDER BY waitstart \gx

-[ RECORD 1 ]-----------------------------------
pid       | 3997906
state     | active
query     | LOCK mytable IN SHARE MODE;
waitstart | 2023-11-13 12:42:23.339828+01
-[ RECORD 2 ]-----------------------------------
pid       | 4001769
state     | active
query     | INSERT INTO mytable VALUES (-1,'xyz');
waitstart | 2023-11-13 13:03:12.876036+01
```

Here we would note that the entry with the earliest waitstart value is the one corresponding to the second session, meaning that it will acquire the lock when the first session ends.

Note also that this previous query is using the @> operator to represent the fact that the array on the right-hand side is a sub-array of the one on the left-hand side. In practice, we select all the rows where one of the blocking sessions is the one whose PID is 3997901.

1. At this point, we can use the first session to add the new column:

   ```
   cookbook=# ALTER TABLE mytable ADD COLUMN col2 timestamptz;
   ```

   NOTE

   If you try to use a third session to read from mytable, your session will be blocked because the DDL lock is still being held by the first session.

2. Then, still in the first session, we can commit the transaction:

   ```
   cookbook=# COMMIT;
   ```

   This causes the first session to release the AccessExclusive lock, and the second session to be granted the Share lock on the same table.

3. Now, as the last step, we can populate the new column, knowing that throughout this operation the table will be readable. We can do it either in a single go, or in multiple batches, depending on how big the table is. For instance:

   ```
   cookbook=# UPDATE mytable SET col2 = clock_timestamp()
   WHERE col2 IS NULL AND id <= 100;
   UPDATE 100

   cookbook=# UPDATE mytable SET col2 = clock_timestamp()
   WHERE col2 IS NULL AND id <= 200;
   UPDATE 100

   cookbook=# UPDATE mytable SET col2 = clock_timestamp()
   WHERE col2 IS NULL AND id <= 300;
   UPDATE 100

   (...)

   cookbook=# COMMIT;
   COMMIT
   ```

Changing the data type of a column

Thankfully, changing column data types is not an everyday task, but we must understand the behavior to ensure we can execute the change without any problem when we need to do it.

Getting ready

Let's start with a simple example of a table, with just one row, as follows:

```
CREATE TABLE birthday
( name      TEXT
, dob       INTEGER);
INSERT INTO birthday VALUES ('simon', 690926);
postgres=# select * from birthday;
```

This gives us the following output:

```
 name  |  dob
-------+--------
 simon | 690926
(1 row)
```

How to do it...

Let's say we want to change the dob column to another data type. Let's try this with a simple example first, as follows:

```
postgres=# ALTER TABLE birthday
postgres-# ALTER COLUMN dob SET DATA TYPE text;
ALTER TABLE
```

This works fine. Let's just change that back to the integer type so that we can try something more complex, such as a date data type:

```
postgres=# ALTER TABLE birthday
postgres-# ALTER COLUMN dob SET DATA TYPE integer;
ERROR:  column "dob" cannot be cast automatically to type integer
HINT:  You might need to specify "USING dob::integer"
```

Oh! What went wrong? Let's try using an explicit conversion with the USING clause, as follows:

```
postgres=# ALTER TABLE birthday
            ALTER COLUMN dob SET DATA TYPE integer
```

```
                USING dob::integer;
ALTER TABLE
```

This works as expected. Now, let's try moving to a date type:

```
postgres=# ALTER TABLE birthday
ALTER COLUMN dob SET DATA TYPE date
USING date(to_date(dob::text, 'YYMMDD') -
     (CASE WHEN dob/10000 BETWEEN 16 AND 69 THEN interval '100
       years'
      ELSE interval '0' END));
```

Now, it gives us what we were hoping to see:

```
postgres=# select * from birthday;
name   |    dob
-------+------------
simon  | 26/09/1969
(1 row)
```

With PostgreSQL, you can also set or drop default expressions, irrespective of whether the NOT NULL constraints are applied:

```
ALTER TABLE foo
ALTER COLUMN col DROP DEFAULT;
ALTER TABLE foo
ALTER COLUMN col SET DEFAULT 'expression';
ALTER TABLE foo
ALTER COLUMN col SET NOT NULL;
ALTER TABLE foo
ALTER COLUMN col DROP NOT NULL;
```

How it works...

Moving from the integer type to the date type uses a complex USING expression. Let's break this down step by step so that we can see why, as follows:

```
postgres=# ALTER TABLE birthday
ALTER COLUMN dob SET DATA TYPE date
USING date(to_date(dob::text, 'YYMMDD') -
     (CASE WHEN dob/10000 > extract('year' from current_date)%100
      THEN interval '100 years'
      ELSE interval '0' END));
```

First, PostgreSQL does not allow a conversion directly from integer to date. We need to convert it into text and then into date. The dob::text statement means *cast to text*.

Once we have text, we can use the to_date() function to move to a date type.

This is not enough; our starting data was 690926, which we presume is a date in the YYMMDD format. PostgreSQL docs say *"In to_date, if the year format specification is less than four digits, such as YYY, and the supplied year is less than four digits, the year will be adjusted to be nearest to the year 2020; for example, 95 becomes 1995."* So, we must add an adjustment factor as well since dates before 1970 will be presumed to be in the future.

It is very strongly recommended that you test this conversion by performing a SELECT first. Converting data types, especially to/from dates, always causes some problems, so don't try to do this quickly. Always take a backup of the data first.

There's more...

The USING clause can also be used to handle complex expressions involving other columns. This could be used for data transformations, which might be useful for DBAs in some circumstances, such as migrating to a new database design on a production database server. Let's put everything together in a full, working example. We will start with the following table, which has to be transformed:

```
postgres=# select * from cust;
 customerid | firstname | lastname | age
------------+-----------+----------+-----
          1 | Philip    | Marlowe  |  38
          2 | Richard   | Hannay   |  42
          3 | Holly     | Martins  |  25
          4 | Harry     | Palmer   |  36
(4 rows)
```

We want to transform it into a table design like the following:

```
postgres=# select * from cust;
 customerid |     custname      | age
------------+-------------------+-----
          1 | Philip Marlowe    |  38
          2 | Richard Hannay    |  42
          3 | Holly Martins     |  25
          4 | Harry Palmer      |  36
(4 rows)
```

We can decide to do this using these simple steps:

```
ALTER TABLE cust ADD COLUMN custname text NOT NULL DEFAULT '';
UPDATE cust SET custname = firstname || ' ' || lastname;
ALTER TABLE cust DROP COLUMN firstname;
ALTER TABLE cust DROP COLUMN lastname;
```

We can also use the SQL commands directly or run them using a tool such as **pgAdmin**. Following those steps may cause problems, as the changes aren't within a transaction, meaning that other users can see the changes when they are only half-finished. Hence, it would be better to do this in a single transaction using BEGIN and COMMIT. Also, those four changes require us to make two passes over the table.

However, we can perform the entire transformation in one pass by using multiple clauses on the ALTER TABLE command. So, instead, we can do the following:

```
BEGIN;
ALTER TABLE cust
  ALTER COLUMN firstname SET DATA TYPE text
        USING firstname || ' ' || lastname,
  ALTER COLUMN firstname SET NOT NULL,
  ALTER COLUMN firstname SET DEFAULT '',
  DROP COLUMN lastname;
ALTER TABLE cust RENAME firstname TO custname;
COMMIT;
```

Some type changes can be performed without actually rewriting rows – for example, if you are casting data from varchar to text, or from NUMERIC(10,2) to NUMERIC(18,2), or simply to NUMERIC. Moreover, foreign key constraints will recognize type changes of this kind on the source table, so it will skip the constraint check whenever it is *safe*.

Note that moving from VARCHAR(128) to VARCHAR(256) is safe, whereas reducing the max length – say, from VARCHAR(256) to VARCHAR(128), is not.

If you are changing from TIMESTAMP to TIMESTAMPTZ, then this is safe *if your session timezone is UTC*. This is a new optimization in Postgres 14.

Changing the definition of an enum data type

PostgreSQL comes with several data types, but users can create custom types to faithfully represent any value. Data type management is mostly, but not exclusively, a developer's job, and data type design goes beyond the scope of this book. This is a quick recipe that only covers the simpler problem of the need to apply a specific change to an existing data type.

Getting ready

Enumerative data types are defined like this:

```
CREATE TYPE satellites_urani AS ENUM ('titania','oberon');
```

The other popular case is composite data types, which are created as follows:

```
CREATE TYPE node AS
( node_name text,
  connstr text,
  standbys text[]);
```

How to do it...

If you misspelled some enumerative values, but you realized it too late, you can fix it like so:

```
ALTER TYPE satellites_urani RENAME VALUE 'titania' TO 'Titania';
ALTER TYPE satellites_urani RENAME VALUE 'oberon' TO 'Oberon';
```

This is very useful if the application expects – and uses – the right names.

A more complicated case is when you are upgrading your database schema to a new version, say because you want to consider some facts that were not available during the initial design, and you need extra values for the enumerative type that we defined in the preceding code. You want to put the new values in a certain position to preserve the correct ordering. For that, you can use the ALTER TYPE syntax, as follows:

```
ALTER TYPE satellites_urani ADD VALUE 'Ariel' BEFORE 'Titania';
ALTER TYPE satellites_urani ADD VALUE 'Umbriel' AFTER 'Ariel';
```

Composite data types can be changed with similar commands. Attributes can be renamed, as shown in the following example:

```
ALTER TYPE node
RENAME ATTRIBUTE replicas TO standbys;
```

And new attributes can be added as follows:

```
ALTER TYPE node
DROP ATTRIBUTE standbys,
ADD ATTRIBUTE async_standbys text[],
ADD ATTRIBUTE sync_standbys text[];
```

This form supports a list of changes, perhaps because composite types are more complex than a list of enumerative values, and can therefore require complicated modifications.

How it works...

Each time you create a table, a composite type is automatically created with the same attribute names, types, and positions. Each ALTER TABLE command that changes the table column definitions will silently issue a corresponding ALTER TYPE statement to keep the type in agreement with *its* table definition.

Enumerative values in PostgreSQL are stored in tables as numbers, which are transparently mapped to strings via the pg_enum catalog table. To be able to insert a new value between two existing ones, enumerative values are indexed by real numbers, which allow decimal points and have the same size in bytes as integer numbers. The motive is to use numeric ordering to encode the order of values that was specified by the user.

In the satellites_urani example, the first two values were Titania and Oberon, which initially got indexed by the real numbers 1 and 2:

```
postgres=# select * from pg_enum where enumtypid = regtype 'satellites_urani';
enumtypid | enumsortorder | enumlabel
-----------+---------------+-----------
    38112 |             1 | Titania
    38112 |             2 | Oberon
(2 rows)
```

When we add a third value before Titania (that is, 1), the number 0 is taken, as you would probably expect:

```
postgres=# ALTER TYPE satellites_urani ADD VALUE 'Ariel' BEFORE 'Titania';
ALTER TYPE
postgres=# select * from pg_enum where enumtypid = regtype 'satellites_urani';
```

```
 enumtypid | enumsortorder | enumlabel
-----------+---------------+-----------
     38112 |             1 | Titania
     38112 |             2 | Oberon
     38112 |             0 | Ariel
(3 rows)
```

And, finally, when adding a fourth value between `Ariel` (0) and `Titania` (1), PostgreSQL can pick the (`real`) value in between, that is, `0.5`:

```
postgres=# ALTER TYPE satellites_urani ADD VALUE 'Umbriel' AFTER 'Ariel';
ALTER TYPE
postgres=# select * from pg_enum where enumtypid = regtype 'satellites_urani';
 enumtypid | enumsortorder | enumlabel
-----------+---------------+-----------
     38112 |             1 | Titania
     38112 |             2 | Oberon
     38112 |             0 | Ariel
     38112 |           0.5 | Umbriel
(4 rows)
```

To test the resulting order, we can build a test table that contains all the possible values, and then sort it:

```
postgres=# CREATE TABLE test(x satellites_urani);
CREATE TABLE
postgres=# INSERT INTO test VALUES ('Ariel'), ('Oberon'), ('Titania'), ('Umbriel');
INSERT 0 4
postgres=# SELECT * FROM test ORDER BY x;
    x
---------
 Ariel
 Umbriel
 Titania
 Oberon
(4 rows)
```

There's more...

When an attribute is removed from a composite data type, the corresponding values will instantly disappear from all the values of that same type that are stored in any database table. What happens is that these values are still inside the tables, but they have become invisible because their attribute is now marked as deleted, and the space they occupy will only be reclaimed when the content of the composite type is parsed again. This can be forced with a query such as the following:

```
UPDATE mycluster SET cnode = cnode :: text :: node;
```

Here, `mycluster` is a table that has a `cnode` column of the `node` type. This query converts the values into the `text` type, displaying only current attribute values, and then back into `node`. You may have noticed that this behavior is very similar to the example of the dropped column in the previous recipe.

Adding a constraint concurrently

A table constraint is a guarantee that must be satisfied by all of the rows in the table. Therefore, adding a constraint to a table is a two-phase procedure – first, the constraint is created, and second, the existing rows are validated. Both happen in the same transaction, and the table will be locked according to the type of constraint for the whole duration.

For example, if we add a Foreign Key to a table, we will lock the table to prevent all write transactions against it. This validation could run for an hour in some cases and prevent writes for all that time.

This recipe demonstrates another case – that it is possible to split those two phases into multiple transactions since this allows validation to occur with a lower lock level than what's required to add the constraint, reducing the effect of locking on the table.

First, we create the constraint and mark it as `NOT VALID` to make it clear that it does not exclude violations, unlike ordinary constraints. Then, we `VALIDATE` all the rows by checking them against the constraint. At this point, the `NOT VALID` mark will be removed from the constraint.

Using the same example we used previously, if we add a `NOT VALID` Foreign Key to a table, we will lock the table to prevent all write transactions against it for a short period. Then, we `VALIDATE` all the rows, which run for 1 hour while holding a lock that does *not* prevent writes.

It is possible to validate the constraint at a later time, for example, when you're allowed by workload or business continuity requirements, which might be a long delay, or in some cases, never.

Getting ready

We'll start this recipe by creating two tables with deliberately inconsistent data so that any attempt to check the existing rows will result in an error message:

```
postgres=# CREATE TABLE ft(fk int PRIMARY KEY, fs text);
CREATE TABLE
postgres=# CREATE TABLE pt(pk int, ftval int);
CREATE TABLE
postgres=# INSERT INTO ft (fk, fs) VALUES (1,'one'), (2,'two');
INSERT 0 2
postgres=# INSERT INTO pt (pk, ftval) VALUES (1, 1), (2, 2), (3, 3);
INSERT 0 3
```

How to do it...

If we attempt to create an ordinary foreign key, we will get an error since the number 3 does not appear in the ft table:

```
postgres=# ALTER TABLE pt ADD CONSTRAINT pt_ft_fkey FOREIGN KEY (ftval)
REFERENCES ft (fk);
ERROR: insert or update on table "pt" violates foreign key constraint pt_ft_fkey"
DETAIL: Key (pk)=(3) is not present in table "ft".
```

However, the same constraint can be successfully created as NOT VALID:

```
postgres=# ALTER TABLE pt ADD CONSTRAINT pt_ft_fkey FOREIGN KEY (ftval)
REFERENCES ft(fk) NOT VALID;
ALTER TABLE
postgres=# \d pt
      Table "public.pt"
Column |  Type   | Modifiers
--------+---------+-----------
pk     | integer |
ftval  | text    |
Foreign-key constraints:
    "pt_ft_fkey" FOREIGN KEY (ftval) REFERENCES ft(fk) NOT VALID
```

NOTE

The invalid state of the foreign key is visible in `psql`.

This violation is detected when we try to transform the NOT VALID constraint into a valid one:

```
postgres=# ALTER TABLE pt VALIDATE CONSTRAINT pt_ft_fkey;
ERROR: insert or update on table "pt" violates foreign key constraint pt_ft_fkey"
DETAIL: Key (ftval)=(3) is not present in table "ft".
```

Validation becomes possible after removing the inconsistency, and the foreign key is upgraded to be fully validated:

```
postgres=# DELETE FROM pt WHERE pk = 3;
DELETE 1
postgres=#
ALTER TABLE
postgres=# \d pt
       Table "public.pt"
 Column |  Type   | Modifiers
--------+---------+-----------
 pk     | integer |
 ftval  | text    |
Foreign-key constraints:
    "pt_ft_fkey" FOREIGN KEY (ftval) REFERENCES ft (fk)
```

How it works...

ALTER TABLE ... ADD CONSTRAINT FOREIGN KEY.. NOT VALID uses ShareRowExclusiveLock, which blocks writes, and VACUUM, yet allows reads on the table to continue. ADD CONSTRAINT CHECK can also be added using the NOT VALID option, but as of Postgres 14, it still takes a full AccessExclusiveLock when it executes, which means it blocks all access to the table, including reads.

The ALTER TABLE ... VALIDATE CONSTRAINT command executes using ShareUpdateExclusiveLock, which allows both reads and writes on the table, yet blocks DDL and VACUUM while it scans the table.

PostgreSQL takes SQL locks according to the ISO standard; that is, locks are taken during the transaction and then released when it ends. This means that algorithms like this one, where there is a short activity requiring stronger locks, followed by a longer activity that needs only lower-strength locks, cannot be implemented within a single transaction.

There's more...

If you want to add `ALTER TABLE ... SET NOT NULL` concurrently, then you need to do it as a three-step process:

1. The first step is as follows:

   ```
   ALTER TABLE pt ADD CONSTRAINT ftval_not_null
   CHECK (ftval IS NOT NULL) NOT VALID;
   ```

2. The second step is as follows:

   ```
   ALTER TABLE pt VALIDATE CONSTRAINT ftval_not_null;
   ```

3. The third step is as follows:

   ```
   ALTER TABLE pt ALTER COLUMN ftval SET NOT NULL;
   ```

The last step is optimized in Postgres 14+ so that it does not have to scan the whole table to validate the `NOT NULL` requirement: instead, it notices the existence of a constraint that proves the validity of that requirement.

Adding/removing schemas

Separating groups of objects is a good way of improving administrative efficiency. You need to know how to create new schemas and remove schemas that are no longer required.

How to do it...

To add a new schema, issue this command:

```
CREATE SCHEMA sharedschema;
```

If you want that schema to be owned by a particular user, then you can add the following option:

```
CREATE SCHEMA sharedschema AUTHORIZATION scarlett;
```

If you want to create a new schema that has the same name as an existing user so that the user becomes the owner, then try this:

```
CREATE SCHEMA AUTHORIZATION scarlett;
```

In many database systems, the schema name is the same as that of the owning user. PostgreSQL allows schemas that are owned by one user to have objects owned by another user within them. This can be especially confusing when you have a schema that has the same name as the owning user. To avoid this, you should have two types of schema: schemas that are named the same as the owning user should be limited to only objects owned by that user. Other general schemas can have shared ownership.

To remove a schema named str, we can issue the following command:

```
DROP SCHEMA str;
```

If you want to ensure that the schema exists in all cases, you can issue the following command:

```
CREATE SCHEMA IF NOT EXISTS str;
```

You need to be careful here because the outcome of the preceding command depends on the previous state of the database, i.e. on whether that schema exists already. As an example, try issuing the following command:

```
CREATE TABLE str.tb (x int);
```

This will generate an error if the str schema contained that table before CREATE SCHEMA IF NOT EXISTS was run. Otherwise, no namespace error will occur.

Irrespective of your PostgreSQL version, there isn't a CREATE OR REPLACE SCHEMA command, so when you want to create a schema, regardless of whether it already exists, you can do the following:

```
DROP SCHEMA IF EXISTS newschema;
CREATE SCHEMA newschema;
```

The DROP SCHEMA command won't work unless the schema is empty or you use the nuclear option:

```
DROP SCHEMA IF EXISTS newschema CASCADE;
```

The nuclear option kills all known germs and all your database objects (*even the good objects*).

There's more...

In the SQL standard, you can also create a schema and the objects it contains in one SQL statement. PostgreSQL accepts the following syntax if you need it:

```
CREATE SCHEMA foo
       CREATE TABLE account
       (id           INTEGER NOT NULL PRIMARY KEY
```

```
    ,balance      NUMERIC(50,2))
CREATE VIEW accountsample AS
SELECT *
FROM account
WHERE random() < 0.1;
```

Mostly, I find this limiting. This syntax exists to allow us to create two or more objects at the same time. This can be achieved more easily using PostgreSQL's ability to allow transactional DDL, which was discussed in the *Writing a script that either succeeds entirely or fails entirely* recipe.

Using schema-level privileges

Privileges can be granted for objects in a schema using the GRANT command, as follows:

```
GRANT SELECT ON ALL TABLES IN SCHEMA sharedschema TO PUBLIC;
```

However, this will only affect tables that already exist. Tables that are created in the future will inherit privileges defined by the ALTER DEFAULT PRIVILEGES command, as follows:

```
ALTER DEFAULT PRIVILEGES IN SCHEMA sharedschema
GRANT SELECT ON TABLES TO PUBLIC;
```

Moving objects between schemas

Once you've created schemas for administration purposes, you'll want to move existing objects to keep things tidy.

How to do it...

To move one table from its current schema to a new schema, use the following command:

```
ALTER TABLE cust
SET SCHEMA anotherschema;
```

If you want to move all objects, you can consider renaming the schema itself by using the following query:

```
ALTER SCHEMA existingschema RENAME TO anotherschema;
```

This only works if another schema with that name does not exist. Otherwise, you'll need to run ALTER TABLE for each table you want to move. You can follow the *Performing actions on many tables* recipe, earlier in this chapter, to achieve that.

Views, sequences, functions, aggregates, and domains can also be moved by ALTER commands with SET SCHEMA options.

How it works...

When you move tables to a new schema, all the indexes, triggers, and rules that have been defined on those tables will also be moved to the new schema. If you've used a SERIAL data type and an implicit sequence has been created, then that also moves to the new schema. Schemas are purely an administrative concept and they do not affect the location of the table's data files. Tablespaces don't work this way, as we will see in later recipes.

Databases, users/roles, languages, and conversions don't exist in a schema. Schemas exist in a particular database. Schemas don't exist within schemas; they are not arranged in a tree or hierarchy. More details can be found in the *Using multiple schemas* recipe of *Chapter 4, Server Control*.

There's more...

PostgreSQL has operators called casts that convert a value from a data type to another one. So, for instance, you have an operation that converts an integer into text. Casts don't exist in schemas, though the data types and functions they reference do exist. These things are not typically something we want to move around.

Adding/removing tablespaces

Tablespaces are a feature in PostgreSQL that allow the user to specify different physical devices on which to store data. We may want to do that for performance or administrative ease, or our database may have run out of disk space.

Getting ready

The very first action, when creating a tablespace in PostgreSQL, is to double-check that you actually need to create a tablespace. This is because tablespaces in PostgreSQL have a different purpose than tablespaces in some other database systems.

The only purpose of having multiple tablespaces is to use a certain directory when storing data files for specific tables or indexes. There are no other implications in terms of crashes or disaster recovery: you need **all the tablespaces** for the database to operate correctly!

If you are thinking of placing some tables on a separate tablespace so they can be recovered separately from the rest of the database, then know this is not how PostgreSQL works. Relational databases provide the notion of global integrity, so they are not really designed to operate correctly if part of the data files are lost.

Conversely, if you have different types of storage, you can benefit from using a separate tablespace for each disk partition, so you can place the right tables on the fastest disks.

Before we can create a useful tablespace, we need the underlying devices in a production-ready form. Think carefully about the speed, volume, and robustness of the disks you are about to use. Make sure that they have been configured correctly. Those decisions will affect your life for the next few months and years!

Disk performance is a subtle issue that most people think can be decided in a few seconds. We recommend reading *Chapter 10, Performance and Concurrency*, of this book as well as additional books on the same topic to learn more.

Once you've done all of that, you can create a directory for your tablespace. The directory must be as follows:

- Empty
- Owned by the PostgreSQL-owning user ID
- Specified with an absolute pathname

On Linux and Unix systems, you shouldn't use a mount point directly. Create a subdirectory and use that instead. This simplifies ownership and avoids some filesystem-specific issues, such as getting `lost+found` directories.

The directory also needs to follow sensible naming conventions so that we can identify which tablespace goes with which server. Do not be tempted to use something simple, such as `data`, because it will make later administration more difficult. Be especially careful that test and development servers do not and cannot get confused with production systems.

How to do it...

Once you've created your directory, adding the tablespace is simple:

```
CREATE TABLESPACE new_tablespace
LOCATION '/usr/local/pgsql/new_tablespace';
```

The command to remove the tablespace is also simple and is as follows:

```
DROP TABLESPACE new_tablespace;
```

Every tablespace has a location assigned to it, except for the `pg_global` and `pg_default` default tablespaces, which are for shared system catalogs and all other objects, respectively. They don't have a separate location because they live in a subdirectory of the `data` directory.

A tablespace can only be dropped when it is empty, so how do you know when a tablespace is empty?

Tablespaces can contain both permanent and temporary objects. Permanent data objects are tables, indexes, and TOAST objects. We don't need to worry too much about TOAST objects because they are created and always live in the same tablespace as their main table, and you cannot manipulate their privileges or ownership.

Indexes can exist in separate tablespaces as a performance option, though that requires explicit specification in the CREATE INDEX statement. The default is to create indexes in the same tablespace as the table that they belong to.

Temporary objects may also exist in a tablespace. These exist when users have explicitly created temporary tables or there may be implicitly created data files when large queries overflow their work_mem settings. These files are created according to the setting of the temp_tablespaces parameter. This might cause an issue because you can't tell what the setting of temp_tablespaces is for each user. Users can change their setting of temp_tablespaces from the default value specified in the postgresql.conf file to something else.

We can identify the tablespace of each user object using the following query:

```
SELECT spcname
      ,relname
      ,CASE WHEN relpersistence = 't' THEN 'temp '
            WHEN relpersistence = 'u' THEN 'unlogged '
       ELSE '' END ||
       CASE
       WHEN relkind = 'r' THEN 'table'
       WHEN relkind = 'p' THEN 'partitioned table'
       WHEN relkind = 'f' THEN 'foreign table'
       WHEN relkind = 't' THEN 'TOAST table'
       WHEN relkind = 'v' THEN 'view'
       WHEN relkind = 'm' THEN 'materialized view'
       WHEN relkind = 'S' THEN 'sequence'
       WHEN relkind = 'c' THEN 'type'
       ELSE 'index' END as objtype
  FROM pg_class c join pg_tablespace ts
    ON (CASE WHEN c.reltablespace = 0 THEN
             (SELECT dattablespace FROM pg_database
```

```
                WHERE datname = current_database())
        ELSE c.reltablespace END) = ts.oid
WHERE relname NOT LIKE 'pg_toast%'
AND relnamespace NOT IN
    (SELECT oid FROM pg_namespace
      WHERE nspname IN ('pg_catalog', 'information_schema'))
;
```

This displays output such as the following:

```
    spcname      |   relname   |  objtype
-----------------+-------------+------------
 new_tablespace  | x           | table
 new_tablespace  | y           | table
 new_tablespace  | z           | temp table
 new_tablespace  | y_val_idx   | index
```

You may also want to look at the spcowner, relowner, relacl, and spcacl columns to determine who owns what and what they're allowed to do. The relacl and spcacl columns refer to the **Access Control List (ACL)** that details the privileges available on those objects. The spcowner and relowner columns record the owners of the tablespace and tables/indexes, respectively.

How it works...

A tablespace is just a directory where we store PostgreSQL data files. We use symbolic links from the data directory to the tablespace.

We exclude TOAST tables because they are always in the same tablespace as their parent tables, but remember that TOAST tables are always in a separate schema. You can exclude TOAST tables using the relkind column, but that would still include the indexes on the TOAST tables. TOAST tables and TOAST indexes both start with pg_toast, so we can exclude those easily from our queries.

The preceding query needs to be complex because the pg_class entry for an object will show reltablespace = 0 when an object is created in the database's default tablespace. So, if you directly join pg_class and pg_tablespace, you end up losing rows.

Note that we can see that a temporary object exists and that we can also see the tablespace that it has created, even though we cannot refer to a temporary object in another user's session.

There's more...

Some more notes on best practices follow.

A tablespace can contain objects from multiple databases, so it's possible to be in a position where no objects are visible in the current database. The tablespace just refuses to go away, giving us the following error:

```
ERROR:  tablespace "old_tablespace" is not empty
```

You can choose to make a separate tablespace for each database to prevent this problem. This can be especially confusing if you have the same schema names and table names in separate databases. However, keep in mind that you will still need all your tablespaces for correct operations, as we noted earlier in this recipe. The general advice is to keep your tablespace configuration simple to reduce the chances of confusion.

You may also wish to consider giving each tablespace a specific owner by using the following query:

```
ALTER TABLESPACE new_tablespace OWNER TO eliza;
```

This may help smooth administration.

You may also wish to set default tablespaces for a user so that tables are automatically created by issuing the following query:

```
ALTER USER eliza SET default_tablespace = 'new_tablespace';
```

Putting pg_wal on a separate device

You might have heard advice about placing the pg_wal directory on a separate device for performance reasons. This is usually a good idea since WAL files have a very different access pattern to data files. There is no explicit command to do this once you have a running database, and files in pg_wal are frequently written. So, if you want pg_wal on a separate device, you must perform the steps outlined in the following example *before* the database is running:

1. Stop the database server:

   ```
   [postgres@myhost ~]$ pg_ctl stop
   ```

2. Move pg_wal to a location that's supported by a different disk device:

   ```
   [postgres@myhost ~]$ mv $PGDATA/pg_wal  /mnt/newdisk/
   ```

3. Create a symbolic link from the old location to the new location:

```
[postgres@myhost ~]$ ln -s /mnt/newdisk/pg_wal    $PGDATA/pg_wal
```

4. Restart the database server:

```
[postgres@myhost ~]$ pg_ctl start
```

5. Verify that everything is working by committing any transaction (preferably, a transaction that does not damage the existing workload):

```
[postgres@myhost ~]$ psql -c 'CREATE TABLE all is ok()'
```

Tablespace-level tuning

Since each tablespace has different I/O characteristics, we may wish to alter the planner cost parameters for each tablespace. These can be set with the following command:

```
ALTER TABLESPACE new_tablespace SET
(seq_page_cost = 0.05, random_page_cost = 0.1);
```

In this example, the settings are roughly appropriate for an SSD drive, and it assumes that the drive is 40 times faster than an HDD for random reads and 20 times faster for sequential reads.

The values that have been provided need more discussion than we have time for here; these are only examples to demonstrate how to change the settings.

Moving objects between tablespaces

At some point, you may need to move data between tablespaces.

Getting ready

First, create your tablespaces. Once the old and new tablespaces exist, we can issue the commands to move the objects inside them.

How to do it...

Tablespaces can contain both permanent and temporary objects.

Permanent data objects include tables, indexes, and `TOAST` objects. We don't need to worry too much about `TOAST` objects because they are created in and always live in the same tablespace as their main table. So, if you alter the tablespace of a table, its `TOAST` objects will also move:

```
ALTER TABLE mytable SET TABLESPACE new_tablespace;
```

Indexes can exist in separate tablespaces, and moving a table leaves the indexes where they are. Don't forget to run `ALTER INDEX` commands as well, one for each index, as follows:

```
ALTER INDEX mytable_val_idx SET TABLESPACE new_tablespace;
```

Temporary objects cannot be explicitly moved to a new tablespace, so we need to ensure they are created somewhere else in the future. To do that, you need to do the following:

1. Edit the `temp_tablespaces` parameter, as shown in the *Updating the parameter file* recipe of *Chapter 3, Server Configuration*.
2. Reload the server configuration file to allow new configuration settings to take effect:

   ```
   SELECT pg_reload_conf()
   ```

How it works...

If you want to move a table and its indexes all in one pass, you can issue all the commands in a single transaction, as follows:

```
BEGIN;
ALTER TABLE mytable SET TABLESPACE new_tablespace;
ALTER INDEX mytable_val1_idx SET TABLESPACE new_tablespace;
ALTER INDEX mytable_val2_idx SET TABLESPACE new_tablespace;
COMMIT;
```

Moving tablespaces means bulk copying data. Copying happens sequentially, block by block. This works well, but there's no way to avoid the fact that the bigger the table, the longer it will take.

The performance will be optimized if archiving or streaming replication is not active, as no WAL will be written in that case.

You should be aware that the table is fully locked (with the `AccessExclusiveLock` lock) while the copy is taking place, so this can cause an effective outage for your application. Be very careful!

If you want to ensure that objects are created in the right place the next time you create them, then you can use the following query:

```
SET default_tablespace = 'new_tablespace';
```

You can run this automatically for all the users that connect to a database using the following query:

```
ALTER DATABASE mydb SET default_tablespace = 'new_tablespace';
```

Ensure that you do not run the following command by mistake, however:

```
ALTER DATABASE mydb SET TABLESPACE new_tablespace;
```

This moves all the objects that do not have an explicitly defined tablespace into new_tablespace. For a large database, this will take a very long time, and your database will be completely locked while it runs, which is not ideal!

There's more...

If you have just discovered that indexes don't get moved when you move a table, then you may want to check whether any indexes are in tablespaces that are different than their parent tables. Run the following code to check this:

```
SELECT i.relname as index_name
     , tsi.spcname as index_tbsp
     , t.relname as table_name
     , tst.spcname as table_tbsp
  FROM ( pg_class t /* tables */
         JOIN pg_tablespace tst
           ON t.reltablespace = tst.oid
           OR ( t.reltablespace = 0
                AND tst.spcname = 'pg_default' )
       )
  JOIN pg_index pgi
    ON pgi.indrelid = t.oid
  JOIN ( pg_class I /* indexes */
         JOIN pg_tablespace tsi
           ON i.reltablespace = tsi.oid
           OR ( i.reltablespace = 0
                AND tsi.spcname =''pg_defaul'' )
       )
    ON pgi.indexrelid = i.oid
 WHERE i.relname NOT LIKE''pg_toast''
   AND i.reltablespace != t.reltablespace
;
```

If we have one table with an index in a separate tablespace, we might see this as a `psql` definition:

```
postgres=# \d y
     Table""public.""
Column | Type | Modifiers
--------+------+-----------
val    | text |
Indexes:
    ""y_val_id"" btree (val), tablespace""new_tablespac""
Tablespace:""new_tablespace""
```

Running the previously presented query gives us the following output:

```
  relname  |    spcname     | relname |    spcname
-----------+----------------+---------+----------------
 y_val_idx | new_tablespace | y       | new_tablespace2
(1 row)
```

In PostgreSQL 16, you can change the tablespace of an index when you run `REINDEX`, so this can be used to resolve these problems using commands like this:

```
REINDEX (TABLESPACE new, CONCURRENTLY) v_val_idx;
```

Accessing objects in other PostgreSQL databases

Sometimes, you may want to access data in other PostgreSQL databases. The reasons for this may be as follows:

- You have more than one database server, and you need to extract data (such as a reference) from one server and load it into the other.
- You want to access data that is in a different database on the same database server, which was split for administrative purposes.
- You want to make some changes that you do not wish to roll back in the event of an error or transaction abortion. These are known as **function side-effects** or **autonomous transactions**.

You may also be considering this because you are exploring the scale-out, sharding, or load-balancing approaches. If so, read the last part of this recipe (the *See also* section) and then skip to *Chapter 12, Replication and Upgrades*.

>
> **NOTE**
>
> PostgreSQL includes two separate mechanisms for accessing external PostgreSQL databases: dblink and the PostgreSQL Foreign Data Wrapper. We focus on the latter, which is now more efficient, at least for exposing remote data.

Getting ready

First of all, let's make a distinction to prevent confusion:

- The **Foreign Data Wrapper** infrastructure, a mechanism that's used to manage the definition of remote connections, servers, and users, is available in all supported PostgreSQL versions. This is like the "driver manager" in JDBC/ODBC.
- The **PostgreSQL Foreign Data Wrapper** is a specific `contrib` extension that uses the Foreign Data Wrapper infrastructure to connect to remote PostgreSQL servers. This is like the driver in JDBC.

Foreign Data Wrapper extensions for other database systems will be discussed in the next recipe, *Accessing objects in other foreign databases*.

How to do it...

Let's use the PostgreSQL foreign data wrapper:

1. The first step is to install the `postgres_fdw` module called `contrib`, which is as simple as this:

    ```
    postgres=# CREATE EXTENSION postgres_fdw;
    ```

2. The result is as follows:

    ```
    CREATE EXTENSION
    ```

3. This extension automatically creates the corresponding Foreign Data Wrapper, as you can check with `psql`'s `\dew` meta-command:

    ```
    postgres=# \dew
                    List of foreign-data wrappers
         Name    | Owner |        Handler        |       Validator
    ------------+-------+-----------------------+----------------------
    ----
    ```

```
postgres_fdw | gianni | postgres_fdw_handler | postgres_fdw_
validator
(1 row)
```

4. We can now define a server:

```
postgres=# CREATE SERVER otherdb
FOREIGN DATA WRAPPER postgres_fdw
OPTIONS (host 'foo', dbname 'otherdb', port '5432');
```

5. This produces the following output:

```
CREATE SERVER
```

6. Then, we can define the user mapping:

```
postgres=# CREATE USER MAPPING FOR PUBLIC SERVER otherdb;
```

7. The output is as follows:

```
CREATE USER MAPPING
```

8. As an example, we will access a portion of a remote table containing (integer, text) pairs:

```
postgres=# CREATE FOREIGN TABLE ft (
num int ,
word text )
SERVER otherdb
OPTIONS (
    schema_name 'public' , table_name 't' );
```

The result is quite laconic:

```
CREATE FOREIGN TABLE
```

This table can now be operated almost like any other table. Let's check whether it is empty:

```
postgres=# select * from ft;
```

9. This is the output:

```
num | word
-----+------
(0 rows)
```

10. We can insert rows as follows:

    ```
    postgres=# insert into ft(num,word) values
    (1,'One'), (2,'Two'),(3,'Three');
    ```

11. This query produces the following output:

    ```
    INSERT 0 3
    ```

12. Then, we can verify that the aforementioned rows have been inserted:

    ```
    postgres=# select * from ft;
    ```

13. This is confirmed by the output:

    ```
    num | word
    ----+------
      1 | One
      2 | Two
      3 | Three
    (3 rows)
    ```

> **NOTE**
>
> You don't have to manage connections or format text strings to assemble your queries. Most of the complexity is handled automatically by the Foreign Data Wrapper.

How it works...

Note that the remote connection persists even across transaction failures and other errors, so there is no need to reconnect.

The postgres_fdw extension can manage connections transparently and efficiently, so if your use case does not involve commands other than SELECT, INSERT, UPDATE, and DELETE, then you should go for it.

Remote data sources look as if they can be treated like tables, and they are represented as such by Foreign Data Wrappers. Ideally, we would like to use foreign tables interchangeably with local tables, with minimum possible performance penalties and maintenance costs, so it is important to know what optimizations work and which ones are still on the wish list.

First, here's the good news: foreign tables can have statistics collected, just like ordinary tables, and they can be used as models to create local tables:

```
CREATE TABLE my_local_copy (LIKE my_foreign_table);
```

This is not supported by dblink because it works on statements instead of managing tables. In general, there is no federated query optimizer. If we join a local table and a remote table with dblink, then data from the remote database is simply pulled through, even if it would have been quicker to send the data and then pull back matching rows. On the other hand, postgres_fdw can share information with the query planner, allowing some optimization, and more improvements are likely to come in the following years now that the infrastructure has been built.

postgres_fdw transparently pushes WHERE clauses to the remote server. Suppose you issue the following command:

```
SELECT * FROM ft WHERE num = 2;
```

Here, only the matching rows will be fetched, using any remote index if available. This is a massive advantage of working with selective queries on large tables. Note that the dblink module cannot automatically send a local WHERE clause to the remote database.

This means that, in general, setting up views of remote data this way isn't very helpful as it encourages users to think that the table location doesn't matter, whereas, from a performance perspective, it does. This isn't any different than other federated or remote access database products.

postgres_fdw can delegate even more activities to the remote node. This includes performing sorts or joins, computing aggregates by carrying out entire UPDATE or DELETE statements, and evaluating the operators or functions provided by suitable extensions.

There's more...

If you are concerned about the overhead of connection time, then you may want to consider using a session pool. This will reserve several database connections, which will allow you to reduce apparent connection time. For more information, look at the *Setting up a connection pool* recipe of *Chapter 4, Server Control*.

Another – and sometimes easier – way of accessing other databases is with a tool named **PL/Proxy**, which is available as a PostgreSQL extension. PL/Proxy allows you to create a local database function that is a proxy for a remote database function. PL/Proxy only works for functions, and some people regard this as a restriction in a way similar to postgres_fdw, which only operates on rows in tables.

Creating a local proxy function is simple:

```
CREATE FUNCTION my_task(VOID)
RETURNS SETOF text AS $$
    CONNECT 'dbname=myremoteserver';
    SELECT my_task();
$$ LANGUAGE plproxy;
```

You need a local function, but you don't need to call a remote function; you can use SQL statements directly. The following example shows a parameterized function:

```
CREATE FUNCTION get_cust_email(p_username text)
RETURNS SETOF text AS $$
    CONNECT 'dbname=myremoteserver';
    SELECT email FROM users WHERE username = p_username;
$$ LANGUAGE plproxy;
```

PL/Proxy is specifically designed to allow more complex architecture for sharding and load balancing. The RUN ON command allows us to dynamically specify the remote database that we will run the SQL statement on. So, the preceding example becomes as follows:

```
CREATE FUNCTION get_cust_email(p_username text)
RETURNS SETOF text AS $$
    CLUSTER 'mycluster';
    RUN ON hashtext(p_username);
    SELECT email FROM users WHERE username = p_username;
$$ LANGUAGE plproxy;
```

You'll likely need to read *Chapter 12, Replication and Upgrades*, before you begin designing application architecture using these concepts.

Accessing objects in other foreign databases

In the previous recipe, you learned how to use objects from a different PostgreSQL database, either with dblink or by using the Foreign Data Wrapper infrastructure. Here, we will explore another variant of the latter – using Foreign Data Wrappers to access databases other than PostgreSQL.

There are many Foreign Data Wrappers for other database systems, all of which are maintained as extensions independently from the PostgreSQL project. The **PostgreSQL Extension Network** (**PGXN**), which we mentioned in *Chapter 3, Server Configuration*, is a good place to see which extensions are available.

Just note this so that you don't get confused: while you can find Foreign Data Wrappers to access several database systems, there are also other wrappers for different types of data sources, such as text files, web services, and so on. There is even postgres_fdw, a backport of the contrib module that we covered in the previous recipe, for users of older PostgreSQL versions who do not have it yet.

NOTE

When evaluating external extensions, I advise you to carefully examine the **README** file in each extension before making stable choices, as the code maturity varies a lot. Some extensions are still development experiments, while others are production-ready extensions, such as oracle_fdw.

Getting ready

For this example, we will use the Oracle Foreign Data Wrapper, oracle_fdw, whose current version is 2.4.0.

You must have obtained and installed the required Oracle software, as specified in the oracle_fdw documentation at https://github.com/laurenz/oracle_fdw/blob/ORACLE_FDW_2_4_0/README.oracle_fdw#L503.

The oracle_fdw wrapper is available in the PostgreSQL Extension Network, so you can follow the straightforward installation procedure described in the *Installing modules from PGXN* section of the *Adding an external module to PostgreSQL* recipe of *Chapter 3, Server Configuration*.

You must have access to an Oracle database server.

How to do it...

Follow these steps to learn how to connect to an Oracle server using oracle_fdw:

1. First, we must ensure that the extension has been loaded:

   ```
   CREATE EXTENSION IF NOT EXISTS oracle_fdw;
   ```

2. Then, we must configure the server and the user mapping:

   ```
   CREATE SERVER myserv
   FOREIGN DATA WRAPPER oracle_fdw
   OPTIONS (dbserver '//myhost/MYDB');
   CREATE USER MAPPING FOR myuser
   SERVER myserv;
   ```

3. Then, we must create a PostgreSQL foreign table with the same column names as the source table in Oracle, and with compatible column types:

```
CREATE FOREIGN TABLE mytab(id bigint, descr text)
SERVER myserv
OPTIONS (user 'scott', password 'tiger');
```

4. Now, we can try to write to the table:

```
INSERT INTO mytab VALUES (-1, 'Minus One');
```

5. Finally, we can read the values that we have inserted:

```
SELECT * FROM mytab WHERE id = -1;
```

This should result in the following output:

```
 id |   descr
----+-----------
 -1 | Minus One
(1 row)
```

How it works...

Our query has a WHERE condition that filters the rows we select from the foreign table. As in the postgres_fdw example of the previous recipe, Foreign Data Wrappers do something clever: the WHERE condition is pushed to the remote server, and only the matching rows are retrieved. Not all Foreign Data Wrappers are able to do this, but many do (as we'll see shortly).

This is good in two ways: first, we delegate some work to another system, and second, we reduce the overall network traffic by not transferring unnecessary data.

Also, note that the WHERE condition is expressed in the PostgreSQL syntax; the Foreign Data Wrapper can translate it into whatever form is required by the remote system.

There's more...

PostgreSQL provides the infrastructure for collecting statistics on foreign tables, so the planner will be able to consider such information, provided that the feature is implemented in the specific Foreign Data Wrapper you are using. For example, statistics are supported by oracle_fdw.

The latest improvements for foreign tables include trigger support, IMPORT FOREIGN SCHEMA, and several improvements to the query planner.

Something particularly useful for database administrators is the `IMPORT FOREIGN SCHEMA` syntax, which can be used to create foreign tables for all the tables and views in a given remote schema with a single statement.

Among the query planner improvements, we wish to mention **Join Pushdown**. In a nutshell, a query that joins some foreign tables that belong to the same server can have the join performed transparently on the remote server. To avoid security issues, this can only happen if these tables are all accessed with the same role.

Open source FDWs are also available for PostgreSQL 14 for the following databases:

- **MySQL** (https://github.com/EnterpriseDB/mysql_fdw): Supports writable FDWs, `SELECT` clauses, `WHERE` clauses, and `JOIN` clause pushdowns, as well as connection pooling
- **MongoDB** (https://github.com/EnterpriseDB/mongo_fdw): Supports writable FDWs and connection pooling
- **HDFS (Apache Hadoop, Apache Spark, Apache Hive)** https://github.com/EnterpriseDB/hdfs_fdw

Making views updatable

PostgreSQL supports the SQL standard `CREATE VIEW` command, which supports automatic `UPDATE`, `INSERT`, and `DELETE` commands, provided they are simple enough.

Note that certain types of updates are forbidden just because they are either impossible or impractical to derive a corresponding list of modifications on the constituent tables. We'll discuss those issues here.

Getting ready

First, you need to consider that only simple views can be made to receive insertions, updates, and deletions easily. The SQL standard differentiates between views that are simple and updatable, and more complex views that cannot be expected to be updatable.

So, before we proceed, we need to understand what a simple updatable view is and what it is not. Let's start with the `cust` table:

```
postgres=# SELECT * FROM cust;
 customerid | firstname | lastname | age
------------+-----------+----------+-----
          1 | Philip    | Marlowe  |  38
          2 | Richard   | Hannay   |  42
```

```
         3 | Holly    | Martins  | 25
         4 | Harry    | Palmer   | 36
         4 | Mark     | Hall     | 47
(5 rows)
```

Let's create a simply updatable view on top of it, as follows:

```
CREATE VIEW cust_view AS
SELECT customerid
      ,firstname
      ,lastname
      ,age
FROM cust;
```

Each row in our view corresponds to one row in a single-source table, and each column is referred to directly without any further processing, except possibly for a column rename. Thus, we expect to be able to make INSERT, UPDATE, and DELETE commands pass through our view into the base table, which is what happens in PostgreSQL.

A view will be automatically updatable if a view has just one table or updatable view in the FROM clause and does not contain functions in the SELECT, WITH, LIMIT DISTINCT, aggregation, window functions, grouping, or sorting clauses.

The following examples are three views where the INSERT, UPDATE, and DELETE commands cannot be made to flow to the base table easily, for the reasons just described:

```
CREATE VIEW cust_avg AS
SELECT avg(age)
FROM cust;
CREATE VIEW cust_above_avg_age AS
SELECT customerid
          ,substr(firstname, 1, 20) as fname
          ,substr(lastname, 1, 20) as lname
          ,age -
          (SELECT avg(age)::integer
          FROM cust) as years_above_avg
FROM cust
WHERE age >
      (SELECT avg(age)
       FROM cust);
```

```
CREATE VIEW potential_spammers AS
SELECT customerid, spam_score(firstname,lastname)
FROM cust
ORDER BY spam_score(firstname,lastname) DESC
LIMIT 100;
```

The first view just shows a single row with the average of a numeric column. Changing an average directly doesn't make much sense. For instance, if we want to raise the average age by 1, should we increase all numbers by 1, resulting in each row that is being updated? Or should we change some rows only, by a larger amount? A user who wants to do this can update the cust table directly.

The second view shows a column called years_above_avg, which is the difference between the age of that customer and the average. Changing that column would be more complex than it seems at first glance: just consider that increasing the age by 10 would not result in increasing years_above_avg by 10, because the average will also be affected.

The third view displays a computed column that can't be updated directly – we can't change the value in the spam_score column without changing the algorithm that's implemented by the spam_score() function.

Now, we can learn how to allow any or all insertions, updates, or deletions to flow from views to base tables since we've clarified whether this makes sense conceptually.

How to do it...

There is nothing to do for simple views – PostgreSQL will propagate modifications to the underlying table automatically.

Conversely, if the view is not simple enough, but you still have a clear idea of how you would like to propagate changes to the underlying table(s), then you can allow updatable views by telling PostgreSQL how to perform **Data Manipulation Language** (**DML**) statements, which in PostgreSQL means INSERT, UPDATE, DELETE, or TRUNCATE.

PostgreSQL supports two mechanisms to achieve updatable views – namely, rewriting rules and INSTEAD OF triggers. The latter provides a mechanism to implement updatable views by creating trigger functions that execute arbitrary code every time a data-modification command is executed on the view.

The INSTEAD OF triggers are part of the SQL standard, and other database systems support them. Conversely, query rewrite rules are specific to PostgreSQL and cannot be found anywhere else in this exact form.

There is no preferable method. On one hand, rules can be more efficient than triggers, while on the other hand, they can be more difficult to understand than triggers and could result in inefficient execution if the code is badly written (although the latter is not an exclusive property of rules, unfortunately).

To explain this point concretely, we will now provide an example of using rules, and then we will re-implement the same example with triggers:

1. We will start with a table of mountains and their heights in meters:

   ```
   CREATE TABLE mountains_m
   ( name text primary key
   , meters int not null
   );
   ```

2. Then, we will create a view that adds a computed column expressing the height in feet, and that displays the data in descending height order:

   ```
   CREATE VIEW mountains AS
   SELECT *, ROUND(meters / 0.3048) AS feet
   FROM mountains_m
   ORDER BY meters DESC;
   ```

3. DML automatically flows to the base table when we insert columns that are not computed:

   ```
   INSERT INTO mountains(name, meters)
   VALUES ('Everest', 8848);
   TABLE mountains;
   name    | meters | feet
   --------+--------+------
   Everest |   8848 | 29029
   (1 row)
   ```

4. However, when we try to insert data with the height specified in feet, we get the following error:

   ```
   INSERT INTO mountains(name, feet)
   VALUES ('K2', 28251);
   ERROR:  cannot insert into column "feet" of view "mountains"
   DETAIL:  View columns that are not columns of their base relation
   are not updatable.
   ```

5. So, we must create a rule that replaces the insert with another query that works all the time:

```
CREATE RULE mountains_ins_rule AS
ON INSERT TO mountains DO INSTEAD
INSERT INTO mountains_m
VALUES (NEW.name, COALESCE (NEW.meters, NEW.feet * 0.3048));
```

6. Now, we can insert both meters and feet:

```
INSERT INTO mountains(name, feet)
VALUES ('K 2', 28251);
INSERT INTO mountains(name, meters)
VALUES ('Kangchenjunga', 8586);
TABLE mountains;
     name      | meters | feet
---------------+--------+------
 Everest       |   8848 | 29029
 K 2           |   8611 | 28251
 Kangchenjunga |   8586 | 28169
(3 rows)
```

7. Updates are also propagated automatically, but only to non-computed columns:

```
UPDATE mountains SET name = 'K2' WHERE name = 'K 2';
TABLE mountains;
     name      | meters | feet
---------------+--------+-------
 Everest       |   8848 | 29029
 K2            |   8611 | 28251
 Kangchenjunga |   8586 | 28169
(3 rows)
UPDATE mountains SET feet = 29064 WHERE name = 'K2';
ERROR:  cannot update column "feet" of view "mountains"
DETAIL:  View columns that are not columns of their base relation
are not updatable.
```

8. If we add another rule that replaces updates with a query that covers all cases, then the last update will succeed and produce the desired effect:

```
CREATE RULE mountains_upd_rule AS
ON UPDATE TO mountains DO INSTEAD
```

```
UPDATE mountains_m
SET name = NEW.name, meters =
CASE
WHEN NEW.meters != OLD.meters
THEN NEW.meters
WHEN NEW.feet != OLD.feet
THEN NEW.feet * 0.3048
ELSE OLD.meters
END
WHERE name = OLD.name;
UPDATE mountains SET feet = 29064 WHERE name = 'K2';
TABLE mountains;
 name          | meters | feet
---------------+--------+-------
 K2            |   8859 | 29065
 Everest       |   8848 | 29029
 Kangchenjunga |   8586 | 28169
(3 rows)
```

9. The query that's used in this rule also covers the simpler case of a non-computed column:

```
UPDATE mountains SET meters = 8611 WHERE name = 'K2';
TABLE mountains;
 name          | meters | feet
---------------+--------+-------
 Everest       |   8848 | 29029
 K2            |   8611 | 28251
 Kangchenjunga |   8586 | 28169
(3 rows)
```

10. The same effect can be achieved by adding the following trigger, which replaces the earlier two rules:

```
CREATE FUNCTION mountains_tf()
RETURNS TRIGGER
LANGUAGE plpgsql
AS $$
BEGIN
IF TG_OP = 'INSERT' THEN
```

```
        INSERT INTO mountains_m VALUES (NEW.name,
        CASE
        WHEN NEW.meters IS NULL
        THEN NEW.feet * 0.3048
        ELSE NEW.meters
        END );
        ELSIF TG_OP = 'UPDATE' THEN
        UPDATE mountains_m
        SET name = NEW.name, meters =
        CASE
        WHEN NEW.meters != OLD.meters
        THEN NEW.meters
        WHEN NEW.feet != OLD.feet
        THEN NEW.feet * 0.3048
        ELSE OLD.meters
        END
        WHERE name = OLD.name;
        END IF;
        RETURN NEW;
        END;
        $$;
        CREATE TRIGGER mountains_tg
        INSTEAD OF INSERT OR UPDATE ON mountains
        FOR EACH ROW
        EXECUTE PROCEDURE mountains_tf();
```

How it works...

In this rule-based example, we used the COALESCE function, which returns the first argument, if it's not null, or the second one otherwise. When the original INSERT statement does not specify a value in meters, then it uses the value in feet divided by 0.3048.

The second rule sets the value in meters to different expressions – if the value in meters was updated, we use the new one; if the value in feet was updated, we use the new value in feet divided by 0.3048; and otherwise, we use the old value in meters (that is, we don't change it).

The logic that's implemented in the trigger function is similar to the previous one; note that we use the TG_OP automatic variable to handle INSERT and UPDATE separately.

We've just scratched the surface of what you can achieve with rules, though I find them too complex for widespread use.

You can do a lot of things with rules, but you need to be careful with them. There are some other important points that I should mention about rules before you dive in and start using them everywhere.

Rules are applied by PostgreSQL once the SQL has been received by the server and parsed for syntax errors, but before the planner tries to optimize the SQL statement.

In the rules in the preceding recipe, we referenced the values of the old or new row, just as we do within trigger functions, using the old and new keywords. Similarly, there are only new values in an INSERT command and only old values in a DELETE command.

One of the major downsides of using rules is that we cannot bulk load data into the table using the COPY command. Also, we cannot transform a stream of inserts into a single COPY command, nor can we perform a COPY operation against the view. Bulk loading requires direct access to the table.

Suppose we have a table and a view, such as the following:

```
CREATE TABLE cust
(customerid BIGINT NOT NULL PRIMARY KEY
,firstname TEXT
,lastname TEXT
,age INTEGER);
CREATE VIEW cust_minor AS
SELECT customerid
,firstname
,lastname
,age
FROM cust
WHERE age < 18;
```

Then, we have some more difficulties. If we wish to update this view, then we might need to read the PostgreSQL manual and understand that we can use a conditional rule by adding a WHERE clause to match the WHERE clause in the view, as follows:

```
CREATE RULE cust_minor_update AS
ON  update TO cust_minor
WHERE new.age < 18
DO INSTEAD
```

```
UPDATE cust SET
firstname = new.firstname
,lastname = new.lastname
,age = new.age
WHERE customerid = old.customerid;
```

This fails, however, as we can see if we try to update `cust_minor`. The fix is to add two rules – one as an unconditional rule that does nothing (literally) and needs to exist for internal reasons, and another to do the work we want:

```
CREATE RULE cust_minor_update_dummy AS ON
update TO cust_minor
DO INSTEAD NOTHING;
CREATE RULE cust_minor_update_conditional AS
ON    update TO cust_minor
WHERE new.age < 18
DO INSTEAD
UPDATE cust SET firstname = new.firstname
,lastname = new.lastname
,age = new.age
WHERE customerid = old.customerid;
```

There's more...

There is yet another question that's posed by updatable views.

As an example, we shall use the `cust_minor` view we just defined, which does not allow you to perform insertions or updates so that the affected rows fall out of the view itself. For instance, consider this query:

```
UPDATE cust_minor SET age = 19 WHERE customerid = 123;
```

The preceding query will not affect any row because of the `WHERE age < 18` conditions in the rule definition.

The `CREATE VIEW` statement has a `WITH CHECK OPTION` clause; if specified, any update that excludes any row from the view will fail.

If a view includes some updatable columns, along with other non-updatable columns (for example expressions, literals, and so on), then updates are allowed if they only change the updatable columns.

Finally, let's show that views are just (empty) tables with a SELECT rule. Let's start by creating an empty table, as follows:

```
CREATE TABLE cust_view AS SELECT * FROM cust WHERE false;
```

With PostgreSQL 15 and earlier versions, the SELECT rule is allowed, provided that it is named _RETURN and the table is empty:

```
postgres # CREATE RULE "_RETURN" AS
    ON SELECT TO cust_view
    DO INSTEAD
    SELECT * FROM cust;
CREATE RULE
postgres=# \d cust_view
               View "public.cust_view"
   Column   |  Type   | Collation | Nullable | Default
------------+---------+-----------+----------+---------
 customerid | bigint  |           |          |
 firstname  | text    |           |          |
 lastname   | text    |           |          |
 age        | integer |           |          |
```

Huh? Didn't we create it as a table?

```
postgres # DROP TABLE cust_view;
ERROR:  "cust_view" is not a table
HINT:  Use DROP VIEW to remove a view
postgres # DROP VIEW cust_view;
DROP VIEW
```

Yes, we had originally created a table, and then added a rule to it. This turned the table into a view.

You might find this confusing. You are not the only one: in PostgreSQL 16 this has been explicitly forbidden, so if you want a view then you just need to use CREATE VIEW:

```
cookbook=# CREATE RULE "_RETURN" AS
    ON SELECT TO cust_view
    DO INSTEAD
    SELECT * FROM cust;
ERROR:  relation "cust_view" cannot have ON SELECT rules
DETAIL:  This operation is not supported for tables.
```

Using materialized views

Every time we select rows from a view, we select from the result of the underlying query. If that query is slow and we need to use it more than once, then it makes sense to run the query once, save its output as a table, and then select the rows from the latter.

This procedure has been available for a long time, and there is a dedicated syntax for it, called `CREATE MATERIALIZED VIEW`, that we will describe in this recipe.

Getting ready

Let's create two randomly populated tables, of which one is large:

```
CREATE TABLE dish
( dish_id SERIAL PRIMARY KEY
, dish_description text
);
CREATE TABLE eater
( eater_id SERIAL
, eating_date date
, dish_id int REFERENCES dish (dish_id)
);
INSERT INTO dish (dish_description)
VALUES ('Lentils'), ('Mango'), ('Plantain'), ('Rice'), ('Tea');
INSERT INTO eater(eating_date, dish_id)
SELECT floor(abs(sin(n)) * 365) :: int + date '2014-01-01'
, ceil(abs(sin(n :: float * n))*5) :: int
FROM generate_series(1,500000) AS rand(n);
```

Notice that the data is not truly random. It is generated by a deterministic procedure, so you can get the same result if you copy the preceding code.

How to do it...

Let's get started:

1. First, create the following view:

    ```
    CREATE VIEW v_dish AS
    SELECT dish_description, count(*)
    FROM dish JOIN eater USING (dish_id)
    GROUP BY dish_description
    ORDER BY 1;
    ```

2. Then, we'll query it:

```
SELECT * FROM v_dish;
```

3. We will obtain the following output:

```
dish_description | count
-----------------+--------
Lentils          |  64236
Mango            |  66512
Plantain         |  74058
Rice             |  90222
Tea              | 204972
(5 rows)
```

4. With a very similar syntax, we will create a materialized view with the same underlying query:

```
CREATE MATERIALIZED VIEW m_dish AS
SELECT dish_description, count(*)
FROM dish JOIN eater USING (dish_id)
GROUP BY dish_description
ORDER BY 1;
```

The corresponding query yields the same output that it did previously:

```
SELECT * FROM m_dish;
```

The materialized version is much faster than the non-materialized version. On my laptop, their execution times are 0.2 milliseconds versus 300 milliseconds.

How it works...

Creating a non-materialized view is the same as creating an empty table with a SELECT rule, as we discovered in the previous recipe. No data is extracted until the view is used.

When creating a materialized view, the default is to run the query immediately and then store its results, as we do for table content.

In short, creating a materialized view is slow, but using it is fast. This is the opposite of standard views, which are created instantly and recomputed at every use.

There's more...

The output of a materialized view is physically stored like a regular table, and the analogy doesn't stop here – you can also create indexes to speed up queries.

A materialized view will not automatically change when its constituent tables change. For that to happen, you must issue the following command:

```
REFRESH MATERIALIZED VIEW m_dish;
```

This replaces all the contents of the view with newly computed ones.

It is possible to quickly create an empty materialized view and populate it later. Just add WITH NO DATA to the end of the CREATE MATERIALIZED VIEW statement. The view cannot be used before it's populated, which you can do with REFRESH MATERIALIZED VIEW, as you just saw.

A materialized view cannot be read while it is being refreshed. For that, you need to use the CONCURRENTLY clause at the expense of a somewhat slower refresh.

As you can see, currently, there is only a partial advantage in using materialized views, compared to previous solutions such as this:

```
CREATE UNLOGGED TABLE m_dish AS SELECT * FROM v_dish;
```

However, when using a declarative language, such as SQL, the same syntax may automatically result in a more efficient algorithm in the case of future improvements to PostgreSQL. For instance, one day, PostgreSQL will be able to perform a faster refresh by simply replacing those rows that changed, instead of recomputing the entire content.

Using GENERATED data columns

You are probably used to the idea that a column can have a default value that's set by a function; this is how we use sequences to set column values in tables. The SQL Standard provides a new syntax for this, which is referred to as GENERATED ... AS IDENTITY. PostgreSQL supports this, but we won't discuss that here.

We can also use views to dynamically calculate new columns as if the data had been stored. PostgreSQL 12+ allows the user to specify that columns can be generated and stored in the table automatically, which is easier and faster than writing a trigger to do this. This is a very important performance and usability feature since we can store data that may take significant time to calculate, so this is much better than just using views. We refer to this feature as GENERATED ALWAYS, which also follows the SQL Standard syntax.

How to do it...

Let's start with an example table:

```
CREATE TABLE example
( id      SERIAL PRIMARY KEY
, descr   TEXT
);
ALTER TABLE example
  ADD COLUMN id2 integer GENERATED ALWAYS AS (id+1) STORED;
```

Note that adding a GENERATED column will always rewrite the table since existing rows need to have a value set for the new column (the ALWAYS keyword in the command means always!).

So, make sure you plan and decide what values you want to generate.

How it works...

The GENERATED value is calculated once on INSERT and then stored. After that, the value is just read from the data block each time it is accessed.

After that, the column can't be updated, so you will get an ERROR message:

```
ERROR:   column "foo" can only be updated to DEFAULT
DETAIL:  Column "foo" is a generated column.
```

So, the value stays just as the table owner intended.

Rows with GENERATED data can be deleted normally.

There's more...

Stored expressions must be IMMUTABLE, meaning they depend solely on the values of other data columns in the same row. This means that adding columns like this seems useful but will just end with an ERROR:

```
ALTER TABLE example
  ADD COLUMN last_access_time timestamp
     GENERATED ALWAYS AS (current_timestamp) STORED
 ,ADD COLUMN last_access_user text
     GENERATED ALWAYS AS (current_user) STORED;
ERROR:   generation expression is not immutable
```

So, if you always want to add dynamically generated values, this still needs to be done using triggers.

Another point that may be confusing is that the SQL syntax for INSERT does allow for a clause called OVERRIDING SYSTEM VALUE, but this only applies to GENERATED ... AS IDENTITY columns.

Using data compression

As data volumes expand, we often think about whether there are ways to compress data to save space. Many patents have been awarded in the realm of data compression, so the development of open source solutions has been slower than normal. PostgreSQL 14 contains some exciting innovations.

Getting ready

Make sure you're running Postgres 14+.

Various types of data compression are available for PostgreSQL:

- Automatic compression of long data values (TOAST)
- Extensions that offer compressed data types (for example, JSON)
- Compression of WAL files
- Dump file compression
- Base backup compression
- SSL compression (this is considered insecure, so it's only used on private networks)
- GiST and SP-GiST index compression
- Btree index compression (also known as deduplication)

Only the first three types of compression will be discussed here, but we focus mainly on the parameters that allow us to control how long data values are automatically compressed.

PostgreSQL will try to compress and/or move longer column values out into an external TOAST table. This is automatic and works optimally for a range of different types of data, but there are a few things we can try if we need to achieve further compression.

How to do it...

Changing to the new compression algorithm is the easiest and most beneficial change. The default for columns with no explicit setting is taken from the value of the default_toast_compression parameter. This only applies to newly inserted data.

If we take an existing table, we can set the compression method explicitly, like this:

```
CREATE TABLE example
( id       SERIAL PRIMARY KEY
, descr    TEXT
);
ALTER TABLE example
  ALTER COLUMN descr SET COMPRESSION lz4;
```

Note that you need to do this separately for each toastable column, because the compression options apply at the column level. A small problem with this is that not all the columns allow this change, so if you try this on an invalid column data type or integer, then you'll get this ERROR, so don't try and just update every column without checking the data type first. Set the compression option for TEXT, JSONB, XML, BYTEA, or GIS data:

```
ALTER TABLE example
  ALTER COLUMN id SET COMPRESSION lz4;
ERROR:  column data type integer does not support compression
```

Setting a new compression method doesn't rewrite the rows into the new compression method, which is good because that would run for a long time. If you want a rewrite to take place and are happy to lock the table while it runs for a long time, just change a column's data type to the same type, a trick we described earlier.

If you're creating a new database, you just need to set this in postgresql.conf once you've created it:

```
default_toast_compression = lz4
```

Since many upgrades use logical replication, we can just set this parameter once and let the rewrite happen automatically during the upgrade process.

How it works...

Above a certain row length, PostgreSQL will attempt to compress and/or move column values out into an external TOAST table. This is done separately for each row, so you will find shorter data columns untouched while longer values from been compressed and/or "toasted." Note that we are using "toast" as both a verb and a noun, describing whether the value has been moved into the toast table.

We can make three different tweaks to this mechanism:

- Change the compression algorithm as we discussed previously.
- Alter the threshold row length at which we consider whether to toast, compress, or do neither.
- Specify whether we don't want to attempt compression and toasting, for special cases.

We can alter the toast threshold using `toast_tuple_target`, which is set separately for each table using the `ALTER TABLE` statement. The default value is 2,040 for an 8 KB block size, though this can be set to anything from 128 bytes to 8,160 bytes. By default, if the total row length is longer than this threshold, then Postgres will attempt to compress and/or toast columns, one at a time, with the longest first until the row is less than this value or it cannot do anything more. This behavior is modified by storage options that can be set for each column. So, what happens on any row depends on the data in all of the columns for that specific row.

PostgreSQL has four different `STORAGE` options, all of which can be set for each column separately:

- `PLAIN`: Inline, uncompressed; for example, the default for `INTEGER`
- `EXTENDED`: External, compressed; for example, the default for `TEXT`
- `EXTERNAL`: External, uncompressed; for example, already compressed data (images and so on)
- `MAIN`: Inline, compressed; for example, medium length `TEXT`, `JSONB`, `XML`, and so on

If you declare a column as using `STORAGE MAIN`, then Postgres will only toast the column value if there is no other way to do this; this column will be at the back of the queue to be toasted. So, if you have some JSONB data that is typically only a few KB, then it might be good to define that column as `MAIN` to reduce the access time for that data.

If the data has already been compressed, then set it to be `EXTERNAL` so that Postgres will not attempt to compress it.

There's more...

An extension called ZSON allows JSONB data to be compressed. This can be used to reduce the size of JSONB data by anything from 0–50%, depending on your data. Unfortunately, this extension is no longer being actively maintained; if your use case would benefit from this approach, please consider spending some effort to compile it for PostgreSQL 16. Open-source software only lives when there are enough users.

Postgres can also be configured to use compression for the WAL transaction log:

```
wal_compression = off (default) | on
```

Learn more on Discord

To join the Discord community for this book – where you can share feedback, ask questions to the author, and learn about new releases – follow the QR code below:

https://discord.gg/pQkghgmgdG

8
Monitoring and Diagnosis

In this chapter, you will find recipes for some common monitoring and diagnosis actions that you will want to perform inside your database. They are meant to answer specific questions that you often face when using PostgreSQL.

In this chapter, we will cover the following recipes:

- Providing PostgreSQL information to monitoring tools
- Monitoring the PostgreSQL message log
- Real-time viewing using pgAdmin
- Checking whether a user is connected
- Checking whether a computer is connected
- Repeatedly executing a query in psql
- Checking which queries are running
- Monitoring the progress of commands and queries
- Checking which queries are active or blocked
- Knowing who is blocking a query
- Killing a specific session
- Knowing whether anybody is using a specific table
- Knowing when a table was last used
- Monitoring I/O statistics
- Usage of disk space by temporary data
- Understanding why queries slow down

- Analyzing the real-time performance of your queries
- Tracking important metrics over time

Databases are not isolated entities. They live on computer hardware using CPUs, RAM, and disk subsystems. Users access databases using networks. Depending on the setup, databases themselves may need network resources to function in any of the following ways: performing some authentication checks when users log in, using disks that are mounted over the network (not generally recommended), replicating data to another database server, backing up over the network, or making remote function calls to other databases.

This means that *monitoring only the database is not enough*. At a minimum, you should also monitor everything directly involved in using the database. This means knowing the following:

- Is the database host available? Does it accept connections?
- How much of the network bandwidth is in use? Have there been network interruptions and dropped connections?
- Is there enough RAM available for the most common tasks? How much of it is left?
- Is there enough disk space available? When will you run out of disk space?
- Is the disk subsystem keeping up? How much more load can it take?
- Can the CPU keep up with the load? How many spare idle cycles do the CPUs have?
- Are other network services the database access depends on (if any) available? For example, if you use Kerberos for authentication, you need to monitor it as well.
- How many context switches are happening when the database is running?
- For most of these things, you are interested in their history – that is, how have things evolved? Was everything mostly the same yesterday or last week?
- When did the disk usage start changing rapidly?

For any larger installation, you probably have something already in place to monitor the health of your hosts and network.

The two aspects of monitoring are *collecting historical data to see how things have evolved* and *getting alerts when things go seriously wrong*.

Tools such as **Munin** and **Prometheus** are quite popular for collecting historical information on all aspects of the servers and presenting this information in an easy-to-follow graphical form. Grafana is a popular tool for this. Real-time monitoring can help when you're trying to figure out why the system is behaving the way it is.

Another aspect of monitoring is getting alerts when something goes wrong and needs (immediate) attention. For alerting, one of the most widely used tools is **Icinga** (a fork of **Nagios**), an established solution. The aforementioned trending tools can integrate with it. check_postgres is a popular Icinga plugin for monitoring many standard aspects of a PostgreSQL database server.

Icinga is a stable and mature solution based on the long-standing approach where each plugin decides whether a given measurement is a cause for alarm, which means that it's more complex to manage and maintain. A more recent tool is the aforementioned **Prometheus**, which is based on a design that separates data collection from the centralized alerting logic. This is covered in more detail next.

Cloud-native monitoring

Prometheus is the tool of choice from the Cloud Native Computing Foundation, so we'll discuss it here. Prometheus is an open source monitoring and alerting toolkit that allows multiple types of systems to feed it monitoring data. An open source Prometheus exporter is available for PostgreSQL, though this is not always needed. For example, EDB's Cloud Native Postgres operator integrates a Prometheus exporter into the Kubernetes operator to provide better security and avoid the need for a separate component in your architecture.

Data from Prometheus is displayed using Grafana. Data from Prometheus can also be stored inside a database and there are various options for storing data inside PostgreSQL or other systems:

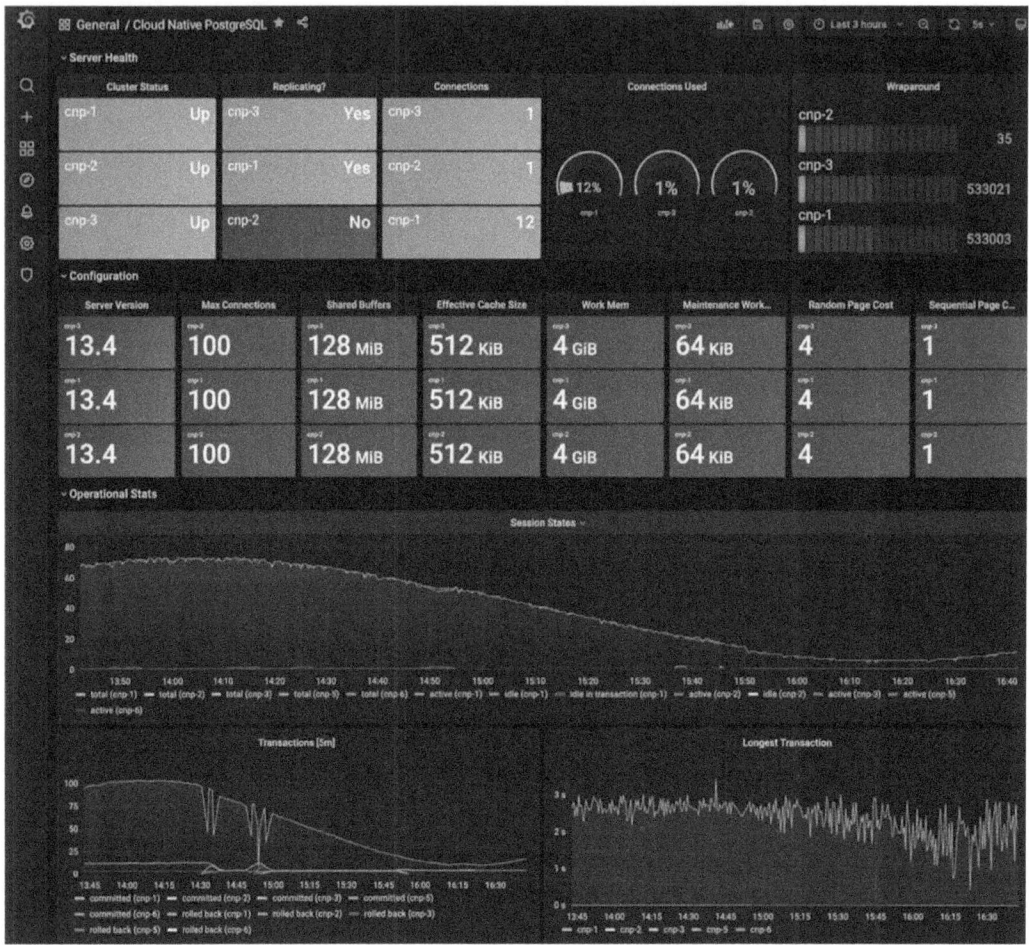

Figure 8.1: Grafana view of PostgreSQL metrics

Remember that the key to successful monitoring is not the tool you use but what information you display with it.

Providing PostgreSQL information to monitoring tools

PostgreSQL exposes a huge amount of information for monitoring. To expose that information securely, make sure your user has the predefined (default) pg_monitor role, which will give you all you need. Some PostgreSQL resources you may encounter may recommend exposing the full contents of pg_stat_activity and similar restricted views, but be careful how and when you do this. Monitoring is important, but so is security.

It's best to use historical monitoring information when all of it is available from the same place and on the same timescale. Most monitoring systems are designed for generic purposes while allowing application and system developers to integrate their specific checks with the monitoring infrastructure. This is possible through a plugin architecture. Adding new kinds of data inputs to them means installing a plugin. Sometimes, you may need to write or develop this plugin, but writing a plugin for a network and performance monitoring framework such as Cacti is easy. You just have to write a script that outputs monitored values in simple text format.

In most common scenarios, the monitoring system is centralized and data is collected directly (and remotely) by the system itself or through some distributed components that are responsible for sending the observed metrics back to the main node.

As far as PostgreSQL is concerned, some useful things to include in graphs are the number of connections, disk usage, number of queries, number of WAL files, most numbers from pg_stat_user_tables and pg_stat_user_indexes, and so on. One *Swiss Army knife* script, which can be used from both Cacti and Nagios/Icinga, is check_postgres. It is available at https://bucardo.org/check_postgres/. It provides ready-made reporting actions for a large array of things that are worth monitoring in PostgreSQL.

We have discussed how to properly monitor a single node. When operating a replication cluster, you need to monitor each node of the cluster plus the replication connections. This is described in detail in the *Monitoring replication* recipe in *Chapter 12, Replication and Upgrades*.

For Munin, there are some PostgreSQL plugins available in the Munin plugin repository at https://github.com/munin-monitoring/contrib/tree/master/plugins/postgresql.

The following screenshot shows a Munin graph about PostgreSQL buffer cache hits for a specific database, where cache hits (the top/blue line) vastly outnumber reads from the disk (the bottom/green line, rarely above zero):

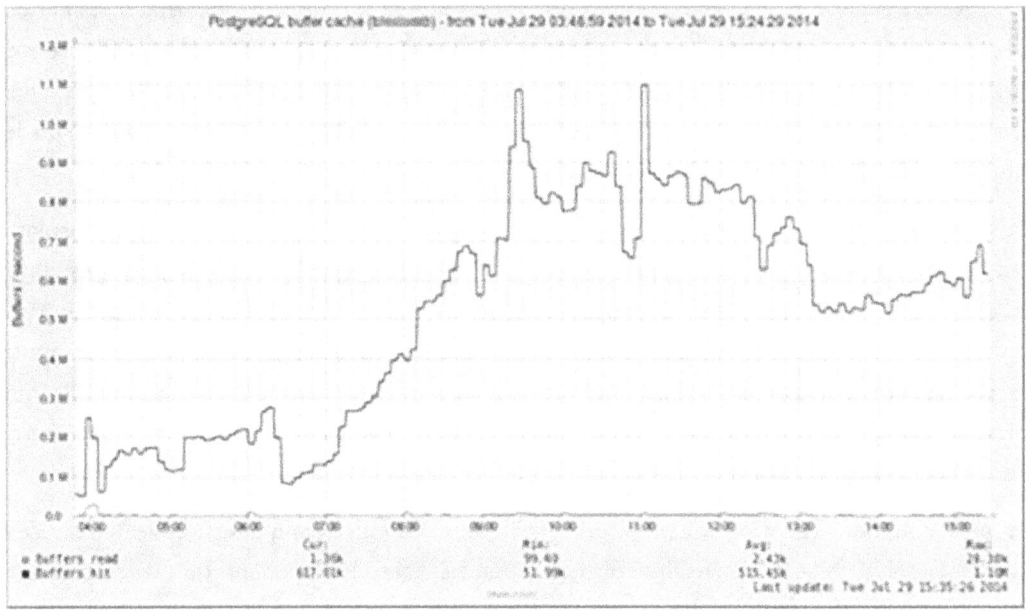

Figure 8.2: Munin graph of buffer cache usage

Finding more information about generic monitoring tools

Setting up the tools themselves is a larger topic, and it is beyond the scope of this book. Each of these tools has more than one book written about them. The basic setup information and the tools themselves can be found at the following URLs:

- RRDtool: http://www.mrtg.org/rrdtool/
- Cacti: https://www.cacti.net/
- Icinga: https://icinga.com/
- Munin: https://munin-monitoring.org/
- Nagios: https://www.nagios.org/
- Zabbix: https://www.zabbix.com/
- Postgres Enterprise Manager: https://www.enterprisedb.com/docs/pem/latest/

Real-time viewing using pgAdmin

You can also use a GUI tool such as pgAdmin, which we discussed for the first time in *Chapter 1, First Steps*, to get a quick view of what is going on in the database.

Getting ready

pgAdmin 4 no longer requires an extension to access PostgreSQL fully, so there is no need to install adminpack, as was required in earlier editions.

How to do it...

This section illustrates the pgAdmin tool.

Once you have connected to the database server, a window similar to the one shown in the following screenshot will be displayed, where you can see a general view, plus information about connections, overall activity, and running transactions:

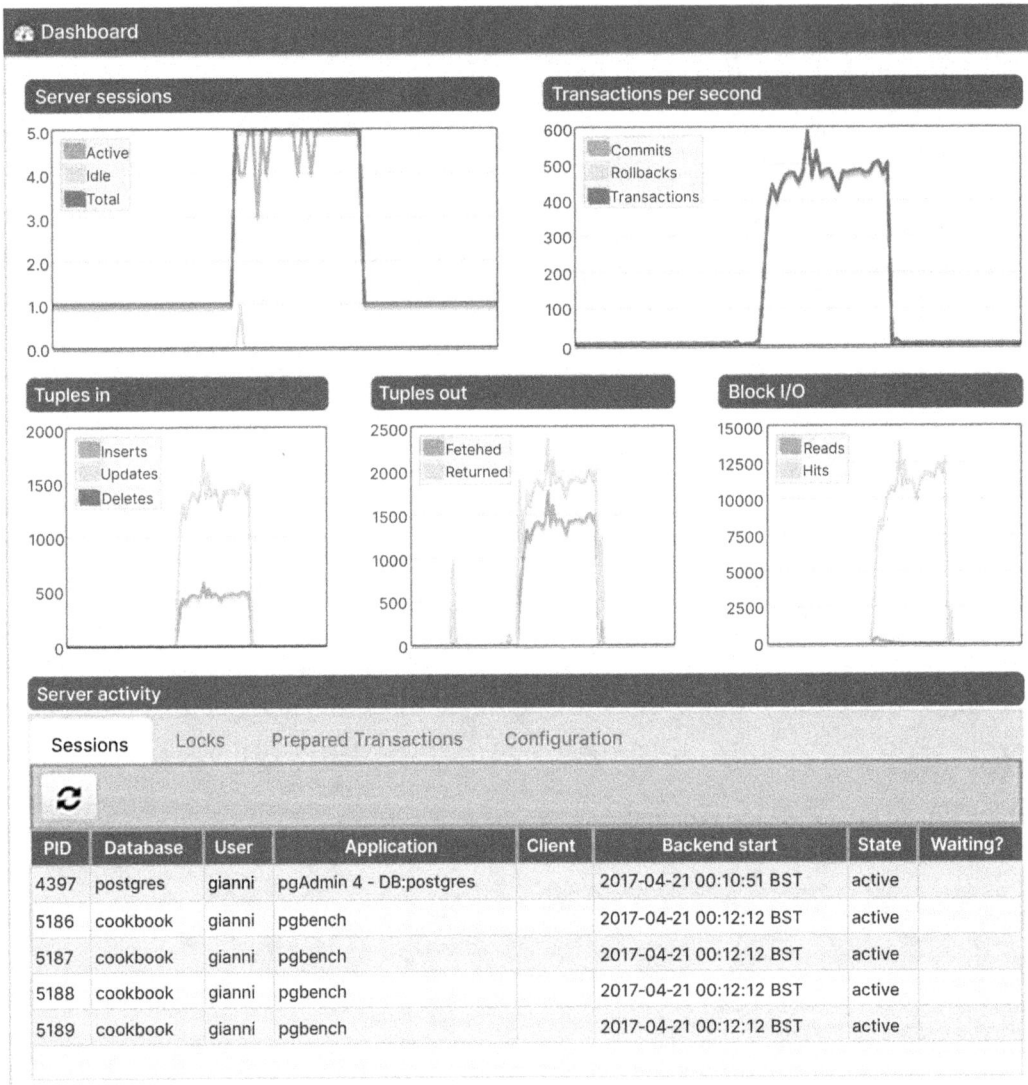

Figure 8.3: pgAdmin dashboard of live usage

Monitoring the PostgreSQL message log

In any production environment, it's always prudent to monitor log files, not only for warning and error detection but also for performance trend analysis. PostgreSQL can output an overwhelming amount of logs per day, but we don't have to scan through every line ourselves – we can use a log analyzer such as pgBadger to generate summaries and reports, which can give us insights into what has happened inside our server for a specific time period.

Getting ready

Most Linux distribution PostgreSQL packages, as well as those from the official community repository, have log file rotation already configured as default. pgBadger can also be installed from your default package manager, either from the community repositories or your distribution's own packages.

We also need to tweak some Postgres settings, such as setting `log_line_prefix` to any format we like, as long as it at least specifies %t (time stamp without milliseconds) and %p (process ID) so that it can be supported by pgBadger, and other parameters that enable it to gather more information, such as:

```
log_checkpoints = on
log_connections = on
log_disconnections = on
log_lock_waits = on
log_temp_files = 0
log_autovacuum_min_duration = 0
log_error_verbosity = default
```

Additionally, in order to reduce the amount of data being output, we can configure `log_min_duration_statement` to the minimum duration over which a statement will get logged. For example, a value of 250ms will mean that we only log statements that take longer than that.

You can find extensive information on how to configure PostgreSQL logging on the relevant documentation pages for Postgres at https://www.postgresql.org/docs/current/runtime-config-logging.html. And for pgBadger, at https://pgbadger.darold.net/documentation.html#POSTGRESQL-CONFIGURATION.

How to do it...

We can call pgBadger to analyze our log files regularly with a cron job (for example, once a day at 4:00 AM). A simple use case is to configure it to produce incremental daily reports and a weekly summary at the end of the week. You can create the following cron job (which needs to all be typed on a single line):

```
0 4 * * * /usr/bin/pgbadger -I -q /var/log/postgresql/postgresql.log.1 -O /var/www/pg_reports/
```

pgBadger can then generate many different kinds of HTML reports with JavaScript charts, which are viewable on your browser.

Here is a sample of the type of reports you can expect. This one highlights the top (N) most time-consuming queries as found in your log files:

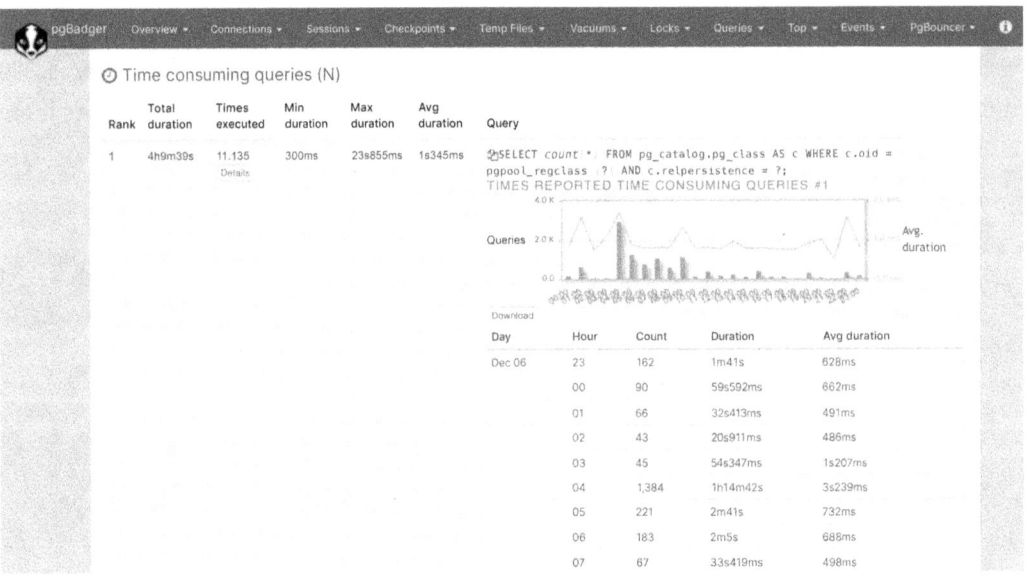

Figure 8.4: pgBadger report showing top time-consuming queries

How it works...

pgBadger also compiles reports on errors and warnings, nicely formatted in HTML, that can be used to determine the most commonly observed errors and any action that needs to be taken. Generally speaking, errors in the log usually point to a significant problem that needs to be addressed either with your application or on your database server, so awareness is essential.

It is recommended that you investigate the root cause of every error you encounter and attempt to eliminate it. Also keep in mind that by letting repeated nuisance errors flood your log files, a more important error that requires immediate remedial action may slip by unnoticed.

There's more...

You can proactively scan your log files regularly even without using pgBadger, by using standard UNIX utilities such as **grep**, which is extremely lightweight and fast. For example, you can filter each log file for errors by running something like this:

```
postgres@dbhost:~$ egrep "FATAL|ERROR" \
/var/log/postgresql/postgresql-16-main.log
```

Checking whether a user is connected

Here, we will show you how to find out whether a certain database user is currently connected to the database.

Getting ready

If you are logged in as a superuser, you will have full access to monitoring information.

How to do it...

Issue the following query to see whether the user bob is connected:

```
SELECT datname FROM pg_stat_activity WHERE usename = 'bob';
```

If this query returns any rows, then that means that bob is connected to the database. The returned value is the name of the database that the user is connected to.

How it works...

PostgreSQL's pg_stat_activity system view keeps track of all running PostgreSQL backends. This includes information such as the query that is being currently executed, the last query that was executed by each backend, who is connected, when the connection, transaction, and/or query were started, and so on.

There's more...

Please spend a few minutes reading the PostgreSQL documentation, which contains more detailed information about pg_stat_activity, available at https://www.postgresql.org/docs/current/monitoring-stats.html#MONITORING-PG-STAT-ACTIVITY-VIEW.

You can find answers to many administration-related questions by analyzing the `pg_stat_activity` view. One common example is outlined in the following recipe.

Checking whether a computer is connected

Often, several different processes may connect as the same database user. In that case, you may want to know whether there is a connection from a specific computer.

How to do it...

You can get this information from the `pg_stat_activity` view as it includes the connected clients' IP address, port, and hostname (where applicable). The port is only needed if you have more than one connection from the same client computer and you need to do further digging to see which process there connects to which database. Run the following command:

```
SELECT datname, usename, client_addr, client_port,
       application_name FROM pg_stat_activity
WHERE backend_type = 'client backend';
```

The `client_addr` and `client_port` parameters help you look up the exact computer and even the process on that computer that has connected to the specific database. You can also retrieve the hostname of the remote computer through the `client_hostname` option (this requires `log_hostname` to be set to on in `postgresql.conf`).

There's more...

I would always recommend including `application_name` in your reports. This field has become widely recognized and honored by third-party application developers (I advise you to do the same with your applications so that it is always clear which application component is performing the action in the database).

For information on how to set the application name for your connections, please refer to *Database Connection Control Functions* in the PostgreSQL documentation at https://www.postgresql.org/docs/current/static/libpq-connect.html.

Repeatedly executing a query in psql

Sometimes, we want to execute a query more than once, repeated at regular intervals; in this recipe, we will look at an interesting `psql` command that does exactly that.

How to do it...

The \watch meta-command allows psql users to automatically (and continuously) re-execute a query. This behavior is similar to the watch utility of some Linux and Unix environments.

In the following example, we will run a simple query on pg_stat_activity and ask psql to repeat it every 5 seconds. You can exit at any time by pressing *Ctrl + C*:

```
postgres=# SELECT count(*) FROM pg_stat_activity;
 count
-------
     6
(1 row)
postgres=# \watch 5
Fri 27 Oct 2023 00:49:06 BST (every 5s)

 count
-------
     6
(1 row)
<snip>
```

There's more...

For further information about the psql utility, please refer to the PostgreSQL documentation at https://www.postgresql.org/docs/current/static/app-psql.html.

Checking which queries are running

In this section, we will show you how to check which query is currently running.

Getting ready

You must make sure that you are logged in as a superuser or as the same database user you want to check. Also, ensure that the track_activities = on parameter is set (which it normally should be, being the default setting). If not, check the *Updating the parameter file* recipe in *Chapter 3, Server Configuration*.

How to do it...

To see which queries connected users are running at this moment, just run the following code:

```
SELECT datname, usename, state, backend_type, query
    FROM pg_stat_activity;
```

This will show normal users as "client backend," but it will also show various PostgreSQL worker processes that you may not want to see. So, you could filter this using `WHERE backend_type = 'client backend'`.

On systems with a lot of users, you may notice that the majority of backends have `state` set to `idle`. This denotes that no query is running, and PostgreSQL is waiting for new commands from the user. The query field shows the statement that was last executed by that particular backend.

If, on the other hand, you are interested in active queries only, limit your selection to those records that have `state` set to `active`:

```
SELECT datname, usename, state, query
    FROM pg_stat_activity
WHERE state = 'active'
  AND backend_type = 'client backend';
```

How it works...

When `track_activities = on` is set, PostgreSQL collects information about all running queries such as their start time, state, query text, which user is running them, and so on. Users with sufficient rights can then view this data using the `pg_stat_activity` system view.

The `pg_stat_activity` view uses a system function named `pg_stat_get_activity (procpid int)`. You can use this function directly to watch for the activity of a specific backend by supplying the process ID as an argument. Giving `NULL` as an argument returns information for all backends.

There's more...

Sometimes, you don't care about getting all the queries that are currently running. You may only be interested in seeing some of these, or you may not like connecting to the database just to see what is running.

Catching queries that only run for a few milliseconds

Since most queries on modern **online transaction processing** (**OLTP**) systems take only a few milliseconds to run, it is often hard to catch the active ones when you're simply probing the `pg_stat_activity` table.

Most likely, you will be able to only see the last executed query for those backends that have state different from active. In some cases, this can be enough.

In general, if you need to perform a deeper analysis, I strongly recommend installing and configuring the pg_stat_statements module, which is described in the *Analyzing the real-time performance of your queries* recipe in this chapter. Another option is to run a post-analysis of log files. Depending on the workload of your system, you may want to restrict the production of highly granular log files (that is, logging all queries) to a short period. For further information on the log analyzer pgBadger, please refer to the *Monitoring the PostgreSQL message log* recipe of this chapter.

Watching the longest queries

Another point of interest that you may want to look for is long-running queries. To get a list of running queries ordered by how long they have been executing, use the following code:

```
SELECT
current_timestamp - query_start AS runtime,
datname, usename, query
FROM pg_stat_activity
WHERE state = 'active'
ORDER BY 1 DESC;
```

This will return currently running queries, with the longest-running queries at the top of the results.

On busy systems, you may want to limit the set of queries that are returned to only the first few queries (add LIMIT 10 at the end) or only the queries that have been running over a certain period. For example, to get a list of queries that have been running for more than 1 minute, use the following query:

```
SELECT
    current_timestamp - query_start AS runtime,
    datname, usename, query
FROM pg_stat_activity
WHERE state = 'active'
    AND current_timestamp - query_start > '1 min'
ORDER BY 1 DESC;
```

Watching queries from ps

If you want, you can also make the queries that are running show up inside the process titles of the ps output by setting the following configuration in the postgresql.conf file:

```
update_process_title = on
```

Although the ps and top outputs are not the best places for watching database queries, they may make sense in some circumstances. In most cases, this is enabled by default, but if you disable it, you will lose visibility of the above.

See also

See PostgreSQL's online documentation, which covers the appropriate settings, at https://www.postgresql.org/docs/current/static/runtime-config-statistics.html.

Monitoring the progress of commands

PostgreSQL 16 now has a growing list of commands that have a "progress bar" – in other words, they provide information to show intermediate progress information for active commands.

Getting ready

Using the earlier recipes, identify the active processes that concern you:

```
SELECT pid, query
FROM pg_stat_activity
WHERE state = 'active';
```

If the query column indicates that they are one of the following actions, then we can look at detailed progress information for them:

- Maintenance commands: ANALYZE, VACUUM, VACUUM FULL/CLUSTER
- Index commands: CREATE INDEX, REINDEX
- Backup/replication: BASE BACKUP
- Data load/unload: COPY

At this time, SELECT statements don't provide detailed progress information.

How to do it...

Each type of command has specific progress information, so you must look in the view that's appropriate to the type of command.

All commands show a pid – the process identifier of the backend running the command.

For each command, consult the appropriate catalog view:

- ANALYZE: pg_stat_progress_analyze
- VACUUM: pg_stat_progress_vacuum
- VACUUM FULL, CLUSTER: pg_stat_progress_cluster
- CREATE INDEX, REINDEX: pg_stat_progress_create_index
- BASE BACKUP: pg_stat_progress_basebackup
- COPY: pg_stat_progress_copy

All types of commands, apart from COPY, show a phase since, in most cases, there are multiple steps involved in processing the command. Each type of command has a specific series of phases (or states) that it will pass through.

We will cover how to monitor and tune a VACUUM in *Chapter 9, Regular Maintenance*.

CREATE INDEX progress is more complex, especially if we are using CONCURRENTLY. The longest phase will be building index since it varies according to the size of the table. For commands with the CONCURRENTLY option, there will also be long index validation phases, which also vary according to the size of the table. At the end of builds with the CONCURRENTLY option, there will be one or more wait phases; if the command stays in this phase for too long, then it will be held up by other running processes, as shown in the current_locker_pid column.

For BASE BACKUP, the longest phase is streaming database files. The backup progress so far is backup_streamed bytes, so the % progress will be as follows:

```
SELECT pid, phase,
100.0*((backup_streamed*1.0)/backup_total) AS "progress%"
FROM pg_stat_progress_basebackup;
```

Although COPY doesn't show the phase, we can calculate the % progress like this:

- COPY FROM % progress will be as follows:

    ```
    SELECT (SELECT relname FROM pg_class WHERE oid = relid),
    100.0*((bytes_processed*1.0)/bytes_total) AS "progress %"
    FROM pg_stat_progress_copy;
    ```

- COPY TO % progress will be as follows:

  ```
  SELECT relname,
  100.0*((tuples_processed*1.0)/(case reltuples WHEN 0 THEN 10 WHEN -1
  THEN 10 ELSE reltuples END))    AS "progress %"
  FROM pg_stat_progress_copy JOIN pg_class on oid = relid;
  ```

All types of commands, apart from BASE BACKUP, show the datid and datname columns, which show the database ID and name, respectively. BASE BACKUP refers to the whole database server, including all databases.

How it works...

When commands run, they update in-memory progress information. By accessing the catalog views, we can see that intermediate progress information.

There's more...

More information is added in each new release, so expect this area to change quickly over time.

Checking which queries are active or blocked

Here, we will show you how to find out whether a query is running or waiting for another query.

Getting ready

Using the predefined (default) pg_monitor role, you will have full access to monitoring information.

How to do it...

Follow these steps to check if a query is waiting for another query:

1. Run the following query:

   ```
   SELECT datname, usename, wait_event_type, wait_event, pid, backend_type, query
   FROM pg_stat_activity
   WHERE wait_event_type IS NOT NULL
   AND wait_event_type NOT IN ('Activity', 'Client');
   ```

2. You will receive the following output:

```
-[ RECORD 1 ]---+-----------------
datname         | postgres
usename         | gianni
wait_event_type | Lock
wait_event      | relation
pid             | 19502
backend_type    | client backend
query           | select * from t;
```

How it works...

The pg_stat_activity system view includes the wait_event_type and wait_event columns, which are set to the kind of wait and to the kind of object that is blocked, respectively. The backend_type column indicates the type of the current backend.

The preceding query uses the wait_event_type field to filter out only those queries that are waiting.

There's more...

PostgreSQL provides a version of the pg_stat_activity view that's capable of capturing many kinds of waits; however, in previous versions, pg_stat_activity could only detect waits on locks such as those placed on SQL objects, via the pg_stat_activity.waiting field.

Although this is the main cause of waiting when using pure SQL, it is possible to write a query in any of PostgreSQL's embedded languages that can wait on other system resources, such as waiting for an HTTP response, for a file write to get completed, or just waiting on a timer.

As an example, you can make your backend sleep for a certain number of seconds using pg_sleep(seconds). While you are monitoring pg_stat_activity, open a new Terminal session with psql and run the following statement in it:

```
db=# SELECT pg_sleep(10);
<it "stops" for 10 seconds here>
 pg_sleep
----------

(1 row)
```

In older versions of Postgres, it will show up as not waiting in the pg_stat_activity view, even though the query is blocked in the timer.

You will see the following output with newer versions of Postgres, where `wait_event_type` is `Timeout` when the server process is waiting for a timeout to expire, and `wait_event` is `PgSleep` when waiting for a process that called `pg_sleep`:

```
-[ RECORD 1 ]---+--------------------
datname         | postgres
usename         | postgres
wait_event_type | Timeout
wait_event      | PgSleep
backend_type    | client backend
query           | SELECT pg_sleep(10);
```

Knowing who is blocking a query

Once you have found out that a query is being blocked, you need to know who or what is blocking it.

Getting ready

If you are logged in as a superuser, you will have full access to monitoring information.

How to do it...

Perform the following steps:

1. Write the following query:

   ```
   SELECT datname, usename, wait_event_type, wait_event, pid, pg_
   blocking_pids(pid) AS blocked_by, backend_type, query
   FROM pg_stat_activity
   WHERE wait_event_type IS NOT NULL
   AND wait_event_type NOT IN ('Activity', 'Client');
   ```

2. You will receive the following output:

   ```
   -[ RECORD 1 ]---+--------------------
   datname         | postgres
   usename         | gianni
   wait_event_type | Lock
   wait_event      | relation
   pid             | 19502
   blocked_by      | {18142}
   backend_type    | client backend
   query           | select * from t;
   ```

This is the query we described in the previous recipe, with the addition of the blocked_by column. Recall that the PID is the unique identifier that's assigned by the operating system to each session; for more details, see *Chapter 4, Server Control*. Here, the PID is used by the pg_blocking_pids(pid) system function to identify blocking sessions.

How it works...

The query is relatively simple: we just introduced the pg_blocking_pids() function, which returns an array composed of the PIDs of all the sessions that were blocking the session with the given PID.

Parallel queries lock via the leader process, so they do not complicate how we monitor locks.

Killing a specific session

Sometimes, the only way to let the system continue as a whole is by *surgically* terminating some offending database sessions. Yes, you read that right: surgically.

In this recipe, you will learn how to intervene, from gracefully canceling a query to brutally killing the actual process from the command line.

How to do it...

Once you have figured out the backend you need to kill, try to use pg_cancel_backend(pid), which cancels the current query, though only if there is one. This can be executed by anyone who is a member of the role whose backend is being canceled.

If that is not enough, then you can use pg_terminate_backend(pid), which kills the backend. This works even for client backends that are idle or idle in a transaction.

You can run these functions as a superuser, or if the calling role is a member of the role whose backend pid is being signed (look for the usename field in the pg_stat_activity view).

You can also grant pg_signal_backend privilege to users to allow this on any user. However, only superusers can cancel superuser backends.

How it works...

When a backend executes these functions, it verifies that the process that's been identified by the pid argument is a PostgreSQL backend. Once we know that, it sends a signal to the process. The backend receiving this signal stops whatever it is doing at the next suitable point in time and terminates it in a controlled way.

If the session is terminated, the client using that backend loses the connection to the database. Depending on how the client application is written, it may silently reconnect, or it may report the error to the user.

There's more...

Killing the session may not always be what you want, so you should consider canceling the statement as well.

Please refer to the *Server Signaling Functions* section in the PostgreSQL documentation at https://www.postgresql.org/docs/current/functions-admin.html#FUNCTIONS-ADMIN-SIGNAL.

Using statement_timeout to clean up queries that take too long to run

Often, you know that you don't have any use for queries that run longer than a given time. Maybe your web frontend just refuses to wait for more than 10 seconds for a query to complete and returns a default answer to users if it takes longer, abandoning the query.

In such a case, it may be a good idea to set statement_timeout = 10s, either in postgresql.conf or as a per-user or per-database setting. Once you do so, queries that are running for too long won't consume precious resources and make other queries fail.

The queries that are terminated by a statement timeout show up in the log, as follows:

```
postgres=# SET statement_timeout TO '3 s';
SET
postgres=# SELECT pg_sleep(10);
ERROR: canceling statement due to statement timeout
```

Killing idle in-transaction sessions

Sometimes, people start a transaction, run some queries, and then just leave, without ending the transaction. This can leave some system resources in a state where some housekeeping processes can't be run. They may even have done something more serious, such as locking a table, thereby causing an immediate *denial of service* for other users who need that table.

You can use the following query to kill all backends that have an open transaction but have been doing nothing for the last 10 minutes:

```
SELECT pg_terminate_backend(pid)
   FROM pg_stat_activity
WHERE state = 'idle in transaction'
    AND current_timestamp - state_change > '10 min';
```

You can even schedule this to run every minute while you are trying to find the specific frontend application that ignores open transactions, or when you have a lazy administrator that leaves a psql connection open, or when a flaky network drops clients without the server noticing it.

Knowing whether anybody is using a specific table

This recipe will help you when you are in doubt about whether an obscure table is being used anymore, or if it has been left over from past use and is just taking up space.

Getting ready

Make sure that you are a superuser, or at least have full rights to the table in question.

How to do it...

Perform the following steps:

1. To see whether a table is currently in active use (that is, whether anyone is using it while you are watching it), run the following query on the database you plan to inspect:

   ```
   CREATE TEMPORARY TABLE tmp_stat_user_tables AS
        SELECT * FROM pg_stat_user_tables;
   ```

2. Then, wait for a while and see what has changed:

   ```
   SELECT * FROM pg_stat_user_tables n
     JOIN tmp_stat_user_tables t
       ON n.relid=t.relid
      AND (n.seq_scan,n.idx_scan,n.n_tup_ins,n.n_tup_upd,n.n_tup_del)
       <> (t.seq_scan,t.idx_scan,t.n_tup_ins,t.n_tup_upd,t.n_tup_del);
   ```

How it works...

The pg_stat_user_tables view shows the current statistics for table usage.

To see whether a table is being used, you can check for changes in its usage counts.

The previous query selects all the tables where any of the usage counts for SELECT or data manipulation have changed.

There's more...

There is a function called pg_stat_reset() that drops a bomb on all usage statistics! This is *NOT* recommended because these statistics are used by autovacuum.

It is often useful to have historical usage statistics for tables when you're trying to solve performance problems or understand usage patterns.

Various tools are available for monitoring table usage, such as EnterpriseDB's **Postgres Enterprise Manager (PEM)**: https://www.enterprisedb.com/products/postgres-enterprise-manager.

You can also collect the data yourself using a table like this:

```
CREATE TABLE backup_stat_user_tables AS
SELECT current_timestamp AS snaptime,*
FROM pg_stat_user_tables
WITH NO DATA;
```

Then, using either a cron or a PostgreSQL-specific scheduler such as pg_agent, you can execute the following query, which adds a snapshot of current usage statistics with a timestamp:

```
INSERT INTO backup_stat_user_tables
SELECT current_timestamp AS snaptime,*
FROM pg_stat_user_tables;
```

Knowing when a table was last used

Once you know that a table is not currently being used, the next question is, *When was it last used?*

Getting ready

You need to use a user with enough monitoring privileges to be able to select from the following views: pg_stat_user_tables, administration functions pg_relation_filenode(), pg_stat_file(), pg_ls_dir(), and catalogs pg_class and pg_namespace.

How to do it...

As we already know, the pg_stat_user_tables view shows current table usage statistics. Since PostgreSQL 16, these include the last time each table was scanned, either in full or using an index (sequential or index scan). These are found respectively in the columns last_seq_scan and last_idx_scan.

To see when a table was last read from, you can run the following query:

```
SELECT last_seq_scan, last_idx_scan
FROM pg_stat_user_tables
WHERE relname=<table name>;
```

And here is the type of output that you can expect:

```
postgres=# SELECT date_trunc('s', last_seq_scan) AS last_seq_scan, date_
trunc('s', last_idx_scan) AS last_idx_scan FROM pg_stat_user_tables WHERE
relname='job_details';
     last_seq_scan      |      last_idx_scan
------------------------+------------------------
 2023-10-27 23:13:19+01 | 2023-10-27 23:12:39+01
(1 row)
```

PostgreSQL does not yet have any built-in information about when a table was last written to, so you have to use other means to figure it out.

If you have set up a cron job to collect usage statistics, as described in the previous recipe, *Knowing whether anybody is using a specific table*, then it is relatively easy to find out the last date of change using a SQL query.

Other than this, there are two possibilities, neither of which give you totally reliable answers.

You can either look at the actual timestamps of the files that the data is stored in, or you can use the xmin and xmax system columns to find out the latest transaction ID that changed the table data.

In this recipe, we will cover the first case and focus on the date information in the table's files.

The following PL/pgSQL function looks for the table's data files to get the value of their last access and modification times:

```
CREATE OR REPLACE FUNCTION table_file_access_info(
    IN schemaname text, IN tablename text,
    OUT last_access timestamp with time zone,
    OUT last_change timestamp with time zone
    ) LANGUAGE plpgsql AS $func$
DECLARE
    tabledir text;
    filenode text;
BEGIN
    SELECT regexp_replace(
        current_setting('data_directory') || '/' || pg_relation_
filepath(c.oid),
        pg_relation_filenode(c.oid) || '$', ''),
      pg_relation_filenode(c.oid)
```

```
            INTO tabledir, filenode
            FROM pg_class c
            JOIN pg_namespace ns
              ON c.relnamespace = ns.oid
             AND c.relname = tablename
             AND ns.nspname = schemaname;
        RAISE NOTICE 'tabledir: % - filenode: %', tabledir, filenode;
        -- find latest access and modification times over all segments
        SELECT max((pg_stat_file(tabledir || filename)).access),
               max((pg_stat_file(tabledir || filename)).modification)
          INTO last_access, last_change
          FROM pg_ls_dir(tabledir) AS filename
          -- only use files matching <basefilename>[.segmentnumber]
         WHERE filename ~ ('^' || filenode || '([.]?[0-9]+)?$');
END;
$func$;
```

Here is the sample output:

```
postgres=# SELECT * FROM table_file_access_info('public','job_status');
NOTICE:  tabledir: /var/lib/postgresql/16/main/base/5/ - filenode: 16850
       last_access       |       last_change
-------------------------+------------------------
 2023-10-28 13:43:49+01  | 2023-10-28 13:47:28+01
```

How it works...

The `table_file_access_info(schemaname, tablename)` function returns the last access and modification times for a given table, using the filesystem as a source of information.

The last query uses this data to get the latest time any of these files were modified or read by PostgreSQL. Beware that this is not a very reliable way to get information about the latest use of any table, but it gives you a rough upper-limit estimate of when it was last modified or read (for example, consider the `autovacuum` process for accessing a table).

Monitoring I/O statistics

Sometimes, in order to understand the performance of your system, especially in relation to the usage of hardware resources, you need to examine the I/O rates of the database and the objects contained within.

Getting ready

The `pg_monitor` role will give you full access to I/O monitoring information.

How to do it...

Since PostgreSQL 16, the `pg_stat_io` view is available to us for retrieving I/O statistics for the entire database server (including all databases). To see the running totals in bytes for each backend type, you can run the following query:

```
SELECT backend_type, reads * op_bytes AS bytes_read, writes * op_bytes AS bytes_written
FROM pg_stat_io;
```

You can expect output similar to this:

```
postgres=# SELECT backend_type, reads * op_bytes AS bytes_read, writes *
op_bytes AS bytes_written FROM pg_stat_io;
     backend_type     | bytes_read | bytes_written
----------------------+------------+---------------
 autovacuum launcher  |          0 |             0
 autovacuum launcher  |       8192 |             0
 autovacuum worker    |          0 |             0
 autovacuum worker    |    1425408 |             0
 autovacuum worker    |          0 |             0
 client backend       |    7086080 |             0
 client backend       |          0 |             0
 client backend       |    5480448 |             0
 client backend       |          0 |             0
 client backend       |          0 |             0
 background worker    |          0 |             0
 background worker    |          0 |             0
 background worker    |          0 |             0
 background worker    |          0 |             0
 background worker    |          0 |             0
 background writer    |            |             0
 checkpointer         |            |     152690688
 standalone backend   |          0 |             0
 standalone backend   |          0 |             0
 standalone backend   |          0 |             0
```

```
 standalone backend |          0 |          0
 startup            |          0 |          0
 startup            |          0 |          0
 startup            |          0 |          0
 startup            |          0 |          0
 walsender          |          0 |          0
 walsender          |          0 |          0
 walsender          |          0 |          0
 walsender          |          0 |          0
 walsender          |          0 |          0
(30 rows)
```

Per-table statistics are also available in the pg_statio_user_tables view, which you can use in the following way:

```
WITH b AS (
    SELECT current_setting('block_size')::int AS blcksz
)
SELECT heap_blks_read * blcksz AS heap_read, heap_blks_hit * blcksz AS
heap_hit, idx_blks_read * blcksz AS idx_read, idx_blks_hit * blcksz AS
idx_hit
FROM pg_statio_user_tables, b
WHERE relname=<table name>;
```

The output will be similar to this sample:

```
postgres=# WITH b AS (SELECT current_setting('block_size')::int AS blcksz)
SELECT heap_blks_read * blcksz AS heap_read, heap_blks_hit * blcksz AS
heap_hit, idx_blks_read * blcksz AS idx_read, idx_blks_hit * blcksz AS
idx_hit FROM pg_statio_user_tables, b WHERE relname='job_details';
 heap_read |  heap_hit  | idx_read |   idx_hit
-----------+------------+----------+------------
      8192 |  841138176 |     8192 | 1637752832
(1 row)
```

How it works...

In the pg_stat_io view, we multiply the number of read and write operations by the op_bytes value to get the total bytes read and written by the PostgreSQL cluster. Similarly, with pg_statio_user_tables, we retrieve the server's data block size in bytes and multiply it by the disk blocks read from tables (heap/read) and indexes (idx/read) and also cached blocks (heap/hit and idx/hit) to get the total bytes.

There's more...

You can find more information on the pg_stat_io and pg_statio_user_tables views in the *The Cumulative Statistics System* section of the PostgreSQL documentation at https://www.postgresql.org/docs/current/monitoring-stats.html.

Usage of disk space by temporary data

In addition to ordinary persistent tables, you can also create temporary tables. Temporary tables have disk files for their data, just as persistent tables do, but those files will be stored in one of the tablespaces listed in the temp_tablespaces parameter or, if not set, in the default tablespace.

PostgreSQL may also use temporary files for query processing for sorts, hash joins, or hold cursors if they are larger than your current work_mem parameter setting.

Getting ready

So, how do you find out how much data is being used by temporary tables and files? You can do this by using any untrusted embedded language or by checking the disk usage directly on the database host. You have to use an untrusted language because trusted languages run in a sandbox, which prohibits them from directly accessing the host filesystem.

How to do it...

Perform the following steps:

1. First, check whether your database defines special tablespaces for temporary files, as follows:

    ```
    SELECT current_setting('temp_tablespaces');
    ```

2. As explained later on in this recipe, if the setting is empty, this means that PostgreSQL is not using temporary tablespaces, and temporary objects will be located in the default tablespace for each database.

3. On the other hand, if temp_tablespaces has one or more tablespaces, then your task is easy because all temporary files, both those used for temporary tables and those used for query processing, are inside the directories of these tablespaces. The following query (which uses WITH queries and string and array functions) demonstrates how to check the space that's being used by temporary tablespaces:

    ```
    WITH temporary_tablespaces AS (SELECT
    unnest(string_to_array(
    ```

```
        current_setting('temp_tablespaces'), ',')
    ) AS temp_tablespace
)
SELECT tt.temp_tablespace,
pg_tablespace_location(t.oid) AS location,
pg_tablespace_size(t.oid) AS size
FROM temporary_tablespaces tt
JOIN pg_tablespace t ON t.spcname = tt.temp_tablespace
ORDER BY 1;
```

The following output shows very limited use of temporary space (I ran the preceding query while I had two open transactions that had just created small, temporary tables using random data through generate_series()):

```
temp_tablespace    |   location    |   size
-------------------+---------------+---------
pgtemp1            | /srv/pgtemp1  | 3633152
pgtemp2            | /srv/pgtemp2  |  376832
(2 rows)
```

Even though you can obtain similar results using different queries or just by checking the disk usage from the filesystem through du (once you know the location of tablespaces), I would like to focus on these functions:

- pg_tablespace_location(oid): This provides the location of the tablespace with the given oid.
- pg_tablespace_size(oid) or pg_tablespace_size(name): This allows you to check the size being used by a named tablespace directly within PostgreSQL.
- In PostgreSQL 12+, you can use pg_ls_tmpdir(oid) to view the file's names, sizes, and last modification time, to allow you to see full details of the temporary file's location(s).

Because the amount of temporary disk space being used can vary a lot in an active system, you may want to repeat the query several times to get a better picture of how the disk usage changes. (With psql, use \watch, as explained in the *Checking whether a user is connected* recipe.)

NOTE

Further information on these functions can be found at https://www.postgresql.org/docs/current/functions-admin.html.

On the other hand, if the `temp_tablespaces` setting is empty, then the temporary tables are stored in the same directory as ordinary tables, and the temporary files that are used for query processing are stored in the `pgsql_tmp` directory inside the main database directory.

Look up the cluster's home directory using the following query:

```
SELECT current_setting('data_directory') || '/base/pgsql_tmp';
```

The size of this directory gives us the total size of current temporary files for query processing.

The total size of the temporary files that are used by a database can be found in the `pg_stat_database` system view, and specifically in the `temp_files` and `temp_bytes` fields. These values are cumulative numbers, not current usage, so expect them to increase over time. The following query returns the cumulative number of temporary files and the space being used by every database since the last reset (`stats_reset`):

```
SELECT datname, temp_files, temp_bytes, stats_reset
  FROM pg_stat_database
WHERE datname is not null;
```

The `pg_stat_database` view holds very important statistics. I recommend that you look at the official documentation at https://www.postgresql.org/docs/current/static/monitoring-stats.html#PG-STAT-DATABASE-VIEW for detailed information and to get further ideas on how to improve your monitoring skills.

How it works...

Because all temporary tables and other larger, temporary on-disk data are stored in files, you can use PostgreSQL's internal tables to find the locations of these files, and then determine the total size of these files.

You can control the max file size by setting the `temp_file_limit` parameter, which is unset by default, noting that this is the total amount of all temporary files, not a limit on just one temporary table. Note that this imposes a limit on all types of temporary files used by queries.

There's more...

While the preceding information about temporary tables is correct, it is not the entire story.

Finding out whether a temporary file is in use anymore

Because temporary files are not as carefully preserved as ordinary tables (this is one of the benefits of temporary tables, as less bookkeeping makes them faster), it may sometimes happen that a system crash leaves a few temporary files, which can (in the worst case) take up a significant amount of disk space. In PostgreSQL 14+, temporary files are removed at restart with the default setting of the remove_temp_files_after_crash = on parameter. In earlier releases, you may need to clean up such files by shutting down the PostgreSQL server and then deleting all files from the pgsql_tmp directory while the database is shut down.

Logging temporary file usage

If you set log_temp_files = 0 or a larger value, then the creation of all temporary files that are larger than this value in kilobytes will be logged to the standard PostgreSQL log.

While monitoring the log and the pg_stat_database view, if you notice an increase in temporary file activity, you should consider increasing work_mem, either globally or (preferably) on a query/session basis. While temporary files don't get synced to disk, they do cause file I/O to occur.

Relevant log entries will look something like this:

```
2023-11-16 21:22:07.471 GMT [903371] movieuser@moviedb LOG:  temporary file: path "base/pgsql_tmp/pgsql_tmp903371.2", size 7700480
2023-11-16 21:22:07.471 GMT [903371] movieuser@moviedb STATEMENT:  select * from movies order by movie;
```

Understanding why queries slow down

In production environments with large databases and high concurrent access, it might happen that queries that used to run in tens of milliseconds suddenly take several seconds.

Likewise, a summary query for a report that used to run in a few seconds may take half an hour to complete.

Here are some ways to find out what is slowing them down.

Getting ready

Any questions of the type *Why is this different today from what it was last week?* are much easier to answer if you have some kind of historical data collection setup.

The tools we mentioned in the *Providing PostgreSQL information to monitoring tools* recipe that can be used to monitor general server characteristics, such as CPU and RAM usage, disk I/O, network traffic, load average, and so on, are very useful for seeing what has changed recently, and for trying to correlate these changes with the observed performance of some database operations.

Also, collecting historical statistics data from pg_stat_* tables, whether daily, hourly, or even every 5 minutes if you have enough disk space, is very useful for detecting possible causes of sudden changes or a gradual degradation in performance.

If you are gathering both of these, then that's even better. If you have none, then the question is actually: *Why is this query slow?*

But don't despair! There are a few things you can do to try to restore performance.

How to do it...

First, analyze your database tables using the following code, for all the tables in your slow query:

```
db_01=# analyze my_table;
ANALYZE
Time: 6231.313 ms
db_01=#
```

This is the first thing you should try as it is usually cheap and is meant to be done quite often anyway. Don't run it on the whole database since that is probably overkill and could take some time.

If this restores the query's performance or at least improves the current performance considerably, then this means that autovacuum is not doing its task well, and the next thing to do is find out why.

You must ensure that the performance improvement is not due to caching the pages that are required by the requested query. Make sure that you repeat your query several times before classifying it as slow. Looking at pg_stat_statements (which will be covered later in this chapter) can help you analyze the impact of a particular query in terms of caching, and is done by inspecting two fields: shared_blks_hit and shared_blks_read.

How it works...

The ANALYZE command updates statistics about data size and data distribution in all tables. If a table's size has changed significantly without its statistics being updated, then PostgreSQL's statistics-based optimizer may choose a bad plan. Manually running the ANALYZE command updates the statistics for all tables.

There's more...

There are a few other common problems, as we will discuss in the following subsections.

Do queries return significantly more data than they did earlier?

If you've initially tested your queries on almost empty tables, you may be querying much more data than you need.

As an example, if you select all users' items and then show the first 10 items, this query runs very fast when the user has 10 or even 50 items, but not so well when they have 50,000.

Ensure that you don't ask for more data than you need. Use the LIMIT clause to return less data to your application (and to give the optimizer at least a chance to select a plan that processes less data when selecting: it may also have a lower startup cost). In some cases, you can evaluate the use of cursors for your applications.

Do queries also run slowly when they run alone?

If you can, try to run the same slow query when the database has no (or very few) other queries running concurrently. If it runs well in this situation, then it may be that the database host is just overloaded (CPU, memory, or disk I/O) or other applications are interfering with PostgreSQL on the same server. Consequently, a plan that works well under a light load is not very good anymore. It may even be that this is not a very good query plan to begin with, and you were fooled by modern computers being fast:

```
db=# select count(*) from t;
  count
---------
 1000000
(1 row)
Time: 329.743 ms
```

As you can see, scanning 1 million rows takes just 0.3 seconds on a laptop that is a few years old if these rows have already been cached.

However, if you have a few such queries running in parallel, and also other queries competing for memory, this query is likely to slow down an order of magnitude or two.

See *Chapter 10*, *Performance and Concurrency*, for general advice on performance tuning.

Is the second run of the same query also slow?

This test is related to the previous test, and it checks whether the slowdown is caused by some of the necessary data not fitting into the memory or because it's being pushed out of memory by other queries.

If the second run of the query is fast, then you probably lack enough memory. Again, see *Chapter 10, Performance and Concurrency*, for details about this.

Table and index bloat

Table bloat is something that can develop over time if some maintenance processes can't be run properly. In other words, due to the way **Multiversion Concurrency Control** (**MVCC**) works, your table will contain a lot of older versions of rows, if these versions can't be removed promptly.

There are several ways this can develop, but all involve lots of updates or deletes and inserts, while autovacuum is prevented from doing its job of getting rid of old tuples. It is possible that, even after the old versions are cleaned up, the table stays at its newly acquired and large size, thanks to visible rows being located at the end of the table and preventing PostgreSQL from shrinking the file. There have been cases where a one-row table has grown to several gigabytes in size.

If you suspect that some tables may be bloated, then run the following query:

```
SELECT pg_relation_size(relid) AS tablesize,schemaname,relname,n_live_tup
FROM pg_stat_user_tables
WHERE relname = <tablename>;
```

Then, see whether the relationship between `tablesize` and `n_live_tup` makes sense. You may also think you need to look at `n_dead_tup`, but even after dead tuples are removed, the bloat they have caused will still be there.

For example, if the table size is tens of megabytes and there are only a small number of rows, then you have bloat, and proper `VACUUM` strategies are necessary (as explained in *Chapter 9, Regular Maintenance*).

It is important to check that the statistics are up to date. You may need to run `ANALYZE` on the table and run the query again.

See also

The following will aid your understanding of this topic:

- You can use the *Tracking important metrics over time* recipe in this chapter to track and chart key performance-related PostgreSQL statistics, which can give you valuable insights into your workload.
- The *Knowing whether anybody is using a specific table* recipe shows one way to collect information on table changes.
- *Chapter 9, Regular Maintenance*.
- *Chapter 10, Performance and Concurrency*.
- The *How many rows are there in a table?* recipe in *Chapter 2, Exploring the Database*, for an introduction to MVCC.
- The auto_explain contrib module, at https://www.postgresql.org/docs/current/auto-explain.html.

Analyzing the real-time performance of your queries

The pg_stat_statements extension adds the capability to track the execution statistics of queries that are run in a database, including the number of calls, total execution time, total number of returned rows, and internal information on memory and I/O access.

It is evident how this approach opens up new opportunities in PostgreSQL performance analysis by allowing database admins to get insights directly from the database through SQL and in real time.

Getting ready

The pg_stat_statements module is available as a contrib module of PostgreSQL. The extension must be installed as a superuser in the desired databases. It also requires administrators to add the library to the postgresql.conf file, as follows:

```
shared_preload_libraries = 'pg_stat_statements'
```

This change requires restarting the PostgreSQL server.

Finally, to use it, the extension must be installed in the desired database through the usual CREATE EXTENSION command (run as a superuser):

```
postgres=# CREATE EXTENSION pg_stat_statements;
CREATE EXTENSION
```

How to do it...

Connect to a database where you have installed the pg_stat_statements extension, preferably as a superuser.

You can start by retrieving a list of the top 10 most frequent queries:

```
SELECT query FROM pg_stat_statements ORDER BY calls DESC LIMIT 10;
```

Alternatively, you can retrieve the queries with the highest average execution time:

```
SELECT query, total_exec_time/calls AS avg, calls
       FROM pg_stat_statements ORDER BY 2 DESC;
```

These are just examples. It is strongly recommended that you look at the PostgreSQL documentation at https://www.postgresql.org/docs/current/pgstatstatements.html for more detailed information on the structure of the pg_stat_statements view.

How it works...

Since the pg_stat_statements shared library has been loaded by the PostgreSQL server, Postgres starts collecting statistics for every database in the instance.

The extension simply installs the pg_stat_statements view and the pg_stat_statements_reset() function in the current database, allowing the database admin to inspect the available statistics.

By default, read access to the pg_stat_statements view is granted to every user who can access the database (even though standard users are only allowed to see the SQL statements of their queries).

The pg_stat_statements_reset() function can be used to discard the statistics that have been collected by the server up to that moment and set all the counters to 0. It requires a superuser to be run.

There's more...

A very important pg_stat_statements feature is normalizing queries that can be planned (SELECT, INSERT, DELETE, and UPDATE). You may notice some ? characters in the query field being returned by the queries we outlined in the previous section. The normalization process intercepts constants in SQL statements run by users and replaces them with a placeholder (identified by a question mark).

Consider the following queries:

```
SELECT * FROM bands WHERE name = 'AC/DC';
SELECT * FROM bands WHERE name = 'Lynyrd Skynyrd';
```

After the normalization process, these two queries appear as one in pg_stat_statements:

```
postgres=# SELECT query, calls FROM pg_stat_statements;
              query                  | calls
-------------------------------------+-------
 SELECT * FROM bands WHERE name = ?; |     2
```

In recent PostgreSQL releases, the pg_stat_statements view has gained a number of notable features, such as the ability to track query planning times (if pg_stat_statements.track_planning is enabled), temporary file block statistics and JIT statistics, all of which enable even deeper query analysis.

The extension comes with a few configuration options, such as which statements to track and the maximum number of queries to be tracked.

Tracking important metrics over time

The pg_statviz extension can take snapshots of important PostgreSQL cumulative and dynamic statistics so you can track them over time, perform analyses, and produce visualizations to aid your understanding of your server's workload. The key benefit of this extension is that it's very lightweight and does not require a module in shared_preload_libraries – therefore, installation does not require a server restart. Finally, it enables this analysis without the overhead of external tools or storage such as Prometheus, Logstash, or Elasticsearch.

Getting ready

You can install the extension by downloading it from the PostgreSQL community repositories with your Linux distribution's package manager by installing the package pg_statviz_extension. Alternatively, you can install it from **PGXN (the PostgreSQL Extension Network)**.

Then the extension must be enabled in the desired database through the CREATE EXTENSION command (run as a superuser):

```
postgres=# CREATE EXTENSION pg_statviz;
CREATE EXTENSION
```

How to do it...

pg_statviz can be used by superusers or any user that has pg_monitor role privileges. To take a snapshot, run the following:

```
SELECT pgstatviz.snapshot();
NOTICE:  created pg_statviz snapshot
           snapshot
-------------------------------

 2023-01-27 11:04:58.055453+00

(1 row)
```

Periodic snapshots can be set up with any job scheduler. Here is an example of how to configure with cron:

```
crontab -e -u postgres
```

Then, inside the postgres user's crontab, add this line to take a snapshot every 10 minutes:

```
*/10 * * * * psql -c "SELECT pgstatviz.snapshot()" >/dev/null 2>&1
```

How it works...

The extension takes timestamped snapshots of Postgres statistics and stores them in tables under the pgstatviz schema. It collects statistics on the background writer, checkpoints, cache hits, connection count, I/O, locks, numbers of tuples read/written, wait events, WAL generation, transactions, and others.

The accompanying visualization utility reads from these tables and can produce charts that enable analysis of this time series data at a glance. It can be installed from the PostgreSQL community repositories as the package pg_statviz. Alternatively, you can install it from **PyPI (the Python Package Index)** as follows:

```
pip install pg_statviz
```

Then, you can create graphs (output to disk as .png images) for a selected time period using all modules at once, or one at a time like this:

```
pg_statviz buf --host localhost -d postgres -U postgres -D 2023-08-25T11:00 2023-01-27
```

Sample output for the tuple read/write rate and transaction rate modules from the same time period on a production database can be found below:

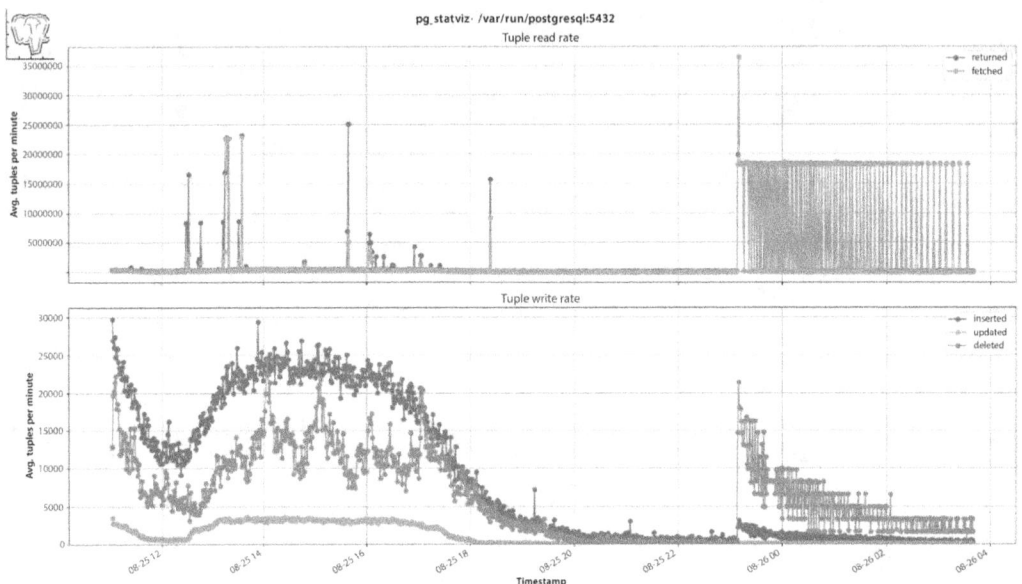

Figure 8.5: pg_statviz tuple read/write rate analysis output

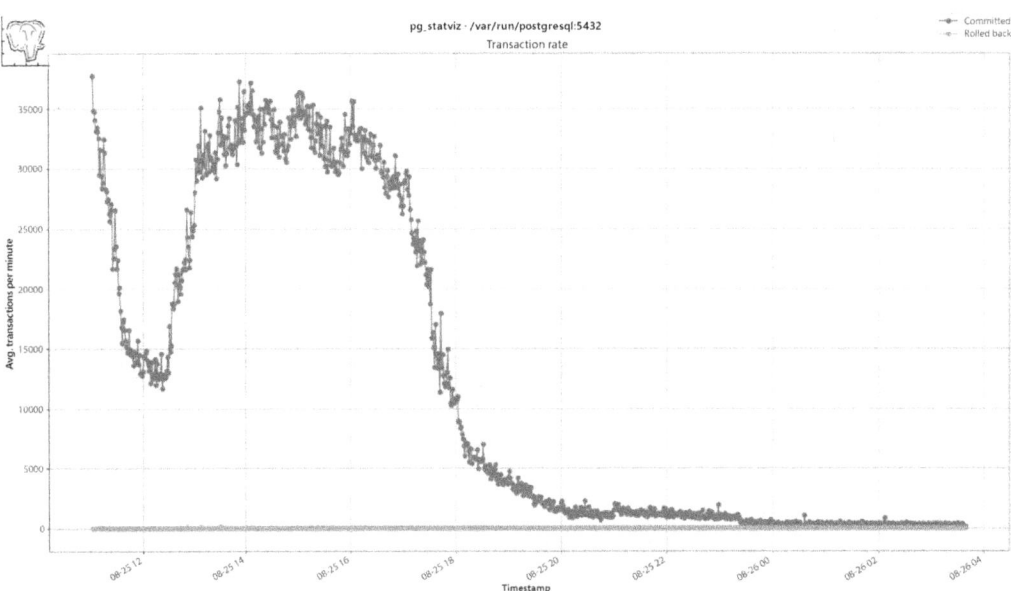

Figure 8.6: pg_statviz transaction rate analysis output

There's more...

The pg_statviz utility accepts most of the standard Postgres command-line tool parameters for connecting to the database, which means that you can use it remotely, and it has a configurable output directory.

You can find the project page at https://github.com/vyruss/pg_statviz.

Learn more on Discord

To join the Discord community for this book – where you can share feedback, ask questions to the author, and learn about new releases – follow the QR code below:

https://discord.gg/pQkghgmgdG

9
Regular Maintenance

In these busy times, many people believe *if it ain't broken, don't fix it*. I believe that too, but it isn't an excuse for not taking action to maintain your database servers and be sure that nothing will break.

Database maintenance is about making your database run smoothly.

PostgreSQL prefers regular maintenance, so please read the *Planning maintenance* recipe in this chapter for more information.

We recognize that you're here for a reason and are looking for a quick solution to your needs. You're probably thinking – *Fix the problem first, and I'll plan later*. So, off we go!

PostgreSQL provides a utility command named VACUUM, which is a reference to a garbage collector that sweeps up all of the bad things and fixes them – or at least most of them. That's the single most important thing you need to remember to do – I say *single* because VACUUM is closely connected to the ANALYZE command, which collects statistics for the SQL optimizer. It's possible to run VACUUM and ANALYZE as a single joint command, VACUUM ANALYZE. These actions are automatically executed for you when appropriate by autovacuum, a special background process that runs as part of the PostgreSQL server.

VACUUM performs a range of cleanup activities, some of which are too complex to describe without a whole sideline into their internals. VACUUM has been heavily optimized over 30 years to take the minimum required lock levels on tables and execute them in the most efficient manner possible, skipping all of the unnecessary work and using L2 cache CPU optimizations when work is required.

Many experienced PostgreSQL database administrators will prefer to execute their VACUUM commands themselves in order to aim for maximum benefit with minimal performance impact. However, autovacuum now provides a fine degree of control, which, if enabled and controlled, can save much of your time. Using both manual and automatic vacuuming gives you control and a safety net.

In this chapter, we will cover the following recipes:

- Controlling automatic database maintenance
- Avoiding auto-freezing
- Removing issues that cause bloat
- Actions for heavy users of temporary tables
- Identifying and fixing bloated tables and indexes
- Monitoring and tuning a vacuum
- Maintaining indexes
- Finding unused indexes
- Carefully removing unwanted indexes
- Planning maintenance

Controlling automatic database maintenance

autovacuum is enabled by default in PostgreSQL and mostly does a great job of maintaining your PostgreSQL database. We say mostly because it doesn't know everything you do regarding the database, such as the best time to perform maintenance actions. Let's explore the settings that can be tuned so that you can use VACUUM commands efficiently.

Getting ready

Exercising control requires some thinking about what you want:

- What are the best times of day to perform maintenance activities? When are system resources more available?
- Which days are quiet, and which are not?
- Which tables are critical to the application, and which are not?

How to do it…

The first thing you must do is make sure that autovacuum is switched on, which is the default. Check that you have the following parameters enabled in your postgresql.conf file:

```
autovacuum = on
track_counts = on
```

PostgreSQL controls autovacuum with more than 40 individually tunable parameters that provide a wide range of options, though this can be a little daunting. The following are the relevant parameters that can be set in postgresql.conf to tune the VACUUM command:

vacuum_buffer_usage_limit
vacuum_cleanup_index_scale_factor
vacuum_cost_delay
vacuum_cost_limit
vacuum_cost_page_dirty
vacuum_cost_page_hit
vacuum_cost_page_miss
vacuum_failsafe_age
vacuum_freeze_min_age
vacuum_freeze_table_age
vacuum_multixact_freeze_min_age
vacuum_multixact_freeze_table_age

There are also postgresql.conf parameters that apply specifically to autovacuum:

autovacuum
autovacuum_analyze_scale_factor
autovacuum_analyze_threshold
autovacuum_freeze_max_age
autovacuum_max_workers
autovacuum_multixact_freeze_max_age
autovacuum_naptime
autovacuum_vacuum_cost_delay
autovacuum_vacuum_cost_limit
autovacuum_vacuum_insert_threshold
autovacuum_vacuum_insert_scale_factor
autovacuum_vacuum_scale_factor
autovacuum_vacuum_threshold
autovacuum_work_mem
log_autovacuum_min_duration

The preceding parameters apply to all tables at once. Individual tables can be controlled by storage parameters, which are set using the following command:

```
ALTER TABLE mytable SET (storage_parameter = value);
```

The storage parameters that relate to maintenance are as follows:

```
autovacuum_enabled
autovacuum_analyze_scale_factor
autovacuum_analyze_threshold
autovacuum_freeze_min_age
autovacuum_freeze_max_age
autovacuum_freeze_table_age
autovacuum_multixact_freeze_max_age
autovacuum_multixact_freeze_min_age
autovacuum_multixact_freeze_table_age
autovacuum_vacuum_cost_delay
autovacuum_vacuum_cost_limit
autovacuum_vacuum_insert_threshold
autovacuum_vacuum_insert_scale_factor
autovacuum_vacuum_scale_factor
autovacuum_vacuum_threshold
vacuum_truncate (no equivalent postgresql.conf parameter)
log_autovacuum_min_duration
```

The toast tables can be controlled with the following parameters. Note that these parameters are *set* on the main table and *not* on the toast table (which gives an error):

```
toast.autovacuum_enabled
toast.autovacuum_analyze_scale_factor
toast.autovacuum_analyze_threshold
toast.autovacuum_freeze_min_age
toast.autovacuum_freeze_max_age
toast.autovacuum_freeze_table_age
toast.autovacuum_multixact_freeze_max_age
toast.autovacuum_multixact_freeze_min_age
toast.autovacuum_multixact_freeze_table_age
toast.autovacuum_vacuum_cost_delay
toast.autovacuum_vacuum_cost_limit
toast.autovacuum_vacuum_insert_threshold
toast.autovacuum_vacuum_insert_scale_factor
toast.autovacuum_vacuum_scale_factor
toast.autovacuum_vacuum_threshold
toast.vacuum_truncate
toast.log_autovacuum_min_duration
```

How it works...

If autovacuum is set, then it will wake up every autovacuum_naptime seconds, and decide whether to run VACUUM, ANALYZE, or both (don't modify that).

There will never be more than autovacuum_max_workers maintenance processes running at any time. As these autovacuum workers perform I/O, they accumulate cost points until they hit the autovacuum_vacuum_cost_limit value, after which they sleep for an autovacuum_vacuum_cost_delay period. This is designed to throttle the resource utilization of autovacuum to prevent it from using all of the available disk I/O bandwidth, which it should never do. So, increasing autovacuum_vacuum_cost_delay will slow down each VACUUM to reduce the impact on user activity, but the general advice is don't do that. The benefits of performing autovacuum outweigh the performance hit incurred in most cases. autovacuum will run ANALYZE when there have been at least autovacuum_analyze_threshold changes and a fraction of the table defined by autovacuum_analyze_scale_factor has been inserted, updated, or deleted.

autovacuum will run VACUUM when there have been at least autovacuum_vacuum_threshold changes, and a fraction of the table defined by autovacuum_vacuum_scale_factor has been updated or deleted.

The autovacuum_* parameters only change VACUUM and ANALYZE operations that are executed by autovacuum. User-initiated VACUUM and ANALYZE commands are affected by vacuum_cost_delay and other vacuum_* parameters.

The log_autovacuum_min_duration setting ensures that any autovacuum process that runs for longer than this value will be logged to the server log, like so:

```
2019-04-19 01:33:55 BST (13130) LOG:  automatic vacuum of table "postgres.public.pgbench_accounts": index scans: 1
        pages: 0 removed, 3279 remain
        tuples: 100000 removed, 100000 remain
        system usage: CPU 0.19s/0.36u sec elapsed 19.01 sec
2019-04-19 01:33:59 BST (13130) LOG:  automatic analyze of table "postgres.public.pgbench_accounts"
        system usage: CPU 0.06s/0.18u sec elapsed 3.66 sec
```

In PostgreSQL 16, the default value is 600,000 milliseconds, so any autovacuum that takes longer than 10 minutes will be logged.

Most of the preceding global parameters can also be set at the table level. For example, the normal autovacuum_cost_delay is 2 ms, but if you want big_table to be vacuumed more quickly, then you can set the following:

```
ALTER TABLE big_table SET (autovacuum_vacuum_cost_delay = 0);
```

It's also possible to set parameters for toast tables. A toast table is where the oversized column values get placed, which the documents refer to as *supplementary storage tables*. If there are no oversized values, then the toast table will occupy little space. Tables with very wide values often have large toast tables. **The Oversized Attribute Storage Technique (TOAST)** is optimized for UPDATE. For example, if you have a heavily updated table, the toast table is often untouched, so it may make sense to turn off autovacuum for the toast table, as follows:

```
ALTER TABLE pgbench_accounts
SET ( toast.autovacuum_enabled = off);
```

NOTE

Autovacuuming the toast table is performed completely separately from the main table, and you can ask for an explicit inclusion or exclusion of the main or toast table yourself when running VACUUM by specifying the PROCESS_MAIN or PROCESS_TOAST option.

Use the following query to display reloptions for tables and their toast tables:

```
postgres=#
SELECT n.nspname
, c.relname
, array_to_string(
    c.reloptions ||
ARRAY(
SELECT 'toast.' || x
FROM unnest(tc.reloptions) AS x
), ', ')
AS relopts
FROM pg_class c
LEFT JOIN pg_class tc    ON c.reltoastrelid = tc.oid
JOIN pg_namespace n ON c.relnamespace  = n.oid
WHERE c.relkind = 'r'
AND nspname NOT IN ('pg_catalog', 'information_schema');
```

An example of the output of this query is shown here:

```
 nspname |     relname      |            relopts
---------+------------------+-------------------------------
 public  | pgbench_accounts | fillfactor=100,
                              autovacuum_enabled=on,
                              autovacuum_vacuum_cost_delay=20
 public  | pgbench_tellers  | fillfactor=100
 public  | pgbench_branches | fillfactor=100
 public  | pgbench_history  |
 public  | text_archive     | toast.autovacuum_enabled=off
```

Managing parameters for many different tables becomes difficult with tens, hundreds, or thousands of tables. We recommend that these parameter settings are used with caution and only when you have good evidence that they are worthwhile. Undocumented parameter settings will cause problems later.

Note that when multiple workers are running, the autovacuum cost delay parameters are "balanced" among all the running workers, so that the total I/O impact on the system is the same regardless of the number of workers running. However, if you set the per-table storage parameters for autovacuum_vacuum_cost_delay or autovacuum_vacuum_cost_limit, then those tables are not considered in the balancing algorithm.

VACUUM allows insertions, updates, and deletions while it runs, but it prevents DDL commands such as ALTER TABLE and CREATE INDEX because they will conflict with the SHARE ACCESS EXCLUSIVE lock it obtains. autovacuum can detect whether a user has requested a conflicting lock on the table while it runs, and it will cancel itself if it is getting in the user's way. VACUUM doesn't cancel itself since we expect that the DBA would not want it to be canceled.

Since PostgreSQL 13, autovacuum can be triggered by insertions, so you may see more VACUUM activity than before in some workloads, but this is likely to be a good thing and nothing to worry about.

Note that VACUUM does not shrink a table when it runs unless there is a large run of space at the end of a table and nobody is accessing the table when we try to shrink it. If you want to avoid trying to shrink a table when we vacuum it, you can turn this off with the following setting:

```
ALTER TABLE pgbench_accounts
SET (vacuum_truncate = off);
```

To shrink a table properly and return the freed space to the filesystem, you'll need VACUUM FULL, but this locks up the whole table for a long time and should be avoided if possible. The VACUUM FULL command will rewrite every row of the table and completely rebuild all indexes. This process is faster than it used to be, though it still takes a long time for larger tables, as well as needing up to twice the current space for the sort and new copy of the table.

There's more...

The postgresql.conf file also allows include directives, which look as follows:

```
include 'autovacuum.conf'
```

These specify another file that will be read at that point, just as if those parameters had been included in the main file.

This can be used to maintain multiple sets of files for the autovacuum configuration. Let's say we have a website that is busy mainly during the daytime, with some occasional nighttime use. We decide to have two profiles – one for daytime when we want less aggressive auto-vacuuming, and another for nighttime, where we can allow more aggressive vacuuming:

1. You need to add the following lines to postgresql.conf:

   ```
   autovacuum = on
   autovacuum_max_workers = 3
   include 'autovacuum.conf'
   ```

2. Remove all other autovacuum parameters.

3. Then, create a file named autovacuum.conf.day that contains the following parameters:

   ```
   autovacuum_analyze_scale_factor = 0.1
   autovacuum_vacuum_cost_delay = 5
   autovacuum_vacuum_scale_factor = 0.2
   ```

4. Then, create another file, named autovacuum.conf.night, that contains the following parameters:

   ```
   autovacuum_analyze_scale_factor = 0.05
   autovacuum_vacuum_cost_delay = 0
   autovacuum_vacuum_scale_factor = 0.1
   ```

5. To swap profiles, simply do the following:

   ```
   $ ln -sf autovacuum.conf.night autovacuum.conf
   $ pg_ctl reload
   ```

The latter command reloads the server configuration, and it must be customized depending on your platform.

This allows us to switch profiles twice per day without needing to edit the configuration files. You can also easily tell which is the active profile simply by looking at the full details of the linked file (using `ls -l`). The exact details of the schedule are up to you. Night and day was just an example, and is unlikely to suit everybody.

See also

The `autovacuum_freeze_max_age` parameter is explained in the next recipe, *Avoiding auto-freezing*, as are the more complex table-level parameters.

Avoiding auto-freezing

In the life cycle of a row, there are two routes that a row can take in PostgreSQL – a row version dies and needs to be removed by VACUUM, or a row version gets old enough and needs to be frozen, a task that is also performed by the VACUUM process. The removal of dead rows is easy to understand, while the second seems strange and surprising because many PostgreSQL users will not be familiar with the concept of freezing. Freezing is necessary for the proper operation of PostgreSQL's **Multiversion Concurrency Control** (**MVCC**) for the following reason.

PostgreSQL uses internal transaction identifiers that are 4 bytes long, so we only have 2^{32} transaction IDs (about 4 billion). PostgreSQL starts again from the beginning when that wraps around, circularly allocating new identifiers. The reason we do this is that moving to an 8-byte identifier has various other negative effects and costs that we would rather not pay for, so we keep the 4-byte transaction identifier. The impact is that we need to do regular sweeps of the entire database to mark tuples as frozen, meaning they are visible to all users – that's why this procedure is known as freezing. Once frozen, they don't need to be touched again, though they can still be updated or deleted later if desired.

How to do it...

Why should we care about preventing auto-freezing at inopportune times? Suppose that we load a table with 100 million rows, and everything is fine. When those rows have been there long enough to begin being frozen, the next VACUUM operation on that table will rewrite all of them to freeze their transaction identifiers. Put another way, autovacuum will wake up and start using lots of I/O to perform the freezing.

The most obvious way to forestall this problem is to explicitly vacuum a table after a major load. Of course, that doesn't remove the problem entirely, because vacuuming doesn't freeze all the rows immediately, so some will remain for later vacuuming.

The knee-jerk reaction for many people is to turn off autovacuum because it keeps waking up at the most inconvenient times. My way of doing this is described in the previous recipe, *Controlling automatic database maintenance*.

Freezing takes place when a transaction identifier on a row becomes more than vacuum_freeze_min_age transactions older than the current next value, measured in xid values, not time. Normal VACUUM operations will perform a small amount of freezing as you go, and in most cases, you won't notice that at all. As explained in the previous example, large transactions leave many rows with the same transaction identifiers, so those might cause problems when it comes to freezing.

The VACUUM command is normally optimized to only look at the chunks of a table that require cleaning, both for normal vacuum and freezing operations.

If you fiddle with the VACUUM parameters to try to forestall heavy VACUUM operations, then you'll notice that the autovacuum_freeze_max_age parameter controls when the table will be scanned by a forced VACUUM command. To put this another way, you can't turn off the need to freeze rows, but you can defer it to a more convenient time. The mistake comes from deferring it completely and then finding that PostgreSQL executes an aggressive, uncancellable vacuum to remedy the lack of freezing. My advice is to control autovacuum, as we described in the previous recipe, or perform explicit VACUUM operations at a time of your choosing, rather than wait for the inevitable emergency freeze operation.

The VACUUM command is also an efficient way to confirm the absence of page corruptions, so it is worth scanning the whole database, block by block, from time to time. To do this, you can run the following command on each of your databases:

```
VACUUM (DISABLE_PAGE_SKIPPING);
```

You can do this table by table as well. There's nothing important about running whole database VACUUM operations anymore; in earlier versions of PostgreSQL, this was important, so you may read that this is a good idea on the web.

You can focus on only the tables that most need freezing by using the vacuumdb utility with the new --min-xid-age and --min-mxid-age options. By setting those options, vacuumdb will skip them if the main table or toast table has a relfrozenxid older than the specified age threshold. If you choose the values carefully, this will skip tables that don't need freezing yet (there is no corresponding option for these in the VACUUM command, as there is in most other cases).

Removing issues that cause bloat

Bloat is disk space previously used by tables that is available for reuse by the database but no longer by the filesystem. It can be caused by long-running queries or long-running write transactions that execute alongside write-heavy workloads. Resolving that is mostly down to understanding the workloads that are running on the server.

Getting ready

Look at the age of the oldest snapshots that are running, like this:

```
postgres=# SELECT now() -
  CASE
  WHEN backend_xid IS NOT NULL
  THEN xact_start
  ELSE query_start END
  AS age
, pid
, backend_xid AS xid
, backend_xmin AS xmin
, state
FROM   pg_stat_activity
WHERE  backend_type = 'client backend'
ORDER BY 1 DESC;
      age         |  pid  |   xid    |   xmin   |       state
------------------+-------+----------+----------+--------------------
 00:00:25.791098  | 27624 |          | 10671262 | active
 00:00:08.018103  | 27591 |          |          | idle in transaction
 00:00:00.002444  | 27630 | 10703641 | 10703639 | active
 00:00:00.001506  | 27631 | 10703642 | 10703640 | active
 00:00:00.000324  | 27632 | 10703643 | 10703641 | active
 00:00:00         | 27379 |          | 10703641 | active
```

The preceding example shows an updated workload of three sessions alongside one session that is waiting in an *idle in transaction* state, plus two other sessions that are only reading data.

How to do it...

If you have sessions stuck in the `idle_in_transaction` state, then you may want to consider setting the `idle_in_transaction_session_timeout` parameter so that transactions in that mode will be canceled. The default for that is 0, meaning there will be no cancellation.

If not, try running shorter transactions or shorter queries.

If that is not an option, then consider setting old_snapshot_threshold. This parameter sets a time delay, after which dead rows are at risk of being removed. If a query attempts to read data that has been removed, then we cancel the query. All queries executing in less time than the old_snapshot_threshold parameter will be safe. This is a very similar concept to the way *hot standby* works (see *Chapter 12, Replication and Upgrades*).

How it works...

VACUUM cannot remove dead rows until they are invisible to all users. The earliest piece of data that's visible to a session is defined by its oldest snapshot's xmin value or, if that is not set, then by the backend's xid value.

There's more...

A session that is not running any query is in the *idle* state if it's outside of a transaction, or in the *idle in transaction* state if it's inside a transaction – that is, between a BEGIN and the corresponding COMMIT. Recall the *Writing a script that either succeeds entirely or fails entirely* recipe in *Chapter 7, Database Administration*, which was about how BEGIN and COMMIT can be used to wrap several commands into one transaction.

The reason to distinguish between these two states is that locks are released at the end of a transaction. Hence, an *idle in transaction* session is not currently doing anything, but it might be preventing other queries, including VACUUM, from accessing some tables.

Actions for heavy users of temporary tables

If you are a heavy user of temporary tables in your applications, then there are some additional actions that you may need to perform.

How to do it...

There are four main things to check, which are as follows:

- Make sure you run VACUUM on system tables or enable autovacuum so that it will do this for you.
- Monitor running queries to see how many temporary files are active and how large they are.
- Tune the memory parameters. Think about increasing the temp_buffers parameter, but be careful not to over-allocate memory.

- Separate the temp table's I/O. In a query-intensive system, you may find that reads/writes to temporary files exceed reads/writes on permanent data tables and indexes because some queries don't fit in working memory and have to spill over into temp files on disk. In this case, you should create a new tablespace(s) on separate disks, and ensure that the temp_tablespaces parameter is configured to use the additional tablespace(s).

How it works...

When we create a temporary table, we insert entries into the pg_class, pg_type, and pg_attribute catalog tables. These catalog tables and their indexes begin to grow and bloat – an issue that will be covered in further recipes. To control that growth, you can either vacuum those tables manually or let autovacuum do its work. You cannot run ALTER TABLE against system tables, so it is not possible to set specific autovacuum settings for any of these tables.

If you vacuum the system catalog tables manually, make sure that you get all of the system tables. You can get the full list of tables to vacuum and a list of their indexes by using the following query:

```
postgres=# SELECT relname, pg_relation_size(oid) FROM pg_class
WHERE relkind in ('i','r') AND relnamespace = 'pg_catalog'::regnamespace
ORDER BY 2 DESC;
```

This results in the following output:

relname	pg_relation_size
pg_proc	450560
pg_depend	344064
pg_attribute	286720
pg_depend_depender_index	204800
pg_depend_reference_index	204800
pg_proc_proname_args_nsp_index	180224
pg_description	172032
pg_attribute_relid_attnam_index	114688
pg_operator	106496
pg_statistic	106496
pg_description_o_c_o_index	98304
pg_attribute_relid_attnum_index	81920
pg_proc_oid_index	73728
pg_rewrite	73728
pg_class	57344

```
    pg_type                        |        57344
    pg_class_relname_nsp_index     |        40960
    ...(partial listing)
```

The preceding values are for a newly created database. These tables can become very large if they're not properly maintained, with values of 11 GB for one index being witnessed in one unlucky installation.

Identifying and fixing bloated tables and indexes

PostgreSQL implements **MVCC**, which allows users to read data at the same time as writers make changes. This is an important feature for concurrency in database applications as it can allow the following:

- Better performance because of fewer locks
- Greatly reduced deadlocking
- Simplified application design and management

Bloated tables and indexes are a natural consequence of MVCC design in PostgreSQL. Bloat is caused mainly by updates, as we must retain both the old and new updates for a certain period. Since these extra row versions are required to provide MVCC, some amount of bloat is normal and acceptable. Tuning to remove bloat completely isn't useful and is probably a waste of time.

Bloating results in increased disk consumption, as well as performance loss – if a table is twice as big as it should be, scanning it takes twice as long. VACUUM is one of the best ways of removing bloat.

Many users execute VACUUM far too frequently, while at the same time complaining about the cost of doing so. This recipe is all about understanding when you need to run VACUUM by estimating the amount of bloat in tables and indexes.

Getting ready

MVCC is a core part of PostgreSQL and cannot be turned off because it's what enables high concurrency with high database performance. The internals of MVCC have some implications for the DBA that need to be understood: each row represents a row version, so it has two system columns – xmin and xmax – indicating the identifiers of the two transactions when the version was created and deleted, respectively. The value of xmax is NULL if that version has not been deleted yet.

The general idea is that, instead of removing row versions, we alter their visibility by changing their xmin and/or xmax values.

To be more precise, when a row is inserted, its xmin value is set to the "XID" or transaction ID of the creating transaction, while xmax is emptied; when a row is deleted, xmax is set to the "XID" of the deleting transaction, without actually removing the row. An UPDATE operation is treated similarly to a DELETE operation, followed by INSERT; the deleted row represents the older version, and the row that's been inserted is the newer version. Finally, when rolling back a transaction, all of its changes are made invisible by marking that transaction ID as aborted.

In this way, we get faster DELETE, UPDATE, and ROLLBACK statements, but the price of these benefits is that the SQL UPDATE command can cause tables and indexes to grow in size because they leave behind dead row versions. The DELETE and aborted INSERT statements take up space, which must be reclaimed by garbage collection. VACUUM is the command we use to reclaim space in a batch operation, though there is another internal feature named **Heap-Only Tuples** (HOT), which allows us to clean data blocks one at a time as we scan each data block if that is possible. HOT also reduces index bloat since not all updates require index maintenance.

How to do it...

The best way to understand this is to look at things the same way that autovacuum does, by using a view that's been created with the following query:

```
CREATE OR REPLACE VIEW av_needed AS
SELECT N.nspname, C.relname
, pg_stat_get_tuples_inserted(C.oid) AS n_tup_ins
, pg_stat_get_tuples_updated(C.oid) AS n_tup_upd
, pg_stat_get_tuples_deleted(C.oid) AS n_tup_del
, CASE WHEN pg_stat_get_tuples_updated(C.oid) > 0
       THEN pg_stat_get_tuples_hot_updated(C.oid)::real
          / pg_stat_get_tuples_updated(C.oid)
       END
  AS HOT_update_ratio
, pg_stat_get_live_tuples(C.oid) AS n_live_tup
, pg_stat_get_dead_tuples(C.oid) AS n_dead_tup
, C.reltuples AS reltuples
, round(COALESCE(threshold.custom, current_setting('autovacuum_vacuum_threshold'))::integer
       + COALESCE(scale_factor.custom, current_setting('autovacuum_vacuum_scale_factor'))::numeric
       * C.reltuples)
  AS av_threshold
```

```
  , date_trunc('minute',
      greatest(pg_stat_get_last_vacuum_time(C.oid),
               pg_stat_get_last_autovacuum_time(C.oid)))
    AS last_vacuum
  , date_trunc('minute',
      greatest(pg_stat_get_last_analyze_time(C.oid),
               pg_stat_get_last_analyze_time(C.oid)))
    AS last_analyze
  , pg_stat_get_dead_tuples(C.oid) >
    round( current_setting('autovacuum_vacuum_threshold')::integer
         + current_setting('autovacuum_vacuum_scale_factor')::numeric
         * C.reltuples)
    AS av_needed
  , CASE WHEN reltuples > 0
         THEN round(100.0 * pg_stat_get_dead_tuples(C.oid) / reltuples)
         ELSE 0 END
    AS pct_dead
FROM pg_class C
LEFT JOIN pg_namespace N ON (N.oid = C.relnamespace)
NATURAL LEFT JOIN LATERAL (
    SELECT (regexp_match(unnest,'^[^=]+=(.+)$'))[1]
    FROM unnest(reloptions)
    WHERE unnest ~ '^autovacuum_vacuum_threshold='
) AS threshold(custom)
NATURAL LEFT JOIN LATERAL (
    SELECT (regexp_match(unnest,'^[^=]+=(.+)$'))[1]
    FROM unnest(reloptions)
    WHERE unnest ~ '^autovacuum_vacuum_scale_factor='
) AS scale_factor(custom)
WHERE C.relkind IN ('r', 't', 'm')
  AND N.nspname NOT IN ('pg_catalog', 'information_schema')
  AND N.nspname NOT LIKE 'pg_toast%'
ORDER BY av_needed DESC, n_dead_tup DESC;
```

We can then use this to look at individual tables, as follows:

```
postgres=# \x
postgres=# SELECT * FROM av_needed WHERE nspname = 'public' AND relname =
'pgbench_accounts';
```

We will get the following output:

```
-[ RECORD 1 ]----+------------------------
nspname          | public
relname          | pgbench_accounts
n_tup_ins        | 100001
n_tup_upd        | 117201
n_tup_del        | 1
hot_update_ratio | 0.123454578032611
n_live_tup       | 100000
n_dead_tup       | 0
reltuples        | 100000
av_threshold     | 20050
last_vacuum      | 2010-04-29 01:33:00+01
last_analyze     | 2010-04-28 15:21:00+01
av_needed        | f
pct_dead         | 0
```

How it works...

We can compare the number of dead row versions, shown as n_dead_tup, against the required threshold, av_threshold.

The preceding query doesn't take into account table-specific autovacuum thresholds. It could do so if you need it, but the main purpose of the query is to give us information to understand what is happening and then set the parameters accordingly – not the other way around.

Notice that the table query shows insertions, updates, and deletions so that you can understand your workload better. There is also something named hot_update_ratio. This shows the fraction of updates that take advantage of the HOT feature, which allows a table to self-vacuum as the table changes. If that ratio is high, then you may afford to avoid VACUUM activities altogether or at least for long periods because of the self-vacuuming optimization. If the ratio is low, then you will need to execute VACUUM commands or autovacuum more frequently. Note that the ratio never reaches 1.0, so if you have it above 0.95, then that is very good and you need not think about it further.

HOT updates take place when the UPDATE statement does not change any of the column values that are indexed by any index, and there is enough free space in the disk page where the updated row is located. If you change even one column that is indexed by just one index, then it will be a non-HOT update, and there will be a performance hit. So, carefully selecting indexes can improve performance and reduce the need for maintenance.

Also, if HOT updates do occur, though not often enough for your liking, you might want to try to decrease the `fillfactor` storage parameter for the table to make more space for them. Remember that this will only be important on your most active tables. Seldom-touched tables don't need much tuning.

To recap, non-HOT updates cause indexes to bloat. The following query is useful in investigating the index size and how it changes over time. It runs fairly quickly and can be used to monitor whether your indexes are changing in size over time:

```
SELECT
nspname,relname,
round(100 * pg_relation_size(indexrelid) /
                pg_relation_size(indrelid)) / 100
            AS index_ratio,
  pg_size_pretty(pg_relation_size(indexrelid))
            AS index_size,
  pg_size_pretty(pg_relation_size(indrelid))
            AS table_size
FROM pg_index I
LEFT JOIN pg_class C ON (C.oid = I.indexrelid)
LEFT JOIN pg_namespace N ON (N.oid = C.relnamespace)
WHERE
  nspname NOT IN ('pg_catalog', 'information_schema', 'pg_toast') AND
  C.relkind='i' AND
  pg_relation_size(indrelid) > 0;
```

Another route is to use the `pgstattuple` contrib extension, which provides very detailed statistics on tables and indexes:

```
CREATE EXTENSION pgstattuple;
```

You can scan tables using `pgstattuple()`, as follows:

```
test=> SELECT * FROM pgstattuple('pg_catalog.pg_proc');
```

The output will look as follows:

```
-[ RECORD 1 ]------+-------
table_len          | 458752
tuple_count        | 1470
tuple_len          | 438896
```

```
 tuple_percent       | 95.67
 dead_tuple_count    | 11
 dead_tuple_len      | 3157
 dead_tuple_percent  | 0.69
 free_space          | 8932
 free_percent        | 1.95
```

The downside of pgstattuple is that it derives exact statistics by scanning the whole table and counting everything. If you have time to scan the table, you may as well vacuum the whole table anyway. So, a better idea is to use pgstattuple_approx(), which is much, much faster, and yet is still fairly accurate. It works by accessing the table's visibility map first and then only scanning the pages that need VACUUM, so I recommend that you use it in all cases for checking tables (there is no equivalent for indexes since they don't have a visibility map):

```
postgres=# select * from pgstattuple_approx('pgbench_accounts');
-[ RECORD 1 ]--------+------------------
table_len            | 268591104
scanned_percent      | 0
approx_tuple_count   | 1001738
approx_tuple_len     | 137442656
approx_tuple_percent | 51.1717082037088
dead_tuple_count     | 0
dead_tuple_len       | 0
dead_tuple_percent   | 0
approx_free_space    | 131148448
approx_free_percent  | 48.8282917962912
```

You can also scan indexes using pgstatindex(), as follows:

```
postgres=> SELECT * FROM pgstatindex('pg_cast_oid_index');
-[ RECORD 1 ]------+------
version            | 2
tree_level         | 0
index_size         | 8192
root_block_no      | 1
internal_pages     | 0
leaf_pages         | 1
empty_pages        | 0
deleted_pages      | 0
```

```
avg_leaf_density      | 50.27
leaf_fragmentation    | 0
```

There's more...

You may want to set up monitoring for the bloated tables and indexes. Look at the Nagios plugin called check_postgres_bloat, which is a part of the check_postgres plugins.

It provides some flexible options to assess bloat. Unfortunately, it's not that well documented, but if you've worked through this recipe, it should make sense. You'll need to play with it to get the thresholding correct anyway, so that shouldn't be a problem.

Also, note that the only way to know the exact bloat of a table or index for certain is to scan the whole relationship. Anything else is just an estimate and may lead to you running maintenance either too early or too late.

Monitoring and tuning a vacuum

This recipe covers both the VACUUM command and autovacuum, which I refer to collectively as vacuums (non-capitalized). It's important to be able to monitor both to know how long they take and what resources they use so you can better tune the behavior and timing of future vacuums.

If you're currently waiting for a long-running vacuum (or autovacuum) to finish, continue to the *How to do it...* section.

If you've just had a long-running vacuum complete, then you may want to think about setting a few parameters for next time, so go straight to the *How it works...* section.

Getting ready

Let's watch what happens when we run a large VACUUM. Don't run VACUUM FULL because, as we mentioned in the recipe *Controlling automatic database maintenance*, it runs for a long time while holding an AccessExclusiveLock on the table. Ouch.

How to do it...

First, locate which process is running this VACUUM by using the pg_stat_activity view to identify the specific pid (34399 is just an example).

Repeatedly execute the following query to see the progress of the VACUUM command, specifying the pid of the process you wish to monitor:

```
postgres=# SELECT * FROM pg_stat_progress_vacuum WHERE pid = 34399;
```

The next section explains what this all means.

How it works...

VACUUM works in various phases:

1. The first phase is *initializing*, but this phase is over so quickly that you'll never see it.
2. The first main phase is *scanning heap*, which performs about 90% of the cleanup of data blocks in the heap. The heap_blks_scanned columns will increase from 0 up to the value of heap_blks_total. The number of blocks that have been vacuumed is shown as heap_blks_vacuumed, and the resulting rows to be removed are shown as num_dead_tuples. During this phase, by default, VACUUM will skip blocks that are currently being pinned by other users – the DISABLE_PAGE_SKIPPING option controls that behavior. If num_dead_tuples reaches max_dead_tuples, then we move straight to the next phase, though we will return later to continue scanning:

   ```
   pid                | 34399
   datid              | 12515
   datname            | postgres
   relid              | 16422
   phase              | scanning heap
   heap_blks_total    | 32787
   heap_blks_scanned  | 25207
   heap_blks_vacuumed | 0
   index_vacuum_count | 0
   max_dead_tuples    | 9541017
   num_dead_tuples    | 537600
   ```

3. After this, we switch to the second main phase, where we start *vacuuming indexes*. We can avoid scanning the indexes altogether, so you may find that vacuuming is faster in this release. You can control whether indexes are vacuumed by setting the vacuum_cleanup_index_scale_factor parameter, which can also be set at the table level if needed, though the default value seems good.

While this phase is happening, the progress data doesn't change until it has vacuumed all of the indexes. This phase can take a long time; more indexes increase the time that is required unless you specify parallelism (more on this later). After this phase, we increment index_vacuum_count. Note that this does not refer to the number of indexes on the table, only how many times we have scanned *all* the indexes:

```
pid                | 3439
datid              | 12515
datname            | postgres
relid              | 16422
phase              | vacuuming indexes
heap_blks_total    | 32787
heap_blks_scanned  | 32787
heap_blks_vacuumed | 0
index_vacuum_count | 0
max_dead_tuples    | 9541017
num_dead_tuples    | 999966
```

4. Once the indexes have been vacuumed, we move on to the third main phase, where we return to *vacuuming heap*. In this phase, we scan through the heap, skipping any blocks that did not have dead tuples, and removing completely any old tuple item pointers.

5. If num_dead_tuples reaches the limit of max_dead_tuples, then we repeat the *"scanning heap"*, *"vacuuming indexes"*, *"vacuuming heap"* phases, until the whole table has been scanned. Each iteration will further increment index_vacuum_count. The value of max_dead_tuples is controlled by the setting of maintenance_work_mem. PostgreSQL needs 6 bytes of memory for each dead row pointer. It's a good idea to set maintenance_work_mem high enough to avoid multiple iterations since these can take lots of extra time:

```
pid                | 34399
datid              | 12515
datname            | postgres
relid              | 16422
phase              | vacuuming heap
heap_blks_total    | 32787
heap_blks_scanned  | 32787
heap_blks_vacuumed | 25051
index_vacuum_count | 1
max_dead_tuples    | 9541017
num_dead_tuples    | 999966
```

6. If the indexes were vacuumed, we then *clean up indexes*, which is a short phase where various pieces of metadata are updated.

7. If there are many empty blocks at the end of the table, VACUUM will attempt to get AccessExclusiveLock on the table. Once acquired, it will truncate the end of the table, showing a phase of *truncating heap*. Truncation does not occur every time because PostgreSQL will only attempt it if the gain is significant and if there's no conflicting lock; if it does, the truncation can often last a long time because it reads from the end of the table backward to find the truncation point. (Note that AccessExclusiveLock is passed through to physical replication standby servers and can cause replication conflicts, so you may wish to avoid it by using the TRUNCATE OFF option. You can also set the vacuum_truncate option on a table to ensure autovacuum doesn't attempt the truncation. However, there is no function to specifically request truncation of a table as an individual action.)

8. Once a table has been vacuumed, we vacuum the TOAST table by default. This behavior is controlled by the TOAST option. This isn't shown as a separate phase in the progress view; vacuuming the TOAST table will be shown as a separate vacuum.

To make VACUUM run in minimal time, maintenance_work_mem should be set to anything up to 1 GB, according to how much memory you can allocate to this task at this time. This will minimize the number of times indexes are scanned. If you avoid running vacuums, then more dead rows will be collected when it runs, which may cause an overflow of max_dead_tuples, thus causing the vacuum to take longer to run.

Using the INDEX_CLEANUP OFF option allows you to request that steps after *"scanning heap"* are skipped, which will then make VACUUM go much faster. This is not an option with autovacuum.

If your filesystem supports it, you may also be able to set maintenance_io_concurrency to tune the number of ANALYZE and VACUUM operations PostgreSQL can expect to be able to run in parallel on your system.

VACUUM can be blocked while waiting for table-level locks by other DDL statements such as a long-running ALTER TABLE or CREATE INDEX. If that happens, the lock waits are not shown in the progress view, so you may also want to look in the pg_stat_activity or pg_locks view. You can request that locked tables be skipped with the SKIP_LOCKED option.

You can request multiple options for a VACUUM command, as shown in these examples, both of which do the same thing:

```
VACUUM (DISABLE_PAGE_SKIPPING, SKIP_LOCKED, VERBOSE) my_table;
VACUUM (DISABLE_PAGE_SKIPPING ON, SKIP_LOCKED ON, VERBOSE ON, ANALYZE OFF) my_table;
```

There's more...

VACUUM doesn't run in parallel on a single table. However, if you have more than one index on a table, the index scanning phases can be conducted in parallel, if specifically requested by the user – autovacuum never does this. To use this feature, add the PARALLEL option and specify the number of workers, which will be limited to the number of indexes, the value of max_parallel_maintenance_workers, and whether we exceed min_parallel_index_scan_size.

If you want to run multiple VACUUM commands at once, you can do this by, for example, running four vacuums, each job with up to two parallel workers to scan indexes, scanning all databases:

```
$ vacuumdb --jobs=4 -parallel=2 --all
```

If you run multiple VACUUM commands at once, you'll use more memory and I/O, so be careful.

Vacuums can be slowed down by raising vacuum_cost_delay or lowering vacuum_cost_limit. Setting vacuum_cost_delay too high is counterproductive. VACUUM is your friend, not your enemy, so delaying it until it doesn't happen at all just makes things worse. Be careful.

Each vacuum sleeps when the work it has performed takes it over its limit, so the processes running VACUUM commands do not all sleep at the same time.

VACUUM commands use the value of vacuum_cost_limit as their limit.

For autovacuum workers, their limit is a share of the total autovacuum_vacuum_cost_limit, so the total amount of work that's done is the same no matter what the setting of autovacuum_max_workers.

autovacuum_max_workers should always be set to more than 2 to ensure that all the tables can begin vacuuming when they need it. Setting it too high may not be very useful, so you need to be careful.

If you need to change the settings to slow down or speed up a running process, then vacuums will pick up any new default settings when you reload the postgresql.conf file.

If you do choose to run VACUUM FULL, the progress for that is available since PostgreSQL 12 via the pg_stat_progress_cluster catalog view, which also covers the CLUSTER command. Note that you can have multiple jobs running VACUUM FULL, but you should not specify parallel workers when using FULL to avoid deadlocks.

PostgreSQL 13+ allows ANALYZE progress reporting via pg_stat_progress_analyze. ANALYZE ignores any parallel workers that have been set.

Maintaining indexes

Just as tables can become bloated, so can indexes. However, reusing space in indexes is much less effective because reclaiming an index page cannot happen if most, but not all, index keys on that page have been deleted. In the *Identifying and fixing bloated tables and indexes* recipe, you saw that non-HOT updates can cause bloated indexes. Non-primary key indexes are also prone to some bloat from normal INSERT commands, as is common in most relational databases. Indexes can become a problem in many database applications that involve a high proportion of INSERT and DELETE commands.

autovacuum does not detect bloated indexes, nor does it do anything to rebuild indexes. Therefore, we need to look at other ways to maintain indexes.

Getting ready

PostgreSQL supports commands that will rebuild indexes for you. The client utility, reindexdb, allows you to execute the REINDEX command conveniently from the operating system:

```
$ reindexdb
```

This executes the SQL REINDEX command on every table in the default database. If you want to reindex all your databases, then use the following command:

```
$ reindexdb -a
```

That's what the manual says, anyway. My experience is that many indexes don't need rebuilding, so you should probably be more selective of what you rebuild.

Also, REINDEX puts a full table lock (AccessExclusiveLock) on the table while it runs, preventing even SELECT statements against the table. You don't want to run that on your whole database!

So, I recommend that you rebuild individual indexes or all the indexes on one table at a time.

Try these steps instead:

1. First, let's create a test table with two indexes – a primary key and an additional index – as follows:

    ```
    DROP TABLE IF EXISTS test; CREATE TABLE test
    (id  INTEGER  PRIMARY  KEY
    ,category  TEXT
    ,  value   TEXT);
    CREATE  INDEX  ON  test  (category);
    ```

2. Now, let's look at the internal identifier of the tables, oid, and the current file number (relfilenodes), as follows:

```
SELECT oid, relname, relfilenode
FROM pg_class
WHERE oid in (SELECT indexrelid
              FROM pg_index
              WHERE indrelid = 'test'::regclass);
  oid  |      relname       | relfilenode
-------+--------------------+-------------
 16639 | test_pkey          |       16639
 16641 | test_category_idx  |       16641
(2 rows)
```

How to do it...

PostgreSQL supports a command known as REINDEX CONCURRENTLY, which builds an index without taking a painful AccessExclusiveLock:

```
REINDEX INDEX CONCURRENTLY test_category_idx;
```

When we check our internal identifiers again, we get the following:

```
SELECT oid, relname, relfilenode
FROM pg_class
WHERE oid in (SELECT indexrelid
              FROM pg_index
              WHERE indrelid = 'test'::regclass);
  oid  |      relname       | relfilenode
-------+--------------------+-------------
 16639 | test_pkey          |       16639
 16642 | test_category_idx  |       16642
(2 rows)
```

Here, we can see that test_category_idx is now a completely new index.

This seems pretty good, and it works on primary keys too.

If you do choose to use the reindexdb tool, make sure that you use these options to reindex one table at a time, concurrently, with some useful output:

```
$ reindexdb --concurrently -t test --verbose
INFO:  index "public.test_category_idx" was reindexed
INFO:  index "public.test_pkey" was reindexed
INFO:  index "pg_toast.pg_toast_16414_index" was reindexed
INFO:  table "public.test" was reindexed
DETAIL:  CPU: user: 0.00 s, system: 0.00 s, elapsed: 0.02 s.
```

How it works...

The REINDEX INDEX CONCURRENTLY statement allows the INSERT, UPDATE, and DELETE commands to be used while the index is being created. It cannot be executed inside another transaction, and only one index per table can be created concurrently at any time.

If you perform REINDEX TABLE CONCURRENTLY, then each index will be recreated one after the other. However, each index can be built in parallel, as discussed shortly.

REINDEX will also work on partitioned tables, from PostgreSQL 14+.

You can also now use REINDEX to change the tablespaces of indexes, as it works.

Also new since PostgreSQL 14+ is the ability to use VACUUM to ignore long-running transactions that execute REINDEX on other tables, making it even more practical to use on production database servers.

There's more...

CREATE INDEX/ REINDEX for B-tree indexes can be run in parallel. The amount of parallelism will be directly controlled by the setting of a table's parallel_workers parameter. Be careful since setting this at the table level affects all queries, not just the index build/rebuild. If the table-level parameter is not set, then the maintenance_work_mem and max_parallel_maintenance_workers parameters will determine how many workers will be used; the default is 64 MB for maintenance_work_mem and 2 MB for max_parallel_maintenance_workers. Increase both to get further gains in performance and/or concurrency. Note that these workers are shared across all users, so be careful not to over-allocate jobs; otherwise, there won't be enough workers to let everybody run in parallel.

If you are fairly new to database systems, you may think that rebuilding indexes for performance is something that only PostgreSQL needs to do. Other DBMSes require this as well – they just don't say so.

Indexes are designed for performance and, in all databases, deleting index entries causes contention and loss of performance. PostgreSQL does not remove index entries for a row when that row is deleted, so an index can be filled with dead entries. PostgreSQL attempts to remove dead entries when a block becomes full, but that doesn't stop a small number of dead entries from accumulating in many data blocks.

Finding unused indexes

Selecting the correct set of indexes for a workload is known to be a hard problem. It usually involves trial and error by developers and DBAs to get a good mix of indexes.

Tools for identifying slow queries exist and many SELECT statements can be improved by adding an index.

What many people forget is to check whether the mix of indexes remains valuable over time, which is something for the DBA to investigate and optimize.

How to do it...

PostgreSQL keeps track of each access against an index. We can view that information and use it to see whether an index is unused, as follows:

```
postgres=# SELECT schemaname, relname, indexrelname, idx_scan
FROM pg_stat_user_indexes ORDER BY idx_scan;
 schemaname |      indexrelname       | idx_scan
------------+-------------------------+----------
 public     | pgbench_accounts_bid_idx |        0
 public     | pgbench_branches_pkey    |    14575
 public     | pgbench_tellers_pkey     |    15350
 public     | pgbench_accounts_pkey    |   114400
(4 rows)
```

As shown in the preceding code, there is one unused index, alongside others that have some usage. You now need to decide whether unused means that you should remove the index. That is a more complex question, so we need to explain how it works.

How it works...

The PostgreSQL statistics accumulate various pieces of useful information. These statistics can be reset to zero using an administrator function. Also, as the data accumulates over time, we usually find that objects that have been there for longer periods have higher apparent usage.

So, if we see a low number for idx_scan, then it may be that the index was newly created (as was the case in my preceding demonstration), or that the index is only used by a part of the application that runs only at certain times of the day, week, month, and so on.

Another important consideration is that the index may be a unique constraint index that exists specifically to safeguard against duplicate INSERT commands. An INSERT operation does not show up as idx_scan, even if the index was used while checking the uniqueness of the newly inserted values, whereas UPDATE or DELETE may show up because they have to locate the row first. So, a table that only has INSERT commands against it will appear to have unused indexes.

Here is an updated version of the preceding query, which excludes unique constraint indexes:

```
SELECT schemaname
, relname
, indexrelname
, idx_scan
FROM pg_stat_user_indexes i
LEFT JOIN pg_constraint c
    ON i.indexrelid = c.conindid
WHERE c.contype IS NULL
ORDER BY idx_scan DESC;
```

Also, some indexes that show usage might be showing historical usage, and there is no further usage. Alternatively, it might be the case that some queries use an index where they could just as easily and almost as cheaply use an alternative index. Those things are for you to explore and understand before you take action. A very common approach is to regularly monitor such numbers to gain knowledge by examining their evolution over time, both on the master database and any replicated hot standby nodes. You have to consider that indexes are replicated to standby nodes and may also be in use on those nodes, or even exclusively on those standbys.

In the end, you may decide that you want to remove an index. If only there was a way to try removing an index and then put it back again quickly in case you cause problems! Rebuilding an index may take hours on a big table, so these decisions can be a little scary. No worries! Just follow the next recipe, *Carefully removing unwanted indexes*.

Carefully removing unwanted indexes

Unwanted indexes are unused or underutilized ones that are not worth the performance cost to update and maintain, so we need to consider carefully removing them. Carefully removing? Do I mean pressing *Enter* gently after typing DROP INDEX? Err, no!

The reasoning is that it takes a long time to build an index and a short time to drop it.

What we want is a way of removing an index so that if we discover that removing it was a mistake, we can put the index back again quickly.

Getting ready

The following query will list all invalid indexes, if any:

```
SELECT ir.relname AS indexname
, it.relname AS tablename
, n.nspname AS schemaname
FROM pg_index i
JOIN pg_class ir ON ir.oid = i.indexrelid
JOIN pg_class it ON it.oid = i.indrelid
JOIN pg_namespace n ON n.oid = it.relnamespace
WHERE NOT i.indisvalid;
```

Take note of these indexes so that you can tell whether a given index is invalid later because we marked it as invalid during this recipe, in which case it can safely be marked as valid, or because it was already invalid for other reasons.

How to do it...

Here, we will describe a procedure that allows us to deactivate an index without actually dropping it so that we can appreciate what its contribution was and possibly reactivate it:

1. First, create the following function:

    ```
    CREATE OR REPLACE FUNCTION trial_drop_index(iname TEXT) RETURNS VOID
    LANGUAGE SQL AS $$ UPDATE pg_index
    SET indisvalid = false
    WHERE indexrelid = $1::regclass;
    $$;
    ```

2. Then run it to perform a trial of dropping the index.
3. If you experience performance issues after dropping the index, then use the following function to undrop the index:

    ```
    CREATE OR REPLACE FUNCTION trial_undrop_index(iname TEXT)
    RETURNS VOID
    ```

```
LANGUAGE  SQL  AS
$$ UPDATE  pg_index
SET  indisvalid  =  true
WHERE  indexrelid  =  $1::regclass;
$$;
```

Be careful to avoid undropping any index that was detected by the query in the *Getting ready* section; if it wasn't marked as `invalid` when applying this recipe, then it may be unusable because it isn't valid.

How it works...

This recipe also uses some inside knowledge. When we create an index using `CREATE INDEX CONCURRENTLY`, it is a two-stage process. The first phase builds the index and then marks it as `invalid`. The `INSERT`, `UPDATE`, and `DELETE` statements now begin maintaining the index, but we perform a further pass over the table to see whether we missed anything before declaring the index valid. User queries don't use the index until it says that it is valid.

Once the index has been built and the `valid` flag has been set, if we set the flag to `invalid`, the index will still be maintained. It's just that it will not be used by queries. This allows us to turn the index off quickly, though with the option to turn it on again if we realize that we do need the index after all. This makes it practical to test whether dropping the index will alter the performance of any of your most important queries.

Planning maintenance

Monitoring systems are not a substitute for good planning. They alert you to unplanned situations that need attention. The more unplanned things you respond to, the greater the chance that you will need to respond to multiple emergencies at once. And when that happens, something will break. Ultimately, that is your fault. If you wish to take your responsibilities seriously, you should plan for this.

How to do it...

This recipe is all about planning, so we'll provide discussion points rather than portions of code. We'll cover the main points that should be addressed and provide a list of points as food for thought, around which the actual implementation should be built:

- **Let's break a rule**: If you don't have a backup, make one now. I mean now – go on, off you go! Then, let's talk some more about planning maintenance. If you already have, well done! It's hard to keep your job as a DBA if you lose data because of missing backups, especially today when everybody's grandmother knows to keep their photos backed up.

- **First, plan your time**: Decide on a regular date to perform certain actions. Don't allow yourself to be a puppet of your monitoring system, running up and down every time the lights change. If you keep getting dragged off on other assignments, then you must understand that you need to get a good handle on the database maintenance to make sure that it doesn't bite you.
- **Don't be scared**: It's easy to worry about what you don't know and either overreact or underreact. Your database probably doesn't need to be inspected daily, but it's never a bad practice.

How it works...

Build a regular cycle of activity around the following tasks:

- **Capacity planning**: Observe long-term trends in system performance and keep track of the growth of database volumes. Plan to schedule any new data feeds and new projects that increase the rates of change. This is best done monthly so that you can monitor what has happened and what will happen.
- **Backups, recovery testing, and emergency planning**: Organize regular reviews of written plans and test scripts. Check the tape rotation, confirm that you still have the password to the off-site backups, and so on. Some sysadmins run a test recovery every night so that they always know that successful recovery is possible.
- **Vacuum and index maintenance**: Do this to reduce bloat, as well as to collect optimizer statistics through the `ANALYZE` command. Also, regularly check index usage, drop unused indexes, and reindex concurrently as needed. Consider `VACUUM` again, with the need to manage the less frequent `freezing` process. This is listed as a separate task so that you don't ignore this and let it bite you later!
- **Server log file analysis**: How many times has the server restarted? Are you sure you know about each incident?
- **Security and intrusion detection**: Has your database already been hacked? What did they do?
- **Understanding usage patterns**: If you don't know much about what your database is used for, then I'll wager it is not very well tuned or maintained. The *Tracking important metrics over time* recipe in *Chapter 8, Monitoring and Diagnosis*, can give you valuable insights.

- **Long-term performance analysis**: It's a common occurrence for me to get asked to come and tune a slow system. Often, what happens is that a database server gets slower over a very long period. Nobody ever noticed any particular day when it got slow – it just got slower over time. Keeping records of response times over time can help you confirm whether everything is as good now as it was months or years earlier. This activity is where you may reconsider current index choices.

Many of these activities are mentioned in this chapter or throughout the rest of this cookbook. Some are not because they aren't very technical and are more about planning and understanding your environment.

There's more...

You may also find time to consider the following:

- **Data quality**: Is the content of the database accurate and meaningful? Could the data be enhanced?
- **Business intelligence**: Is the data being used for everything that can bring value to the organization?

Learn more on Discord

To join the Discord community for this book – where you can share feedback, ask questions to the author, and learn about new releases – follow the QR code below:

https://discord.gg/pQkghgmgdG

10
Performance and Concurrency

Performance and concurrency are two problems that are often tightly coupled—when concurrency problems are encountered, performance usually degrades, in some cases by a lot. If you take care of concurrency problems, you will achieve better performance.

In this chapter, you will see how to find slow queries and how to find queries that make other queries slow.

Performance tuning, unfortunately, is still not an exact science, so you may also encounter a performance problem that's not covered by any of the given methods.

We will also see how to get help in the final recipe, *Reporting performance problems*, if none of the other recipes that are covered here work.

In this chapter, we will cover the following recipes:

- Finding slow SQL statements
- Finding out what makes SQL slow
- Reducing the number of rows returned
- Simplifying complex SQL queries
- Speeding up queries without rewriting them
- Discovering why a query is not using an index
- Forcing a query to use an index
- Using parallel query
- Using Just-In-Time (JIT) compilation
- Creating time-series tables using partitioning
- Using optimistic locking to avoid long lock waits

- Reporting performance problems

Finding slow SQL statements

Two main kinds of slowness can manifest themselves in a database.

The first kind is a single query that can be too slow to be really usable, such as a customer information query in a **customer relationship management** (**CRM**) system running for minutes, a password check query running in tens of seconds, or a daily data aggregation query running for more than a day. These can be found by logging queries that take over a certain amount of time, either at the client end or in the database.

The second kind is a query that is run frequently (say, a few thousand times a second) and used to run in single-digit **milliseconds** (**ms**) but is now running in several tens or even hundreds of milliseconds, hence slowing the system down.

Here, we will show you several ways to find statements that are either slow or cause the database as a whole to slow down (although they are not slow by themselves).

Getting ready

Connect to the database as the user whose statements you want to investigate or as a superuser to investigate all users' queries:

1. Check that you have the pg_stat_statements extension installed:

    ```
    postgres=# \x
    postgres=# \dx pg_stat_statements
    ```

2. Here is a list of our installed extensions:

    ```
    -[ RECORD 1 ]-------------------------------------------------------
    Name        | pg_stat_statements
    Version     | 1.10
    Schema      | public
    Description | track execution statistics of all SQL statements
    executed
    ```

3. If you can't see them, then issue the following command:

    ```
    postgres=# CREATE EXTENSION pg_stat_statements;
    postgres=# ALTER SYSTEM
               SET shared_preload_libraries = 'pg_stat_statements';
    ```

4. Then, restart the server, or refer to the *Using an installed module/extension* and *Managing installed extensions* recipes in *Chapter 3, Server Configuration*, for more details.

How to do it...

Run this query to look at the top 10 highest workloads on your server side:

```
SELECT calls, total_exec_time, query
FROM pg_stat_statements
ORDER BY total_exec_time DESC LIMIT 10;
```

The output is ordered by total_exec_time, so it doesn't matter whether it was a single query or thousands of smaller queries.

Many additional columns are useful in tracking down further information about particular entries:

```
postgres=# \d pg_stat_statements
          View "public.pg_stat_statements"
        Column         |       Type       | Modifiers
-----------------------+------------------+-----------
 userid                | oid              |
 dbid                  | oid              |
 toplevel              | bool             |
Unique identifier for SQL
 queryid               | bigint           |
The SQL being executed
 query                 | text             |
Number of times planned and timings
 plans                 | bigint           |
 total_plan_time       | double precision |
 min_plan_time         | double precision |
 max_plan_time         | double precision |
 mean_plan_time        | double precision |
 stddev_plan_time      | double precision |
Number of times executed and timings
 calls                 | bigint           |
 total_exec_time       | double precision |
 min_exec_time         | double precision |
 max_exec_time         | double precision |
 mean_exec_time        | double precision |
```

```
 stddev_exec_time      | double precision |
Number of rows returned by query
 rows                  | bigint           |
Columns related to tables that all users can access
 shared_blks_hit       | bigint           |
 shared_blks_read      | bigint           |
 shared_blks_dirtied   | bigint           |
 shared_blks_written   | bigint           |
Columns related to session-specific temporary tables
 local_blks_hit        | bigint           |
 local_blks_read       | bigint           |
 local_blks_dirtied    | bigint           |
 local_blks_written    | bigint           |
Columns related to temporary files
 temp_blks_read        | bigint           |
 temp_blks_written     | bigint           |
I/O timing
 blk_read_time         | double precision |
 blk_write_time        | double precision |
Columns related to WAL usage
 wal_records           | bigint           |
 wal_fpi               | bigint           |
 wal_bytes             | numeric          |
JIT compile/generation and timing
  jit_functions         | bigint           |        |        |
  jit_generation_time   | double precision |        |        |
JIT inlining and time
  jit_inlining_count    | bigint           |        |        |
  jit_inlining_time     | double precision |        |        |
JIT optimization and time
  jit_optimization_count | bigint          |        |        |
  jit_optimization_time  | double precision |        |        |
JIT statement emission and time.
  jit_emission_count    | bigint           |        |        |
  jit_emission_time     | double precision |        |        |
```

How it works...

pg_stat_statements collects data on all running queries by accumulating data in memory, with low overheads.

Similar SQL statements are normalized so that the constants and parameters that are used for execution are removed. This allows you to see all similar SQL statements in one line of the report, rather than seeing thousands of lines, which would be fairly useless. While useful, it can sometimes mean that it's hard to work out which parameter values are actually causing the problem.

There's more...

Another way to find slow queries is to set up PostgreSQL to log them to the server log. For example, if you decide to monitor any query that takes over 10 seconds, then use the following command:

```
postgres=# ALTER SYSTEM
          SET log_min_duration_statement = 10000;
```

Remember that the duration is in ms. After doing this, reload PostgreSQL. All queries whose duration exceeds the threshold will be logged. You should pick a threshold that is above 99% of queries so that you only get the worst outliers logged. As you progressively tune your system, you can reduce the threshold over time.

PostgreSQL log files are usually located together with other log files; for example, on Debian/Ubuntu Linux, they are in the /var/log/postgresql/ directory.

You can also find the location of the log file by using the following command:

```
postgres=# SELECT pg_current_logfile();
  pg_current_logfile
------------------------
log/postgresql-Fri.log
(1 row)
```

The above output shows that the log file name is postgresql-Fri.log and the directory is log. If the directory name doesn't start with /, then it indicates that it's a relative directory located inside PGDATA.

If you set log_min_duration_statement = 0, then all queries would be logged, which will typically swamp the log file, causing more performance problems itself, and thus this is not recommended. A better idea would be to use the log_min_duration_sample parameter, available in PostgreSQL 13+, to set a limit for sampling queries. The two settings are designed to work together:

- Any query elapsed time less than log_min_duration_sample is not logged at all.
- Any query elapsed time higher than log_min_duration_statement is always logged.
- For any query elapsed time that falls between the two settings, we sample the queries and log them at a rate set by log_statement_sample_rate (default 1.0 = all). Note that the sampling is blind—it is not stratified/weighted, so rare queries may not show up at all in the log.

Query logging will show the parameters that are being used for the slow query, even when pg_stat_statements does not.

Finding out what makes SQL slow

An SQL statement can be slow for a lot of reasons. Here, we will provide a short list of these reasons, with at least one way of recognizing each.

Getting ready

If the SQL statement is still running, look at *Chapter 8*, *Monitoring and Diagnosis*.

How to do it...

The core issues are likely to be the following:

- You're asking the SQL statement to do too much work.
- Something is stopping the SQL statement from doing the work.

This might not sound that helpful at first, but it's good to know that there's nothing really magical going on that you can't understand if you look.

In more detail, the main reasons/issues are these:

- Returning too much data.
- Processing too much data.
- Index needed.
- The wrong plan for other reasons—for example, poor estimates.
- Locking problems.

- Cache or **input/output (I/O)** problems. It's possible the system itself has bottlenecks such as single-core, slow **central processing units (CPUs)**, insufficient memory, or reduced I/O throughput. Those issues may be outside the scope of this book—here, we discuss just the database issues.

The first issue can be handled as described in the *Reducing the number of rows returned* recipe. The rest of the preceding reasons can be investigated from two perspectives: the SQL itself and the objects that the SQL touches. Let's start by looking at the SQL itself by running the query with EXPLAIN ANALYZE. We're going to use the optional form, as follows:

```
postgres=# EXPLAIN (ANALYZE, BUFFERS) ...SQL...
```

The EXPLAIN command provides output to describe the execution plan of the SQL, showing access paths and costs (in abstract units). The ANALYZE option causes the statement to be executed (be careful), with instrumentation to show the number of rows accessed and the timings for that part of the plan. The BUFFERS option provides information about the number of database buffers read and the number of buffers that were hit in the cache. Taken together, we have everything we need to diagnose whether the SQL performance is reduced by one of the earlier-mentioned issues:

```
postgres=# EXPLAIN (ANALYZE, BUFFERS) SELECT count(*) FROM t;
                                         QUERY PLAN
------------------------------------------------------------------------
 Finalize Aggregate  (cost=11430.22..11430.23 rows=1 width=8) (actual time=401.669..404.667 rows=1 loops=1)
   Buffers: shared hit=2144 read=2336
   ->  Gather  (cost=11430.00..11430.21 rows=2 width=8) (actual time=401.656..404.658 rows=3 loops=1)
         Workers Planned: 2
         Workers Launched: 2
         Buffers: shared hit=2144 read=2336
         ->  Partial Aggregate  (cost=10430.00..10430.01 rows=1 width=8) (actual time=390.315..390.317 rows=1 loops=3)
               Buffers: shared hit=2144 read=2336
               ->  Parallel Seq Scan on t  (cost=0.00..9240.00 rows=476000 width=0) (actual time=0.033..199.622 rows=333333 loops=3)
                     Buffers: shared hit=2144 read=2336
 Planning Time: 0.047 ms
 Execution Time: 404.693 ms
(12 rows)
```

Let's use this technique to look at an SQL statement that would benefit from an index.

For example, if you want to get the three latest rows in a 1 million rows table, run the following query:

```
SELECT * FROM events ORDER BY id DESC LIMIT 3;
```

You can either read through just three rows using an index on the id SERIAL column or you can perform a sequential scan of all rows followed by a sort, as shown in the following code snippet. Your choice depends on whether you have a usable index on the field from which you want to get the top three rows:

```
postgres=# EXPLAIN (ANALYZE)
           SELECT * FROM events ORDER BY id DESC LIMIT 3;

                                                                   QUERY PLAN
-------------------------------------------------------------------------------
-------------------------------------------------------------
 Limit  (cost=16203.26..16203.61 rows=3 width=4) (actual time=420.839..424.983 rows=3 loops=1)
   ->  Gather Merge  (cost=16203.26..125913.99 rows=940312 width=4) (actual time=420.836..424.977 rows=3 loops=1)
         Workers Planned: 2
         Workers Launched: 2
         ->  Sort  (cost=15203.24..16378.63 rows=470156 width=4) (actual time=405.781..405.784 rows=2 loops=3)
               Sort Key: id DESC
               Sort Method: top-N heapsort  Memory: 25kB
               Worker 0:  Sort Method: top-N heapsort  Memory: 25kB
               Worker 1:  Sort Method: top-N heapsort  Memory: 25kB
               ->  Parallel Seq Scan on events  (cost=0.00..9126.56 rows=470156 width=4) (actual time=0.020..207.075 rows=333333 loops=3)
 Planning Time: 0.110 ms
 Execution Time: 425.013 ms
(12 rows)
postgres=# CREATE INDEX events_id_ndx ON events(id);
CREATE INDEX
postgres=# EXPLAIN (ANALYZE)
           SELECT * FROM events ORDER BY id DESC LIMIT 3;
                                                                   QUERY
```

```
 PLAN
-----------------------------------------------------------------------------
-----------------------------------------------------------------
 Limit  (cost=0.42..0.50 rows=3 width=4) (actual time=0.046..0.052 rows=3
loops=1)
   ->  Index Only Scan Backward using events_id_ndx on events
(cost=0.42..25980.42 rows=1000000 width=4) (actual time=0.044..0.046
rows=3 loops=1)
         Heap Fetches: 0
 Planning Time: 0.170 ms
 Execution Time: 0.067 ms
(5 rows)
```

This produces a huge difference in query runtime, even when all of the data is in the cache.

If you run the same analysis using EXPLAIN (ANALYZE, BUFFERS) on your production system, you'll be able to see the cache effects as well. Databases work well if the "active set" of data blocks in a database can be cached in **random-access memory (RAM)**. The active set, also known as the working set, is a subset of the data that is accessed by queries on a regular basis. Each new index you add will increase the pressure on the cache, so it is possible to have too many indexes.

You can also look at the statistics for objects touched by queries, as mentioned in the *Knowing whether anybody is using a specific table* recipe from *Chapter 8, Monitoring and Diagnosis*. In pg_stat_user_tables, the fast growth of seq_tup_read means that there are lots of sequential scans occurring. The ratio of seq_tup_read to seq_scan shows how many tuples each seqscan reads. Similarly, the idx_scan and idx_tup_fetch columns show whether indexes are being used and how effective they are.

There's more...

If not enough of the data fits in the shared buffers, lots of rereading of the same data happens, causing performance issues. In pg_statio_user_tables, watch the heap_blks_hit and heap_blks_read fields, or the equivalent ones for index and toast relations. They give you a fairly good idea of how much of your data is found in PostgreSQL's shared buffers (heap_blks_hit) and how much had to be fetched from the disk (heap_blks_read). If you see large numbers of blocks being read from the disk continuously, you may want to tune those queries; if you determine that the disk reads were justified, you can make the configured shared_buffers value bigger.

If your shared_buffers parameter is tuned properly and you can't rewrite the query to perform less block I/O, you might need a bigger server.

You can find a lot of resources on the web that explain how shared buffers work and how to set them based on your available hardware and your expected data access patterns. Our professional advice is to always test your database servers and perform benchmarks before you deploy them in production. Information on the shared_buffers configuration parameter can be found at http://www.postgresql.org/docs/current/static/runtime-config-resource.html.

Locking problems

Thanks to its **multi-version concurrency control** (**MVCC**) design, PostgreSQL does not suffer from most locking problems, such as writers locking out readers or readers locking out writers, but it still has to take locks when more than one process wants to update the same row. Also, it has to hold the write lock until the current writer's transaction finishes.

So, if you have a database design where many queries update the same record, you can have a locking problem. Running **Data Definition Language** (**DDL**) will also require stronger locks that may interrupt applications.

Refer to the *Knowing who is blocking a query* recipe of *Chapter 8*, *Monitoring and Diagnosis*, for more detailed information.

To diagnose locking problems retrospectively, use the log_lock_waits parameter to generate log output for locks that are held for a long time.

EXPLAIN options

Use the FORMAT option to retrieve the output of EXPLAIN in a different format, such as **JavaScript Object Notation** (**JSON**), **Extensible Markup Language** (**XML**), and **YAML Ain't Markup Language** (**YAML**). This could allow us to write programs to manipulate the outputs.

The following command is an example of this:

```
EXPLAIN (ANALYZE, BUFFERS, FORMAT JSON) SELECT count(*) FROM t;
```

Not enough CPU power or disk I/O capacity for the current load

These issues are usually caused by suboptimal query plans but, sometimes, your computer is just not powerful enough.

In this case, top is your friend. For quick checks, run the following code from the command line:

```
user@host:~$ top
```

First, watch the percentage of idle CPU from top. If this is in low single digits most of the time, you probably have problems with the CPU's power.

If you have a high load average with a lot of CPU idle left, you are probably out of disk bandwidth. In this case, you should also have lots of Postgres processes in the D status, meaning that the process is in an uninterruptible state (usually waiting for I/O).

See also

For further information on the syntax of the EXPLAIN SQL command, refer to the PostgreSQL documentation at http://www.postgresql.org/docs/current/static/sql-explain.html.

Reducing the number of rows returned

Although the problem often produces too many rows in the first place, it is made worse by returning all unnecessary rows to the client. This is especially true if the client and server are not on the same host.

Here are some ways to reduce the traffic between the client and server.

How to do it...

Consider the following scenario: a full-text search returns 10,000 documents, but only the first 20 are displayed to users. In this case, order the documents by rank on the server, and return only the top 20 that actually need to be displayed:

```
SELECT title, ts_rank_cd(body_tsv, query, 20) AS text_rank
FROM articles, plainto_tsquery('spicy potatoes') AS query
WHERE body_tsv @@ query
ORDER BY rank DESC
LIMIT 20
;
```

The ORDER BY clause ensures the rows are ranked, and then the LIMIT 20 returns only the top 20.

If you need the next 20 documents, don't just query with a limit of 40 and throw away the first 20. Instead, use OFFSET 20 LIMIT 20 to return the next 20 documents.

The SQL optimizer understands the LIMIT clause and will change the execution plan accordingly.

To gain some stability so that documents with the same rank still come out in the same order when using OFFSET 20, add a unique field (such as the id column of the articles table) to ORDER BY in both queries:

```
SELECT title, ts_rank_cd(body_tsv, query, 20) AS text_rank
FROM articles, plainto_tsquery('spicy potatoes') AS query
```

```
WHERE body_tsv @@ query
ORDER BY rank DESC, articles.id
OFFSET 20 LIMIT 20;
```

Another use case is an application that requests all products of a branch office so that it can run a complex calculation over them. In such a case, try to do as much data analysis as possible inside the database.

There is no need to run the following:

```
SELECT * FROM accounts WHERE branch_id = 7;
```

Also, instead of counting and summing the rows on the client side, you can run this:

```
SELECT count(*), sum(balance) FROM accounts WHERE branch_id = 7;
```

With some research on SQL, you can carry out an amazingly large portion of your computation using plain SQL (for example, do not underestimate the power of window functions).

If SQL is not enough, you can use **Procedural Language/PostgreSQL** (**PL/pgSQL**) or any other embedded procedural language supported by PostgreSQL for even more flexibility.

There's more...

Consider one more scenario: an application runs a huge number of small lookup queries. This can easily happen with modern **object-relational mappers** (**ORMs**) and other toolkits that do a lot of work for the programmer but, at the same time, hide a lot of what is happening.

For example, if you define a **HyperText Markup Language** (**HTML**) report over a query in a templating language and then define a lookup function to resolve an **identifier** (**ID**) inside the template, you may end up with a form that performs a separate, small lookup for each row displayed, even when most of the values looked up are the same. This doesn't usually pose a big problem for the database, as queries of the `SELECT name FROM departments WHERE id = 7` form are really fast when the row for `id = 7` is in shared buffers. However, repeating this query thousands of times still takes seconds due to network latency, process scheduling for each request, and other factors.

The two proposed solutions are as follows:

- Make sure that the value is cached by your ORM
- Perform the lookup inside the query that gets the main data so that it can be displayed directly

Exactly how to carry out these solutions depends on the toolkit, but they are both worth investigating as they really can make a difference in speed and resource usage.

PostgreSQL 9.5 introduced the TABLESAMPLE clause into SQL. This allows you to run commands much faster by using a sample of a table's rows, this, of course, only gives an approximate answer. In certain cases though, this can be just as useful as the most accurate answer:

```
postgres=# SELECT avg(id) FROM events;
         avg
--------------------
 500000.500
(1 row)
postgres=# SELECT avg(id) FROM events TABLESAMPLE system(1);
         avg
--------------------
 507434.635
(1 row)
postgres=# EXPLAIN (ANALYZE, BUFFERS) SELECT avg(id) FROM events;
                            QUERY PLAN
-------------------------------------------------------------------
 Aggregate  (cost=16925.00..16925.01 rows=1 width=32) (actual time=204.841..204.841 rows=1 loops=1)
   Buffers: shared hit=96 read=4329
   ->  Seq Scan on events  (cost=0.00..14425.00 rows=1000000 width=4) (actual time=1.272..105.452 rows=1000000 loops=1)
         Buffers: shared hit=96 read=4329
 Planning time: 0.059 ms
 Execution time: 204.912 ms
(6 rows)
postgres=# EXPLAIN (ANALYZE, BUFFERS)
           SELECT avg(id) FROM events TABLESAMPLE system(1);
                            QUERY PLAN
-------------------------------------------------------------------
 Aggregate  (cost=301.00..301.01 rows=1 width=32) (actual time=4.627..4.627 rows=1 loops=1)
   Buffers: shared hit=1 read=46
   ->  Sample Scan on events  (cost=0.00..276.00 rows=10000 width=4) (actual time=0.074..2.833 rows=10622 loops=1)
         Sampling: system ('1'::real)
```

```
              Buffers: shared hit=1 read=46
 Planning time: 0.066 ms
 Execution time: 4.702 ms
 (7 rows)
```

Simplifying complex SQL queries

There are two types of complexity that you can encounter in SQL queries.

First, the complexity can be directly visible in the query if it has hundreds—or even thousands—of rows of SQL code in a single query. This can cause both maintenance headaches and slow execution.

This complexity can also be hidden in subviews, so the SQL code of the query may seem simple but it uses other views and/or functions to do part of the work, which can, in turn, use others. This is much better for maintenance, but it can still cause performance problems.

Both types of queries can either be written manually by programmers or data analysts or emerge as a result of a query generator.

Getting ready

First, verify that you really have a complex query.

A query that simply returns lots of database fields is not complex in itself. In order to be complex, the query has to join lots of tables in complex ways.

The easiest way to find out whether a query is complex is to look at the output of EXPLAIN. If it has lots of rows, the query is complex, and it's not just that there is a lot of text that makes it so.

All of the examples in this recipe have been written with a very typical use case in mind: sales.

What follows is a description of a fictitious model that's used in this recipe. The most important fact is the sale event, stored in the sale table (I specifically used the word *fact*, as this is the right term to use in a *data warehousing* context). Every sale takes place at a point of sale (the salespoint table) at a specific time and involves an item. That item is stored in a warehouse (see the item and warehouse tables, as well as the item_in_wh link table).

Both warehouse and salespoint are located in a geographical area (the location table). This is important, for example, to study the provenance of a transaction.

Here is a simplified **entity-relationship model (ERM)**, which is useful for understanding all of the joins that occur in the following queries:

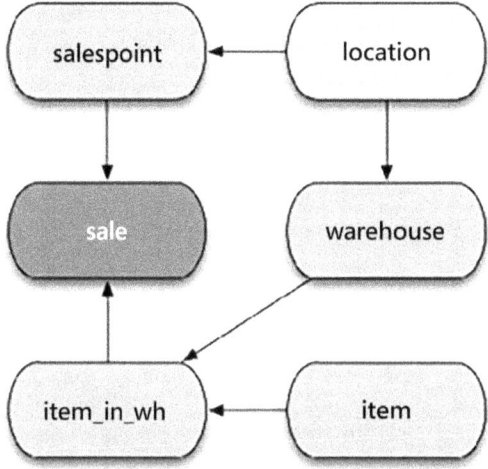

Figure 10.1: Data model for the example code

How to do it...

Simplifying a query usually means restructuring it so that parts of it can be defined separately and then used by other parts.

We'll illustrate these possibilities by rewriting the following query in several ways.

The complex query in our example case is a so-called **pivot** or **cross-tab** query. This query retrieves the quarterly profit for non-local sales from all shops, as shown in the following code snippet:

```
SELECT shop.sp_name AS shop_name,
       q1_nloc_profit.profit AS q1_profit,
       q2_nloc_profit.profit AS q2_profit,
       q3_nloc_profit.profit AS q3_profit,
       q4_nloc_profit.profit AS q4_profit,
       year_nloc_profit.profit AS year_profit
  FROM (SELECT * FROM salespoint ORDER BY sp_name) AS shop
  LEFT JOIN (
    SELECT
      spoint_id,
      sum(sale_price) - sum(cost) AS profit,
      count(*) AS nr_of_sales
    FROM sale s
    JOIN item_in_wh iw ON s.item_in_wh_id=iw.id
    JOIN item i ON iw.item_id = i.id
```

```
                JOIN salespoint sp ON s.spoint_id = sp.id
                JOIN location sploc ON sp.loc_id = sploc.id
                JOIN warehouse wh ON iw.whouse_id = wh.id
                JOIN location whloc ON wh.loc_id = whloc.id
            WHERE sale_time >= '2013-01-01'
                AND sale_time < '2013-04-01'
                AND sploc.id != whloc.id
            GROUP BY 1
        ) AS q1_nloc_profit
        ON shop.id = Q1_NLOC_PROFIT.spoint_id
    LEFT JOIN (
< similar subquery for 2nd quarter >
        ) AS q2_nloc_profit
        ON shop.id = q2_nloc_profit.spoint_id
    LEFT JOIN (
< similar subquery for 3rd quarter >
        ) AS q3_nloc_profit
        ON shop.id = q3_nloc_profit.spoint_id
    LEFT JOIN (
< similar subquery for 4th quarter >
        ) AS q4_nloc_profit
        ON shop.id = q4_nloc_profit.spoint_id
    LEFT JOIN (
< similar subquery for full year >
        ) AS year_nloc_profit
        ON shop.id = year_nloc_profit.spoint_id
ORDER BY 1;
```

Since the preceding query has an almost identical repeating part for finding the sales for a period (the four quarters of 2013, in this case), it makes sense to move it to a separate view (for the whole year) and then use that view in the main reporting query, as follows:

```
CREATE VIEW non_local_quarterly_profit_2013 AS
    SELECT
        spoint_id,
        extract('quarter' from sale_time) as sale_quarter,
        sum(sale_price) - sum(cost) AS profit,
        count(*) AS nr_of_sales
```

```
            FROM sale s
            JOIN item_in_wh iw ON s.item_in_wh_id=iw.id
            JOIN item i ON iw.item_id = i.id
            JOIN salespoint sp ON s.spoint_id = sp.id
            JOIN location sploc ON sp.loc_id = sploc.id
            JOIN warehouse wh ON iw.whouse_id = wh.id
            JOIN location whloc ON wh.loc_id = whloc.id
        WHERE sale_time >= '2013-01-01'
          AND sale_time < '2014-01-01'
          AND sploc.id != whloc.id
        GROUP BY 1,2;
SELECT shop.sp_name AS shop_name,
       q1_nloc_profit.profit as q1_profit,
       q2_nloc_profit.profit as q2_profit,
       q3_nloc_profit.profit as q3_profit,
       q4_nloc_profit.profit as q4_profit,
       year_nloc_profit.profit as year_profit
  FROM (SELECT * FROM salespoint ORDER BY sp_name) AS shop
  LEFT JOIN non_local_quarterly_profit_2013 AS q1_nloc_profit
      ON shop.id = Q1_NLOC_PROFIT.spoint_id
   AND q1_nloc_profit.sale_quarter = 1
  LEFT JOIN non_local_quarterly_profit_2013 AS q2_nloc_profit
      ON shop.id = Q2_NLOC_PROFIT.spoint_id
   AND q2_nloc_profit.sale_quarter = 2
  LEFT JOIN non_local_quarterly_profit_2013 AS q3_nloc_profit
      ON shop.id = Q3_NLOC_PROFIT.spoint_id
   AND q3_nloc_profit.sale_quarter = 3
  LEFT JOIN non_local_quarterly_profit_2013 AS q4_nloc_profit
      ON shop.id = Q4_NLOC_PROFIT.spoint_id
   AND q4_nloc_profit.sale_quarter = 4
  LEFT JOIN (
         SELECT spoint_id, sum(profit) AS profit
           FROM non_local_quarterly_profit_2013 GROUP BY 1
       ) AS year_nloc_profit
      ON shop.id = year_nloc_profit.spoint_id
 ORDER BY 1;
```

Moving the subquery to a view has not only made the query shorter but also easier to understand and maintain.

You might want to consider **materialized views**—more on this later.

Before that, we will be using common table expressions (also known as WITH queries) instead of a separate view. Starting with PostgreSQL version 8.4, you can use a WITH statement to define a view in line, as follows:

```
WITH nlqp AS (
      SELECT
          spoint_id,
          extract('quarter' from sale_time) as sale_quarter,
          sum(sale_price) - sum(cost) AS profit,
          count(*) AS nr_of_sales
        FROM sale s
        JOIN item_in_wh iw ON s.item_in_wh_id=iw.id
        JOIN item i ON iw.item_id = i.id
        JOIN salespoint sp ON s.spoint_id = sp.id
        JOIN location sploc ON sp.loc_id = sploc.id
        JOIN warehouse wh ON iw.whouse_id = wh.id
        JOIN location whloc ON wh.loc_id = whloc.id
       WHERE sale_time >= '2013-01-01'
         AND sale_time < '2014-01-01'
         AND sploc.id != whloc.id
       GROUP BY 1,2
)
SELECT shop.sp_name AS shop_name,
       q1_nloc_profit.profit as q1_profit,
       q2_nloc_profit.profit as q2_profit,
       q3_nloc_profit.profit as q3_profit,
       q4_nloc_profit.profit as q4_profit,
       year_nloc_profit.profit as year_profit
  FROM (SELECT * FROM salespoint ORDER BY sp_name) AS shop
  LEFT JOIN nlqp AS q1_nloc_profit
       ON shop.id = Q1_NLOC_PROFIT.spoint_id
      AND q1_nloc_profit.sale_quarter = 1
  LEFT JOIN nlqp AS q2_nloc_profit
       ON shop.id = Q2_NLOC_PROFIT.spoint_id
```

```
        AND q2_nloc_profit.sale_quarter = 2
    LEFT JOIN nlqp AS q3_nloc_profit
            ON shop.id = Q3_NLOC_PROFIT.spoint_id
        AND q3_nloc_profit.sale_quarter = 3
    LEFT JOIN nlqp AS q4_nloc_profit
            ON shop.id = Q4_NLOC_PROFIT.spoint_id
        AND q4_nloc_profit.sale_quarter = 4
    LEFT JOIN (
            SELECT spoint_id, sum(profit) AS profit
              FROM nlqp GROUP BY 1
        ) AS year_nloc_profit
            ON shop.id = year_nloc_profit.spoint_id
ORDER BY 1;
```

For more information on WITH queries (also known as **Common Table Expressions (CTEs)**), read the official documentation at http://www.postgresql.org/docs/current/static/queries-with.html.

There's more...

Another ace in the hole is represented by temporary tables that are used for parts of a query. By default, a temporary table is dropped at the end of a Postgres session, but the behavior can be changed at the time of creation.

PostgreSQL itself can choose to materialize parts of a query during the query optimization phase but, sometimes, it fails to make the best choice for the query plan, either due to insufficient statistics or because—as can happen for large query plans where **Genetic Query Optimization (GEQO)** is used—it may have just overlooked some possible query plans.

If you think that materializing (separately preparing) some parts of a query is a good idea, you can do this by using a temporary table, simply by running CREATE TEMPORARY TABLE my_temptable01 AS <the part of the query you want to materialize> and then using my_temptable01 in the main query, instead of the materialized part.

You can even create indexes on a temporary table for PostgreSQL to use in the main query:

```
BEGIN;
CREATE TEMPORARY TABLE nlqp_temp ON COMMIT DROP
AS
        SELECT
```

```
            spoint_id,
            extract('quarter' from sale_time) as sale_quarter,
            sum(sale_price) - sum(cost) AS profit,
            count(*) AS nr_of_sales
         FROM sale s
         JOIN item_in_wh iw ON s.item_in_wh_id=iw.id
         JOIN item i ON iw.item_id = i.id
         JOIN salespoint sp ON s.spoint_id = sp.id
         JOIN location sploc ON sp.loc_id = sploc.id
         JOIN warehouse wh ON iw.whouse_id = wh.id
         JOIN location whloc ON wh.loc_id = whloc.id
         WHERE sale_time >= '2013-01-01'
           AND sale_time <  '2014-01-01'
           AND sploc.id != whloc.id
         GROUP BY 1,2
;
```

You can create indexes on a table and analyze the temporary table here:

```
SELECT shop.sp_name AS shop_name,
       q1_NLP.profit as q1_profit,
       q2_NLP.profit as q2_profit,
       q3_NLP.profit as q3_profit,
       q4_NLP.profit as q4_profit,
       year_NLP.profit as year_profit
  FROM (SELECT * FROM salespoint ORDER BY sp_name) AS shop
  LEFT JOIN nlqp_temp AS q1_NLP
      ON shop.id = Q1_NLP.spoint_id AND q1_NLP.sale_quarter = 1
  LEFT JOIN nlqp_temp AS q2_NLP
      ON shop.id = Q2_NLP.spoint_id AND q2_NLP.sale_quarter = 2
  LEFT JOIN nlqp_temp AS q3_NLP
      ON shop.id = Q3_NLP.spoint_id AND q3_NLP.sale_quarter = 3
  LEFT JOIN nlqp_temp AS q4_NLP
      ON shop.id = Q4_NLP.spoint_id AND q4_NLP.sale_quarter = 4
  LEFT JOIN (
         select spoint_id, sum(profit) AS profit FROM nlqp_temp GROUP BY 1
       ) AS year_NLP
      ON shop.id = year_NLP.spoint_id
```

```
    ORDER BY 1
    ;
    COMMIT;  -- here the temp table goes away
```

Using materialized views

If the part you put in the temporary table is large, does not change very often, and/or is hard to compute, then you may be able to do it less often for each query by using a technique named **materialized views**.

Materialized views are views that are prepared before they are used (similar to a cached table). They are either fully regenerated as underlying data changes or, in some cases, can update only those rows that depend on the changed data.

PostgreSQL natively supports materialized views through the CREATE MATERIALIZED VIEW, ALTER MATERIALIZED VIEW, REFRESH MATERIALIZED VIEW, and DROP MATERIALIZED VIEW commands. At the time of writing, PostgreSQL only supports full regeneration of materialized tables using REFRESH MATERIALIZED VIEW CONCURRENTLY, though this uses a parallel query to execute very quickly.

A fundamental aspect of materialized views is that they can have their own indexes, as with any other table. See http://www.postgresql.org/docs/current/static/sql-creatematerializedview.html for more information on creating materialized views.

For instance, you can rewrite the example in the previous recipe using a materialized view instead of a temporary table:

```
    CREATE MATERIALIZED VIEW nlqp_temp AS
        SELECT spoint_id,
            extract('quarter' from sale_time) as sale_quarter,
            sum(sale_price) - sum(cost) AS profit,
            count(*) AS nr_of_sales
        FROM sale s
        JOIN item_in_wh iw ON s.item_in_wh_id=iw.id
        JOIN item i ON iw.item_id = i.id
        JOIN salespoint sp ON s.spoint_id = sp.id
        JOIN location sploc ON sp.loc_id = sploc.id
        JOIN warehouse wh ON iw.whouse_id = wh.id
        JOIN location whloc ON wh.loc_id = whloc.id
            WHERE sale_time >= '2013-01-01'
```

```
            AND sale_time < '2014-01-01'
            AND sploc.id != whloc.id
        GROUP BY 1,2
```

Using set-returning functions for some parts of queries

Another possibility for achieving similar results to temporary tables and/or materialized views is by using a **set-returning function** for some parts of the query.

It is easy to have a materialized view freshness check inside a function. However, detailed analysis and an overview of these techniques go beyond the goals of this book, as they require a deep understanding of the PL/pgSQL procedural language.

Speeding up queries without rewriting them

Often, you either can't or don't want to rewrite a query. However, you can still try and speed it up through any of the techniques we will discuss here.

How to do it...

By now, we assume that you've looked at various problems already, so the following are more advanced ideas for you to try.

Increasing work_mem

For queries involving large sorts or for join queries, it may be useful to increase the amount of working memory that can be used for query execution. Try setting the following:

```
SET work_mem = '1TB';
```

Then, run EXPLAIN (not EXPLAIN ANALYZE). If EXPLAIN changes for the query, then it may benefit from more memory. I'm guessing that you don't have access to 1 **terabyte** (**TB**) of RAM; the previous setting was only used to prove that the query plan is dependent on available memory. Now, issue the following command:

```
RESET work_mem;
```

Now, choose a more appropriate value for production use, such as the following:

```
SET work_mem = '128MB';
```

Remember to increase maintenace_work_mem when creating indexes or adding **foreign keys** (**FKs**), rather than work_mem.

Setting recursive_worktable_factor

The `recursive_worktable_factor` parameter plays a pivotal role in enhancing the performance of recursive queries. It aids the query planner by providing a refined estimate of the working table's size.

A misjudgment in estimating the working table's size can cause the planner to opt for a less efficient query plan, hampering performance. Specifically:

- Underestimating the size might lead the planner to choose a plan demanding multiple data passes or utilize an inadequately sized temporary table
- Overestimating the size can make the planner opt for an unnecessarily intricate and costly plan, like using an oversized temporary table or employing a less efficient algorithm for joining the working table with other tables in the query

By correctly adjusting the `recursive_worktable_factor` parameter, you equip the planner to select the most efficient query plan for your recursive query. Such adjustments can bring about substantial performance boosts, particularly for intricate queries or those handling vast data quantities.

Below is an example of how setting the `recursive_worktable_factor` parameter can improve the performance of a recursive query:

```
postgres=# SET recursive_worktable_factor = 1;
SET
postgres=# EXPLAIN ANALYZE WITH RECURSIVE direct_reports (id, manager_id)
AS (
  SELECT id, manager_id
  FROM employees
  WHERE manager_id = 3
  UNION ALL
  SELECT e.id, e.manager_id
  FROM employees e
  JOIN direct_reports dr ON e.manager_id = dr.id
)
SELECT id
FROM direct_reports
ORDER BY id;

QUERY PLAN
```

```
------------------------------------------------------------------
------------------------------------------------------------------
----------
 Sort  (cost=31027062111.26..31277447865.09 rows=100154301533 width=4)
(actual time=11828.285..12404.234 rows=1000001 loops=1)
   Sort Key: direct_reports.id
   Sort Method: external merge  Disk: 11768kB
   CTE direct_reports
     ->  Recursive Union  (cost=0.00..2508283266.29 rows=100154301533
width=8) (actual time=85.956..10062.950 rows=1000001 loops=1)
           ->  Seq Scan on employees  (cost=0.00..37739.00 rows=1001533
width=8) (actual time=85.954..722.830 rows=1000001 loops=1)
                 Filter: (manager_id = 3)
                 Rows Removed by Filter: 999999
           ->  Merge Join  (cost=430293.53..150670251.20 rows=10015330000
width=8) (actual time=8200.796..8200.804 rows=0 loops=1)
                 Merge Cond: (dr.id = e.manager_id)
                 ->  Sort  (cost=133547.85..136051.68 rows=1001533
width=4) (actual time=1173.078..1173.081 rows=1 loops=1)
                       Sort Key: dr.id
                       Sort Method: external merge  Disk: 11768kB
                       ->  WorkTable Scan on direct_reports dr
(cost=0.00..20030.66 rows=1001533 width=4) (actual time=17.385..614.639
rows=1000001 loops=1)
                 ->  Materialize  (cost=296745.69..306745.69 rows=2000000
width=8) (actual time=2499.394..5884.538 rows=1999999 loops=1)
                       ->  Sort  (cost=296745.69..301745.69 rows=2000000
width=8) (actual time=2499.389..3695.102 rows=1999999 loops=1)
                             Sort Key: e.manager_id
                             Sort Method: external merge  Disk: 35248kB
                             ->  Seq Scan on employees e
(cost=0.00..32739.00 rows=2000000 width=8) (actual time=0.019..1152.668
rows=2000000 loops=1)
   ->  CTE Scan on direct_reports  (cost=0.00..2003086030.66
rows=100154301533 width=4) (actual time=85.963..11268.729 rows=1000001
loops=1)
 Planning Time: 0.133 ms
 JIT:
   Functions: 15
```

     Options: Inlining true, Optimization true, Expressions true, Deforming true
   Timing: Generation 0.714 ms, Inlining 4.864 ms, Optimization 59.807 ms, Emission 38.585 ms, Total 103.970 ms
 Execution Time: 12934.389 ms
(26 rows)

postgres=# SET recursive_worktable_factor = 10;
SET

postgres=# EXPLAIN ANALYZE WITH RECURSIVE direct_reports (id, manager_id) AS (
  SELECT id, manager_id
  FROM employees
  WHERE manager_id = 3
  UNION ALL
  SELECT e.id, e.manager_id
  FROM employees e
  JOIN direct_reports dr ON e.manager_id = dr.id
)
SELECT id
FROM direct_reports
ORDER BY id;

QUERY PLAN

 Sort (cost=326878354279.47..329382189283.30 rows=1001534001533 width=4) (actual time=9567.752..10143.355 rows=1000001 loops=1)
   Sort Key: direct_reports.id
   Sort Method: external merge Disk: 11768kB
   CTE direct_reports
     -> Recursive Union (cost=0.00..25058077949.25 rows=1001534001533 width=8) (actual time=97.913..7801.959 rows=1000001 loops=1)
           -> Seq Scan on employees (cost=0.00..37739.00 rows=1001533 width=8) (actual time=97.910..735.372 rows=1000001 loops=1)
                 Filter: (manager_id = 3)

```
                    Rows Removed by Filter: 999999
              ->  Merge Join  (cost=1935481.17..1504270019.49
rows=100153300000 width=8) (actual time=5926.530..5926.537 rows=0 loops=1)
                    Merge Cond: (e.manager_id = dr.id)
                    ->  Sort  (cost=296745.69..301745.69 rows=2000000
width=8) (actual time=2501.454..3668.316 rows=1999999 loops=1)
                          Sort Key: e.manager_id
                          Sort Method: external merge  Disk: 35248kB
                          ->  Seq Scan on employees e  (cost=0.00..32739.00
rows=2000000 width=8) (actual time=0.031..1152.541 rows=2000000 loops=1)
                    ->  Materialize  (cost=1638735.48..1688812.13
rows=10015330 width=4) (actual time=1169.809..1169.813 rows=1 loops=1)
                          ->  Sort  (cost=1638735.48..1663773.81
rows=10015330 width=4) (actual time=1169.803..1169.805 rows=1 loops=1)
                                Sort Key: dr.id
                                Sort Method: external merge  Disk: 11768kB
                                ->  WorkTable Scan on direct_reports dr
(cost=0.00..200306.60 rows=10015330 width=4) (actual time=17.372..614.035
rows=1000001 loops=1)
  ->  CTE Scan on direct_reports  (cost=0.00..20030680030.66
rows=1001534001533 width=4) (actual time=97.921..9007.778 rows=1000001
loops=1)
 Planning Time: 0.221 ms
 JIT:
   Functions: 15
   Options: Inlining true, Optimization true, Expressions true, Deforming
true
   Timing: Generation 0.948 ms, Inlining 5.361 ms, Optimization 68.912 ms,
Emission 40.936 ms, Total 116.157 ms
 Execution Time: 10673.570 ms
(26 rows)
```

More ideas with indexes

Try to add a multicolumn index that is specifically tuned for that query.

If you have a query that, for example, selects rows from the t1 table on the a column and sorts on the b column, then creating the following index enables PostgreSQL to do it all in one index scan:

```
CREATE INDEX t1_a_b_idx ON t1(a, b);
```

PostgreSQL 9.2 introduced a new plan type: **index-only scans**. This allows you to utilize a technique known as **covering indexes**. If all of the columns requested by the SELECT list of a query are available in an index, that particular index is a covering index for that query. This technique allows PostgreSQL to fetch valid rows directly from the index, without accessing the table (**heap**), so performance improves significantly. If the index is non-unique, you can just add columns onto the end of the index, like so – however, please be aware that this only works for non-unique indexes:

```
CREATE INDEX t1_a_b_c_idx ON t1(a, b, c);
```

PostgreSQL 11+ provides syntax to identify covering index columns in a way that works for both unique and non-unique indexes, like this:

```
CREATE INDEX t1_a_b_cov_idx ON t1(a, b) INCLUDE (c);
```

Another often underestimated (or unknown) feature of PostgreSQL is **partial indexes**. If you use SELECT on a condition, especially if this condition only selects a small number of rows, you can use a conditional index on that expression, like this:

```
CREATE INDEX t1_proc_ndx ON t1(i1)
WHERE needs_processing = TRUE;
```

The index will be used by queries that have a WHERE clause that includes the index clause, like so:

```
SELECT id, ... WHERE needs_processing AND i1 = 5;
```

There are many types of indexes in PostgreSQL so you may find that there are multiple types of indexes that can be used for a particular task and many options to choose from:

- **ID data**: BTREE and HASH
- **Categorical data**: BTREE
- **Text data**: GIST and GIN
- **JSONB or XML data**: GIN, plus selective use of btree
- **Time-range data**: BRIN (and partitioning)
- **Geographical data**: GIST, SP-GIST, and BRIN

Performance gains in Postgres can also be obtained with another technique: **clustering tables on specific indexes**. However, index access may still not be very efficient if the values that are accessed by the index are distributed randomly, all over the table. If you know that some fields are likely to be accessed together, then cluster the table on an index defined on those fields. For a multicolumn index, you can use the following command:

```
CLUSTER t1_a_b_ndx ON t1;
```

Clustering a table on an index rewrites the whole table in index order. This can lock the table for a long time, so don't do it on a busy system. Also, CLUSTER is a one-time command. New rows do not get inserted in cluster order, and to keep the performance gains, you may need to cluster the table every now and then.

Once a table has been clustered on an index, you don't need to specify the index name in any cluster commands that follow. It is enough to type this:

```
CLUSTER t1;
```

It still takes time to rewrite the entire table, though it is probably a little faster once most of the table is in index order.

There's more...

We will complete this recipe by listing four examples of query performance issues that can be addressed with a specific solution.

Time-series partitioning

Refer to the *Creating time-series tables* recipe for more information on this.

Using a view that contains TABLESAMPLE

Where some queries access a table, replace that with a view that retrieves fewer rows using a TABLESAMPLE clause. In this example, we are using a sampling method that produces a sample of the table using a scan lasting no longer than 5 seconds; if the table is small enough, the answer is exact; otherwise, progressive sampling is used to ensure that we meet our time objective:

```
CREATE EXTENSION tsm_system_time;
CREATE SCHEMA fast_access_schema;
CREATE VIEW fast_access_schema.tablename AS
SELECT *
FROM data_schema.tablename TABLESAMPLE system_time(5000); --5 secs
SET search_path = 'fast_access_schema, data_schema';
```

So, the application can use the new table without changing the SQL. Be careful, as some answers can change when you're accessing fewer rows (for example, sum()), making this particular idea somewhat restricted; the overall idea of using views is still useful.

In case of many updates, set fillfactor on the table

If you often update only some tables and can arrange your query/queries so that you don't change any indexed fields, then setting `fillfactor` to a lower value than the default of 100 for those tables enables PostgreSQL to use **heap-only tuples** (**HOT**) updates, which can be an **order of magnitude** (**OOM**) faster than ordinary updates. HOT updates not only avoid creating new index entries but can also perform a fast mini-vacuum inside the page to make room for new rows:

```
ALTER TABLE t1 SET (fillfactor = 70);
```

This tells PostgreSQL to fill only 70% of each page in the t1 table when performing insertions so that 30% is left for use by in-page (HOT) updates.

Rewriting the schema – a more radical approach

In some cases, it may make sense to rewrite the database schema and provide an old view for unchanged queries using views, triggers, rules, and functions.

One such case occurs when refactoring the database, and you would want old queries to keep running while changes are made.

Another case is an external application that is unusable with the provided schema but can be made to perform OK with a different distribution of data between tables.

Discovering why a query is not using an index

This recipe explains what to do if you think your query should use an index, but it isn't.

There could be several reasons for this but, most often, the reason is that the optimizer believes that, based on the available distribution statistics, it is cheaper and faster to use a query plan that does not use that specific index.

Getting ready

First, check that your index exists, and ensure that the table has been analyzed. If there is any doubt, rerun it to be sure—though it's better to do this only on specific tables:

```
postgres=# ANALYZE;
ANALYZE
```

How to do it...

Force index usage and compare plan costs with an index and without, as follows:

```
postgres=# EXPLAIN ANALYZE SELECT count(*) FROM itable WHERE id > 500;
                    QUERY PLAN
-----------------------------------------------------------------
  Aggregate  (cost=188.75..188.76 rows=1 width=0)
             (actual time=37.958..37.959 rows=1 loops=1)
    ->  Seq Scan on itable  (cost=0.00..165.00 rows=9500 width=0)
             (actual time=0.290..18.792 rows=9500 loops=1)
          Filter: (id > 500)
Total runtime: 38.027 ms
(4 rows)
postgres=# SET enable_seqscan TO false;
SET
postgres=# EXPLAIN ANALYZE SELECT count(*) FROM itable WHERE id > 500;
                    QUERY PLAN
-----------------------------------------------------------------
  Aggregate  (cost=323.25..323.26 rows=1 width=0)
             (actual time=44.467..44.469 rows=1 loops=1)
    ->  Index Scan using itable_pkey on itable
             (cost=0.00..299.50 rows=9500 width=0)
             (actual time=0.100..23.240 rows=9500 loops=1)
          Index Cond: (id > 500)
Total runtime: 44.556 ms
(4 rows)
```

Note that you must use EXPLAIN ANALYZE rather than just EXPLAIN. EXPLAIN ANALYZE shows you how much data is being requested and measures the actual execution time, while EXPLAIN only shows what the optimizer thinks will happen. EXPLAIN ANALYZE is slower, but it gives an accurate picture of what is happening.

In PostgreSQL 16, please use these EXPLAIN (ANALYZE ON, SETTINGS ON, BUFFERS ON, WAL ON) options rather than just using EXPLAIN ANALYZE. Note that SETTINGS will give you information about any non-default options, while BUFFERS and WAL will give you more information about the data access for read/write.

How it works...

By setting the enable_seqscan parameter to off, we greatly increase the cost of sequential scans for the query. This setting is never recommended for production use—only use it for testing because this setting affects the whole query, not just the part of it you would like to change.

This allows us to generate two different plans, one with SeqScan and one without. The optimizer works by selecting the lowest-cost option available. In the preceding example, the cost of SeqScan is 188.75 and the cost of IndexScan is 323.25, so for this specific case, IndexScan will not be used.

Remember that each case is different and always relates to the exact data distribution.

There's more...

Be sure that the WHERE clause you are using can be used with the type of index you have. For example, the abs(val) < 2 WHERE clause won't use an index because you're performing a function on the column, while val BETWEEN -2 AND 2 could use the index. With more advanced operators and data types, it's easy to get confused as to the type of clause that will work, so check the documentation for the data type carefully.

In PostgreSQL 10, join statistics were also improved by the use of FKs since they can be used in some queries to prove that joins on those keys return exactly one row.

Forcing a query to use an index

Often, we think we know better than the database optimizer. Most of the time, your expectations are wrong, and if you look carefully, you'll see that. So, recheck everything and come back later.

It is a classic error to try to get the database optimizer to use indexes when the database has very little data in it. Put some genuine data in the database first, then worry about it. Better yet, load some data on a test server first, rather than doing this in production.

Sometimes, the optimizer gets it wrong. You feel elated—and possibly angry—that the database optimizer doesn't see what you see. Please bear in mind that the data distributions within your database change over time, and this causes the optimizer to change its plans over time as well.

If you have found a case where the optimizer is wrong, this can sometimes change over time as the data changes. It might have been correct last week and will be correct again next week, or it correctly calculated that a change of plan was required, but it made that change slightly ahead of time or slightly too late. Again, trying to force the optimizer to do the right thing *now* might prevent it from doing the right thing *later*, when the plan changes again. So hinting fixes things in the short term, but in the longer term can cause problems to resurface.

In the long run, it is not recommended to try to force the use of a particular index.

Getting ready

Still here? Oh well.

If you really feel this is necessary, then your starting point is to run an EXPLAIN command for your query, so please read the previous recipe first.

How to do it...

The most common problem is selecting too much data.

A typical point of confusion comes from data that has a few very common values among a larger group. Requesting data for very common values costs more because we need to bring back more rows. As we bring back more rows, the cost of using the index increases. Therefore, it is possible that we won't use the index for very common values, whereas we would use the index for less common values. To use an index effectively, make sure you're using the LIMIT clause to reduce the number of rows that are returned.

Since different index values might return more or less data, it is common for execution times to vary depending on the exact input parameters. This could cause a problem if we are using prepared statements—the first five executions of a prepared statement are made using "custom plans" that vary according to the exact input parameters. From the sixth execution onward, the optimizer decides whether to use a "generic plan" or not, if it thinks the cost will be lower on average. Custom plans are more accurate, but the planning overhead makes them less efficient than generic plans. This heuristic can go wrong at times and you might need to override it using plan_cache_mode = force_generic_plan or force_custom_plan.

Another technique for making indexes more usable is **partial indexes**. Instead of indexing all of the values in a column, you might choose to index only a set of rows that are frequently accessed—for example, by excluding NULL or other unwanted data. By making the index smaller, it will be cheaper to access and will fit within the cache better, preventing pointless work by targeting the index at only the important data. Data statistics are kept for such indexes, so it can also improve the accuracy of query planning. Let's look at an example:

```
CREATE INDEX ON customer(id)
WHERE blocked = false AND subscription_status = 'paid';
```

Another common problem is that the optimizer may make errors in its estimation of the number of rows returned, causing the plan to be incorrect. Some optimizer estimation errors can be corrected using CREATE STATISTICS. If the optimizer is making errors, it can be because the WHERE clause contains multiple columns. For example, queries that mention related columns such as state and phone_area_code or city and zip_code will have poor estimates because those pairs of columns have data values that are correlated.

You can define additional statistics that will be collected when you next analyze the table:

```
CREATE STATISTICS cust_stat1 ON state, area_code FROM cust;
```

The execution time of ANALYZE will increase to collect the additional stats information, plus there is a small increase in query planning time, so use this sparingly when you can confirm this will make a difference. If there is no benefit, use DROP STATISTICS to remove them again. By default, multiple types of statistics will be collected—you can fine-tune this by specifying just a few types of statistics if you know what you are doing.

Unfortunately, the statistics command doesn't automatically generate names, so include the table name in the statistics you create since the name is unique within the database and cannot be repeated on different tables. In future releases, we may also add cross-table statistics.

Additionally, you cannot collect statistics on individual fields within JSON documents at the moment, nor collect dependency information between them; this command only applies to whole column values at this time.

Another nudge toward using indexes is to set random_page_cost to a lower value—maybe even equal to seq_page_cost. This makes PostgreSQL prefer index scans on more occasions, but it still does not produce entirely unreasonable plans, at least for cases where data is mostly cached in shared buffers or system disk caches, or underlying disks are **solid-state drives** (**SSDs**).

The default values for these parameters are provided here:

```
random_page_cost = 4;
seq_page_cost = 1;
```

Try setting this:

```
set random_page_cost = 2;
```

See if it helps; if not, you can try setting it to 1.

Changing random_page_cost allows you to react to whether data is on disk or in memory. Letting the optimizer know that more of an index is in the cache will help it to understand that using the index is actually cheaper.

Index scan performance for larger scans can also be improved by allowing multiple asynchronous I/O operations by increasing effective_io_concurrency. Both random_page_cost and effective_io_concurrency can be set for specific tablespaces or for individual queries.

There's more...

PostgreSQL does not directly support hints, but they are available via an extension.

If you absolutely, positively have to use the index, then you'll want to know about an extension called pg_hint_plan. It is available for PostgreSQL 9.1 and later versions. For more information and to download it, go to http://pghintplan.sourceforge.jp/. Hints can be added to your application SQL using a special comment added to the start of a query, like this:

```
/*+ IndexScan(tablename indexname) */ SELECT …
```

It works, but, as I said previously, try to avoid fixing things now and causing yourself pain later when the data distribution changes.

EnterpriseDB (EDB) and **Postgres Advanced Server (EPAS)** also support hints in an Oracle-style syntax to allow you to select a specific index, like this:

```
SELECT /*+ INDEX(tablename indexname) */ … rest of query …
```

EPAS has many compatibility features such as this for migrating application logic from Oracle. See https://www.enterprisedb.com/docs/epas/latest/epas_compat_ora_dev_guide/05_optimizer_hints/ for more information on this.

Using parallel query

PostgreSQL now has an increasingly effective parallel query feature.

Response times from long-running queries can be improved by the use of parallel processing. The concept is that if we divide a large task up into multiple smaller pieces then we get the answer faster, but we use more resources to do that.

Very short queries won't get faster by using parallel query, so if you have lots of them you'll gain more by thinking about better indexing strategies. Parallel query is aimed at making very large tasks faster, so it is useful for reporting and **business intelligence (BI)** queries.

How to do it...

Take a query that needs to do a big chunk of work, such as the following:

```
\timing
SET max_parallel_workers_per_gather = 0;
SELECT count(*) FROM big;
count
---------
1000000
(1 row)
Time: 46.399 ms
SET max_parallel_workers_per_gather = 2;
SELECT count(*) FROM big;
count
---------
1000000
(1 row)
Time: 29.085 ms
```

By setting the max_parallel_workers_per_gather parameter, we've improved performance using parallel query. Note that we didn't need to change the query at all. (The preceding queries were executed multiple times to remove any cache effects.)

In PostgreSQL 9.6 and 10, parallel query only works for read-only queries, so only SELECT statements that do not contain the FOR clause (for example, SELECT ... FOR UPDATE). In addition, a parallel query can only use functions or aggregates that are marked as PARALLEL SAFE. No user-defined functions are marked PARALLEL SAFE by default, so read the docs carefully to see whether your functions can be enabled for parallelism for the current release.

How it works...

The plan for our earlier example of parallel query looks like this:

```
postgres=# EXPLAIN ANALYZE
SELECT count(*) FROM big;
                        QUERY PLAN
-------------------------------------------------------------------
Finalize Aggregate  (cost=11614.55..11614.56 rows=1 width=8) (actual
time=59.810..62.074 rows=1 loops=1)
```

```
         -> Gather  (cost=11614.33..11614.54 rows=2 width=8) (actual
time=59.709..62.067 rows=3 loops=1)
               Workers Planned: 2
               Workers Launched: 2
                     -> Partial Aggregate  (cost=10614.33..10614.34 rows=1 width=8)
(actual time=56.298..56.299 rows=1 loops=3)
                           -> Parallel Seq Scan on big  (cost=0.00..9572.67
rows=416667 width=0) (actual time=0.009..32.138 rows=333333 loops=3)
 Planning Time: 0.056 ms
 Execution Time: 62.110 ms
(8 rows)
```

By default, a query will use only one process. Parallel query is enabled by setting max_parallel_workers_per_gather to a value higher than zero (the default is 2). This parameter specifies the maximum number of **additional** processes that are available if needed. So, a setting of 1 will mean you have the leader process plus one additional worker process, so two processes in total.

The query optimizer will decide whether parallel query is a useful plan based on cost, just as with other aspects of the optimizer. Importantly, it will decide how many parallel workers to use in its plan, up to the maximum you specify.

Note that the performance increase from adding more workers isn't linear for anything other than simple plans, so there are diminishing returns from using too many workers. The biggest gains are from adding the first few extra processes.

PostgreSQL will assign a number of workers according to the size of the table compared to the min_parallel_table_scan_size value, using the logarithm (base 3) of the ratio. With default values, this means:

Size of Table	Number of Parallel Workers
<24 megabytes (MB)	1
24 MB+	2
216 MB+	4
1.9 gigabytes (GB)	6
17 GB	8
1.4 TB	12
114 TB	16

Table 10.1: Table size and assigned parallel workers

Decreasing min_parallel_table_scan_size will increase the number of workers assigned.

Across the whole server, the maximum number of worker processes available is specified by the `max_parallel_workers` parameter and is set at server start only.

At execution time, the query will use its planned number of worker processes if that many are available. If worker processes aren't available, the query will run with fewer worker processes. As a result, it pays to not be too greedy, since if all concurrent users specify more workers than are available, you'll end up with variable performance as the number of concurrent parallel queries changes.

Using Just-In-Time (JIT) compilation

PostgreSQL's **Just-In-Time** (JIT) compilation, introduced by default in PostgreSQL 11, translates SQL expressions into native machine code at runtime and gives performance improvements, especially for complex operations or large datasets. However, its benefits are most noticeable for long-running, CPU-intensive queries, typically analytical ones. On the other hand, the overhead for JIT might negate its advantages for shorter queries.

Key JIT optimizations in PostgreSQL include:

- **Expression evaluation**: Primarily for WHERE clauses, aggregates, and projections. This optimization generates code for specific scenarios.
- **Tuple deforming**: Converts on-disk tuples into in-memory formats, accelerated using functions suited for specific table layouts.

Getting ready

JIT is triggered based on the estimated query cost. If this surpasses the `jit_above_cost` setting, JIT is engaged. Depending on the cost, specific optimizations may be applied, such as inlining short functions (controlled by `jit_inline_above_cost`) or employing advanced optimizations (governed by `jit_optimize_above_cost`).

It's important to note that these cost evaluations occur during the planning phase, not during execution. Thus, for prepared statements, JIT's behavior is determined by the configuration parameters set during preparation, not those during execution.

How it works...

Below is an example that shows the benefit of using JIT:

```
postgres=# CREATE TABLE large_data (a INTEGER, b INTEGER);
CREATE TABLE
```

```
postgres=# INSERT INTO large_data
           SELECT x, x % 1000
           FROM generate_series(1, 10000000) AS x;
INSERT 0 10000000

postgres=# SET JIT TO off;
SET
postgres=# EXPLAIN ANALYZE
SELECT b, AVG(a)
FROM large_data
GROUP BY b
ORDER BY b;

QUERY PLAN
--------------------------------------------------------------------------
--------------------------------------------------------------------------
----
 Finalize GroupAggregate  (cost=107807.71..108063.56 rows=1000 width=36)
(actual time=4199.428..4205.325 rows=1000 loops=1)
   Group Key: b
   ->  Gather Merge  (cost=107807.71..108041.06 rows=2000 width=36)
(actual time=4199.412..4202.464 rows=3000 loops=1)
         Workers Planned: 2
         Workers Launched: 2
         ->  Sort  (cost=106807.68..106810.18 rows=1000 width=36) (actual
time=4186.897..4187.478 rows=1000 loops=3)
               Sort Key: b
               Sort Method: quicksort  Memory: 95kB
               Worker 0:  Sort Method: quicksort  Memory: 95kB
               Worker 1:  Sort Method: quicksort  Memory: 95kB
               ->  Partial HashAggregate  (cost=106747.86..106757.86
rows=1000 width=36) (actual time=4185.534..4186.172 rows=1000 loops=3)
                     Group Key: b
                     Batches: 1  Memory Usage: 321kB
                     Worker 0:  Batches: 1  Memory Usage: 321kB
                     Worker 1:  Batches: 1  Memory Usage: 321kB
                     ->  Parallel Seq Scan on large_data
```

```
(cost=0.00..85914.57 rows=4166657 width=8) (actual time=0.163..1934.249
rows=3333333 loops=3)
 Planning Time: 0.074 ms
 Execution Time: 4205.909 ms
(18 rows)

postgres=# SET JIT TO on;
SET
postgres=# EXPLAIN ANALYZE
SELECT b, AVG(a)
FROM large_data
GROUP BY b
ORDER BY b;

QUERY PLAN
----------------------------------------------------------------------------------------------------------------------------------
 Finalize GroupAggregate  (cost=107807.71..108063.56 rows=1000 width=36)
(actual time=4184.608..4190.889 rows=1000 loops=1)
   Group Key: b
   ->  Gather Merge  (cost=107807.71..108041.06 rows=2000 width=36)
(actual time=4184.578..4188.087 rows=3000 loops=1)
         Workers Planned: 2
         Workers Launched: 2
         ->  Sort  (cost=106807.68..106810.18 rows=1000 width=36) (actual
time=4171.068..4171.650 rows=1000 loops=3)
               Sort Key: b
               Sort Method: quicksort  Memory: 95kB
               Worker 0:  Sort Method: quicksort  Memory: 95kB
               Worker 1:  Sort Method: quicksort  Memory: 95kB
               ->  Partial HashAggregate  (cost=106747.86..106757.86
rows=1000 width=36) (actual time=4169.693..4170.329 rows=1000 loops=3)
                     Group Key: b
                     Batches: 1  Memory Usage: 321kB
                     Worker 0:  Batches: 1  Memory Usage: 321kB
                     Worker 1:  Batches: 1  Memory Usage: 321kB
                     ->  Parallel Seq Scan on large_data
```

```
 (cost=0.00..85914.57 rows=4166657 width=8) (actual time=0.031..1955.915
rows=3333333 loops=3)
 Planning Time: 0.067 ms
 JIT:
   Functions: 24
   Options: Inlining false, Optimization false, Expressions true,
Deforming true
   Timing: Generation 1.462 ms, Inlining 0.000 ms, Optimization 0.854 ms,
Emission 17.770 ms, Total 20.086 ms
 Execution Time: 4192.145 ms
(22 rows)
```

In short, JIT compilation can significantly improve the performance of PostgreSQL queries and other operations, especially for those that involve complex expressions or large amounts of data.

However, it is important to note that JIT compilation can add overhead to the startup time of PostgreSQL, and it may not always be beneficial.

Creating time-series tables using partitioning

In many applications, we need to store data in time series. There are various mechanisms in PostgreSQL that are designed to support this.

How to do it...

If you have a huge table and a query to select only a subset of that table, then you may wish to use a **block range index (BRIN index)**. These indexes give performance improvements when the data is naturally ordered as it is added to the table, such as logtime columns or a naturally ascending OrderId column. Adding a BRIN index is fast and very easy, and works well for the use case of time-series data logging, though it works less well under intensive updates, even with the new BRIN features in PostgreSQL 16. INSERT commands into BRIN indexes are specifically designed to not slow down as the table gets bigger, so they perform much better than B-tree indexes for write-heavy applications. B-trees do have faster retrieval performance but require more resources. To try BRIN, just add an index, like so:

```
CREATE TABLE measurement (
     logtime      TIMESTAMP WITH TIME ZONE NOT NULL,
     measures     JSONB NOT NULL);
CREATE INDEX ON measurement USING BRIN (logtime);
```

Partitioning syntax was introduced in PostgreSQL 10. Over the last five releases, partitioning has been very heavily tuned and extended to make it suitable for time-series logging, BI, and fast **Online Transaction Processing (OLTP)** SELECT, UPDATE, or DELETE commands.

The best reason to use partitioning is to allow you to drop old data quickly. For example, if you are only allowed to keep data for 30 days, it might make sense to store data in 30 partitions. Each day, you would add one new empty partition and detach/drop the last partition in the time series.

For example, to create a table for time-series data, you may want something like this:

```
CREATE TABLE measurement (
    logtime    TIMESTAMP WITH TIME ZONE NOT NULL,
    measures   JSONB NOT NULL
    ) PARTITION BY RANGE (logtime);
CREATE TABLE measurement_week1 PARTITION OF measurement
    FOR VALUES FROM ('2019-03-01') TO ('2019-04-01');
CREATE INDEX ON measurement_week1 USING BRIN (logtime);
CREATE TABLE measurement_week2 PARTITION OF measurement
    FOR VALUES FROM ('2019-04-01') TO ('2019-05-01');
CREATE INDEX ON measurement_week2 USING BRIN (logtime);
```

For some applications, the time taken to SELECT/UPDATE/DELETE from the table will increase with the number of partitions, so if you are thinking you might need more than 100 partitions, you should benchmark carefully with fully loaded partitions to check this works for your application.

You can use both BRIN indexes and partitioning at the same time so that there is less need to have a huge number of partitions. As a guide, partition size should not be larger than shared buffers, to allow the whole current partition to sit within shared buffers.

For more details on partitioning, check out https://www.postgresql.org/docs/current/ddl-partitioning.html.

How it works...

Each partition is actually a normal table, so you can refer to partitions directly in queries. A partitioned table is somewhat similar to a view, since it links all of the partitions under it together. The partition key defines which data goes into which partition so that each row lives in exactly one partition. Partitioning can also be defined with multiple levels—so, a single top-level partitioned table, then with each sub-table also having sub-sub-partitions.

B-tree performance degrades very slowly as tables get bigger, so having single tables larger than a few hundred GB may no longer be optimal. Using partitions and limiting the size of each partition will prevent any bad news as data volumes climb over time. Let me repeat the "very slowly" part—so, no need to rush around changing all of your tables when you get to 101 GB.

As of PostgreSQL 16, adding and detaching partitions are both now optimized to hold a lower level of lock, allowing SELECT statements to continue while those activities occur. Adding a new partition with a reduced lock level just uses the syntax shown previously. Simply dropping a partition will hold an AccessExclusiveLock—or, in other words, will be blocked by SELECT statements and will block them while it runs. Dropping a partition using a reduced lock level should be done in two steps, like this:

```
ALTER TABLE measurement
    DETACH PARTITION measurement_week2 CONCURRENTLY;
DROP TABLE measurement_week2;
```

Note that you cannot run those two commands in one transaction. If the ALTER TABLE command is interrupted, then you will need to run FINALIZE to complete the operation, like this:

```
ALTER TABLE measurement
    DETACH PARTITION measurement_week2 FINALIZE;
```

Partitioned tables also support default partitions, but I recommend against using them because of the way table locking works with that feature. If you add a new partition that partially overlaps the default partition, it will lock the default partition, scan it, and then move data to the new partition. That activity can lock out the table for some time and should be avoided on production systems. Note also that you can't use concurrent detach if you have a default partition.

There's more...

The ability to do a "partition-wise join" can be very useful for large queries when joining two partitioned tables. The join must contain all columns of the partition key and be the same data type, with a 1:1 match between the partitions. If you have multiple partitioned tables in your application, you may wish to enable the enable_partitionwise_join = on optimizer parameter, which defaults to off.

If you do large aggregates on a partitioned table, you may also want to enable another optimizer parameter, enable_partitionwise_aggregate = on, which defaults to off.

PostgreSQL 11 adds the ability to have **primary keys** (**PKs**) defined over a partitioned table, enforcing uniqueness across partitions. This requires that the partition key is the same or a subset of the columns of the PK. Unfortunately, you cannot have a unique index across an arbitrary set of columns of a partitioned table because multi-table indexes are not yet supported—and it would be very large if you did.

You can define references from a partitioned table to normal tables to enforce FK constraints. References to a partitioned table are possible in PostgreSQL 12+.

Partition tables can have before-and-after row triggers.

Partitioned tables can be used in publications and subscriptions, as well as in **EDB Postgres Distributed** (**PGD**).

Using optimistic locking to avoid long lock waits

If you perform work in one long transaction, the database will lock rows for long periods of time. Long lock times often result in application performance issues because of long lock waits:

```
BEGIN;
SELECT * FROM accounts WHERE holder_name ='BOB' FOR UPDATE;
<do some calculations here>
UPDATE accounts SET balance = 42.00 WHERE holder_name ='BOB';
COMMIT;
```

If that is happening, then you may gain some performance benefits by moving from explicit locking (SELECT ... FOR UPDATE) to optimistic locking.

Optimistic locking assumes that others don't update the same record, and checks this at update time, instead of locking the record for the time it takes to process the information on the client side.

How to do it...

Rewrite your application so that the SQL is transformed into two separate transactions, with a double-check to ensure that the rows haven't changed (pay attention to the placeholders):

```
SELECT A.*, (A.*::text) AS old_acc_info
FROM accounts a WHERE holder_name ='BOB';
<do some calculations here>
UPDATE accounts SET balance = 42.00
WHERE holder_name ='BOB'
AND (A.*::text) = <old_acc_info from select above>;
```

Then, check whether the UPDATE operation really did update one row in your application code. If it did not, then the account for BOB was modified between SELECT and UPDATE, and you probably need to rerun your entire operation (both transactions).

How it works...

Instead of locking Bob's row for the time that the data from the first SELECT command is processed in the client, PostgreSQL queries the old state of Bob's account record in the old_acc_info variable and then uses this value to check that the record has not changed when we eventually update.

You can also save all fields individually and then check them all in the UPDATE query; if you have an automatic last_change field, then you can use that instead. Alternatively, if you only care about a few fields changing—such as balance—and are fine ignoring others—such as email—then you only need to check the relevant fields in the UPDATE statement.

There's more...

You can also use the serializable transaction isolation level when you need to be absolutely sure that the data you are looking at is not affected by other user changes.

The default transaction isolation level in PostgreSQL is read-committed, but you can choose from two more levels—repeatable read and serializable—if you require stricter control over the visibility of data within a transaction. See http://www.postgresql.org/docs/current/static/transaction-iso.html for more information.

Another design pattern that's available in some cases is to use a single statement for the UPDATE clause and return data to the user via the RETURNING clause, as in the following example:

```
UPDATE accounts
SET balance = balance - i_amount
WHERE username = i_username
AND balance - i_amount > - max_credit
RETURNING balance;
```

In some cases, moving the entire computation to the database function is a very good idea. If you can pass all of the necessary information to the database for processing as a database function, it will run even faster, as you save several round-trips to the database. If you use a PL/pgSQL function, you also benefit from automatically saving query plans on the first call in a session and using saved plans in subsequent calls.

Therefore, the preceding transaction is replaced by a function in the database, like so:

```
CREATE OR REPLACE FUNCTION consume_balance
( i_username text
, i_amount numeric(10,2)
, max_credit numeric(10,2)
, OUT success boolean
, OUT remaining_balance numeric(10,2)
) AS
$$
BEGIN
  UPDATE accounts SET balance = balance - i_amount
  WHERE username = i_username
  AND balance - i_amount > - max_credit
  RETURNING balance
  INTO remaining_balance;
  IF NOT FOUND THEN
    success := FALSE;
    SELECT balance
    FROM accounts
    WHERE username = i_username
    INTO remaining_balance;
  ELSE
    success := TRUE;
  END IF;
END;
$$ LANGUAGE plpgsql;
```

You can call it by simply running the following line of code from your client:

```
SELECT * FROM consume_balance ('bob', 7, 0);
```

The output will return the success variable. It tells you whether there was a sufficient balance in Bob's account. The output will also return a number, telling you the balance bob has left after this operation.

Reporting performance problems

Sometimes, you face performance issues and feel lost, but you should never feel alone when working with one of the most successful open source projects ever.

How to do it...

If you need to get some advice on your performance problems, then the right place to do so is the performance mailing list at http://archives.postgresql.org/pgsql-performance/.

First, you may want to ensure that it is not a well-known problem by searching the mailing-list archives.

A very good description of what to include in your performance problem report is available at http://wiki.postgresql.org/wiki/Guide_to_reporting_problems.

There's more...

More performance-related information can be found at http://wiki.postgresql.org/wiki/Performance_Optimization.

Learn more on Discord

To join the Discord community for this book – where you can share feedback, ask questions to the author, and learn about new releases – follow the QR code below:

https://discord.gg/pQkghgmgdG

11
Backup and Recovery

Most people admit that backups are essential, though they also devote a very small amount of time to thinking about the topic.

The first recipe in this chapter is about understanding and controlling crash recovery. You need to understand what happens if a database server crashes so that you can understand whether you need to perform a recovery operation.

The next recipe is all about planning. That's really the best place to start before you perform backups.

The physical backup mechanisms were initially written by Simon Riggs (one of the authors of this book) for PostgreSQL 8.0 in 2004, and have been supported by him ever since, with ever-increasing help from the community as Postgres's popularity grew. 2ndQuadrant and EDB have also been providing database recovery services since 2004, and regrettably, many people have needed them as a result of missing or damaged backups.

It is important to note that, in the last few years, the native streaming replication protocol has become more and more relevant in PostgreSQL. It can be used for backup purposes too – not only to make a base backup with `pg_basebackup` but also to stream **Write-Ahead Log** (**WAL**) files using `pg_receivewal`. Given that some of the recipes in this chapter will use streaming replication, we will assume that you have a basic familiarity with it and refer you to the next chapter for more details.

As a final note, all of the examples regarding physical backup and recovery in this chapter are thoroughly explained so that you understand what is happening behind the scenes. However, unless you have very specific requirements dictating otherwise, we highly recommend that when in production, you use Barman (our open source backup and recovery tool) or a similar product that is specialized in this area. We will introduce Barman in the relevant recipe later in this chapter.

In this chapter, we will cover the following recipes:

- Understanding and controlling crash recovery
- Planning your backups
- Hot logical backup of one database
- Hot logical backup of all databases
- Backup of database object definitions
- A standalone hot physical backup
- Hot physical backup with Barman
- Recovery of all databases
- Recovery to a point in time
- Recovery of a dropped/damaged table
- Recovery of a dropped/damaged database
- Extracting a logical backup from a physical one
- Improving the performance of logical backup/recovery
- Improving the performance of physical backup/recovery
- Validating backups

Understanding and controlling crash recovery

Crash recovery is the PostgreSQL subsystem that saves us should the server crash or fail as part of a system crash.

It's good to understand a little about it and what we can do to control it in our favor.

How to do it...

If PostgreSQL crashes, there will be a message in the server log with the severity level set to PANIC. PostgreSQL will immediately restart and attempt to recover using the transaction log or the WAL.

The WAL consists of a series of files written to the pg_wal subdirectory of the PostgreSQL data directory. Each change made to the database is recorded first in WAL, hence the name write-ahead log, which is a synonym for a *transaction log*. Note that the former is probably more accurate, since, in the WAL, there are also changes not related to transactions. When a transaction commits, the default (and safe) behavior is to force the WAL records to disk. Should PostgreSQL crash, the WAL will be replayed, which returns the database to the point of the last committed transaction, ensuring the durability of any database changes.

Database changes themselves aren't written to disk at transaction commit. On a well-tuned server, those changes are written to disk sometime later by asynchronous processes, such as the background writer or the checkpointer.

Crash recovery replays the WAL, but from what point does it start to recover? Recovery starts from points in the WAL known as **checkpoints**. The duration of a crash recovery depends on the number of changes in the transaction log since the last checkpoint. A checkpoint is a known safe starting point for recovery, since it guarantees that all of the previous changes to the database have already been written to disk.

A checkpoint can become a performance bottleneck on busy database servers because of the number of writes required. We're about to look at a number of ways to fine-tune that, but you must also understand the effect that those tuning options may have on crash recovery.

A checkpoint can be either immediate or scheduled. Immediate checkpoints are triggered by the action of a superuser, such as the CHECKPOINT command, and they are performed at full speed, so an immediate checkpoint will complete as soon as possible. Scheduled checkpoints are decided automatically by PostgreSQL, and their speed is throttled to spread the load over a longer period of time and reduce the impact on other activities, such as queries or replication.

Two parameters control the occurrence of scheduled checkpoints. The first is checkpoint_timeout, which is the number of seconds until the next checkpoint. While this parameter is time-based, the second parameter, max_wal_size, influences the amount of WAL data that will be written before a checkpoint is triggered; the actual limit is computed from that parameter, taking into account the fact that WAL files can only be deleted after one checkpoint (two in older releases). A checkpoint is called whenever either of these two limits is reached.

If checkpoints are too frequent, then the amount of I/O will increase, so it's tempting to banish checkpoints as much as possible by setting the following parameters:

```
max_wal_size = 20GB
checkpoint_timeout = '1 day'
```

However, if you do this, you should give some thought to how long crash recovery will take and whether you want that; you must also consider how many changes will accumulate before the next checkpoint and, more importantly, how much I/O the checkpoint will generate due to those changes. Also, if you are using replication, then you might not care about the recovery time because if the primary crashes, you can fail over to a standby without waiting for crash recovery to complete.

Also, you should make sure that the pg_wal directory is mounted on disks with enough disk space. By default, max_wal_size is set to 1 GB. The amount of disk space required by pg_wal might also be influenced by the following:

- Unexpected spikes in workload
- Failures in continuous archiving (see archive_command in the *Hot physical backup with Barman* recipe)
- The wal_keep_size setting (you will need at least wal_keep_size MB of space)

In contrast to max_wal_size, with min_wal_size, you can control the minimum size allotted to WAL storage, meaning that PostgreSQL will recycle existing WAL files instead of removing them.

How it works...

In this recipe, we saw how to control crash recovery. Recovery continues until the end of the transaction log is reached. As WAL data is being written continually, there is no defined endpoint; it is literally the last correct record. Each WAL record is individually checked with a **Cyclic Redundancy Check (CRC)** so that we know whether a record is complete and valid before trying to process it. Each record contains a pointer to the previous record, so we can tell that the record forms a valid link in the chain of actions recorded in the WAL. As a result of that, recovery always ends with some kind of error in reading the next WAL record. That is normal and means *the next record does not exist (yet)*.

Recovery performance can be very fast, though its speed does depend on the actions being recovered. The best way to test recovery performance is to set up a standby replication server, as described in *Chapter 12, Replication and Upgrades*, which is actually implemented as a variant of crash recovery.

There's more...

It's possible that a problem with replaying the transaction log can cause the database server to fail to start.

Some people's response to this is to use a utility named pg_resetwal, which removes the current transaction log files and tidies up after that operation has taken place.

The pg_resetwal utility destroys data changes, which means you are going to face data loss. If you do decide to run that utility, make sure that you make a backup of the pg_wal directory first. Our advice is to seek immediate assistance rather than do this. You don't know for certain that doing this will fix a problem, but once you've done it, going back will be hard.

When discussing `min_wal_size`, we mentioned that WAL files are recycled; what this actually means is that older WAL files are renamed so that they are ready to be reused as future WAL files. This reduces commit latency in case of heavy write workloads because creating a new file is slower than writing into an existing one.

Planning your backups

This recipe is all about thinking ahead and planning. If you're reading this recipe before you've decided to make a backup, well done!

The key thing to understand is that you should plan your recovery, not your backup. The type of backup you make influences the type of recovery that is possible, so you must give some thought to what you are trying to achieve beforehand.

If you want to plan your recovery, then you need to consider the different types of failure that can occur. What type of recovery do you wish to perform?

You need to consider the following main aspects:

- Full or partial database?
- Everything or just object definitions?
- Is **Point-in-Time Recovery (PITR)** going to be needed?
- What are the requirements for restore performance?

We need to look at the characteristics of the utilities to understand what our backup and recovery options are. It's often beneficial to have multiple types of backups to cover the different possible types of failure.

Your main backup options are the following:

- Logical backup, using `pg_dump`
- Physical backup, which is a filesystem backup

The `pg_dump` utility comes in two main flavors – `pg_dump` and `pg_dumpall`. The `pg_dump` utility has the `-F` option for producing backups in various file formats. The file format is very important when it comes to restoring from backups, so you need to pay close attention to it.

As far as physical backup is concerned, in this chapter, we will focus on filesystem backup using `pg_backup_start()` and `pg_backup_stop()`. However, it is important to note that PostgreSQL has its own built-in application for physical base backups, `pg_basebackup`, which relies on the native streaming replication protocol.

In order to distribute the content more evenly, we have decided to cover `pg_basebackup` and streaming replication in the next chapter, *Chapter 12, Replication and Upgrades*.

How to do it...

The following table shows the features that are available, depending on the backup technique selected. The details of these techniques are covered in the remaining recipes in this chapter:

	SQL dump to an archive file: `pg_dump -F c`	SQL dump to a script file: `pg_dump -F p` or `pg_dumpall`	Filesystem backup using `pg_backup_start` and `pg_backup_stop`
Backup type	Logical	Logical	Physical
Recover to point in time	No	No	Yes
Zero data loss	No	No	Yes (see note 6)
Back up all databases	One at a time	Yes (`pg_dumpall`)	Yes
All databases backed up at the same time	No	No	Yes
Selective backup	Yes	Yes	No (see note 2)
Incremental backup	No	No	Possible (see note 3)
Selective restore	Yes	Possible (see note 1)	No (see note 4)
`DROP TABLE` recovery	Yes	Yes	Possible (see note 4)
Compressed backup files	Yes	Yes	Yes
Backup in multiple files	No	No	Yes
Parallel backup possible	No (but see note 8)	No	Yes
Parallel restore possible	Yes	No	Yes
Restore to later release	Yes	Yes	No (but see note 7)
Standalone backup	Yes	Yes	Yes (see note 6)
Allows DDL during backup	No	No	Yes

Table 11.1: Different backup techniques compared in terms of available features

The following notes were mentioned in the preceding table:

1. If you've generated a script with pg_dump or pg_dumpall and need to restore just a single object, then you will need to go deeper. You will need to write a Perl script (or similar) to read the file and extract the parts you want. This is messy and time-consuming but probably faster than restoring the whole thing to a second server and then extracting just the parts you need with another pg_dump.
2. Selective backup with a physical backup is possible but will cause problems later when you try to restore.
3. See the *A standalone hot physical backup* recipe.
4. Selective restore with a physical backup isn't possible with the currently supplied utilities; however, please see the *Recovery of a dropped/damaged table* and *Extracting a logical backup from a physical one* recipes for partial recovery.
5. See the A *standalone hot physical database backup* recipe.
6. See the *Hot physical backup with Barman* recipe. Barman 3.x fully supports synchronous WAL streaming, allowing you to achieve a **Recovery Point Objective** (**RPO**) equal to 0, meaning *zero data loss*.
7. A physical backup cannot be directly restored to a different PostgreSQL major version. However, it is possible to restore it to the same PostgreSQL major version and then follow the procedure described in the *Major upgrades in-place* recipe, in *Chapter 12, Replication and Upgrades*, to upgrade restored files to a newer major version.
8. The directory format (-F d), however, allows parallelization.

There's more...

Choosing physical backups is a safer approach: if you can take a logical backup, then you can also take a physical backup and then extract the same logical backup from it, while the opposite is not possible. For more details, refer to the *Extracting a logical backup from a physical one* recipe later in this chapter.

Hot logical backup of one database

Logical backup makes a copy of the data in the database by dumping the content of each table, as well as object definitions for that same database (such as schemas, tables, indexes, views, privileges, triggers, and constraints).

How to do it...

The command to do this is simple. The following is an example of doing this when using a database called pgbench:

```
pg_dump -F c pgbench > dumpfile
```

Alternatively, you can use the following command:

```
pg_dump -F c -f dumpfile pgbench
```

Finally, note that you can also run pg_dump via the **pgAdmin 4** GUI by selecting the backup source from the tree control and right-clicking to open the context menu that contains the **Backup...** option. Selecting that will open the **Backup** dialog, as shown in the following screenshot:

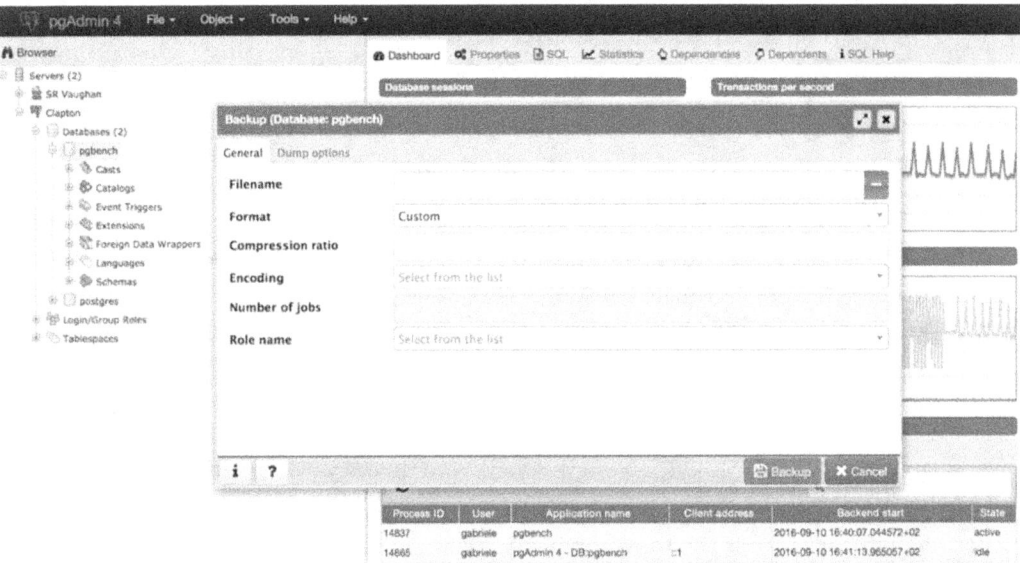

Figure 11.1: Using the pgAdmin 4 GUI

How it works...

The pg_dump utility produces a single output file. This output file can use the standard UNIX split command to separate the file into multiple pieces if required.

The pg_dump archive file, also known as the **custom format**, is lightly compressed by default. Compression can be disabled or made more aggressive.

Even though, by default, pg_dump writes a SQL script directly to standard output, it is recommended to use the archive file instead by enabling the custom format through the -F c option. As we will cover later in this chapter, backing up in the form of archive files gives you more flexibility and versatility when restoring. Archive files must be used with a tool called pg_restore.

The pg_dump utility runs by executing SQL statements against the database to unload data. When PostgreSQL runs a SQL statement, we take a *snapshot* of transactions that are currently running, which freezes our viewpoint of the database. The pg_dump utility can take a parallel dump of a single database by using the **snapshot export** feature of Postgres.

We can't (yet) share that snapshot across sessions connected to more than one database, so we cannot run an exactly consistent pg_dump in parallel across multiple databases. The time of the snapshot is the only moment we can recover to – we can't recover to a time either before or after. Note that the snapshot time is the start of the backup, not the end.

When pg_dump runs, it holds the very lowest kind of lock on the tables being dumped. Those are designed to prevent DDL from running against the tables while the dump takes place. If a dump is run at the point at which other DDLs are already running, then the dump will sit and wait. If you want to limit the waiting time, you can do so by setting the --lock-wait-timeout option.

Since pg_dump runs SQL queries to extract data, it will have some performance impact. This must be taken into account when executing on a live server.

The pg_dump utility allows you to make a selective backup of tables. The -t option also allows you to specify views and sequences. There's no way to dump other object types individually using pg_dump. You can use some supplied functions to extract individual snippets of information from the catalog.

More details on these functions are available at this URL: https://www.postgresql.org/docs/16/functions-info.html#FUNCTIONS-INFO-CATALOG-TABLE.

The pg_dump utility works against earlier releases of PostgreSQL, so it can be used to migrate data between releases.

When migrating your database from an earlier version, it is generally recommended to use pg_dump of the same version as the target PostgreSQL. For example, if you are migrating a PostgreSQL 12.7 database to PostgreSQL 13, you should use pg_dump v13 to remotely connect to the 12.7 server and back up the database.

As far as extensions are concerned, pg_dump is aware of any objects (namely, tables and functions) that have been installed as part of an additional package, such as PostGIS or Slony. Thanks to that, they can be recreated by issuing appropriate CREATE EXTENSION commands instead of dumping and restoring them together with the other database objects. Extension support removes such difficulties when restoring from a logical backup, maintaining the list of additional tables that have been created as part of the software installation process. Refer to the *Managing installed extensions* recipe in *Chapter 3*, *Server Configuration*, for more details.

There's more…

What time was pg_dump taken? The snapshot for pg_dump is taken at the beginning of a backup. The file modification time will tell you when the dump is finished. The dump is consistent at the time of the snapshot, so you may need to know that time.

If you are making a script dump, you can do a verbose dump; continuing the previous example, you just need to add -v, as follows:

```
pg_dump -F c -f dumpfile pgbench -v
```

This adds the time to the top of the script. Custom dumps store the start time as well, and that can be accessed using the following command:

```
pg_restore --schema-only -v dumpfile -f - 2>/dev/null | head | grep Started
-- Started on 2023-10-03 09:05:46 BST
```

See also

Note that pg_dump does not dump global objects that belong to all databases such as roles (e.g., users and groups), privilege grants, and tablespaces. Those are only dumped by pg_dumpall; see the following recipes for more detailed descriptions.

Hot logical backup of all databases

If you have more than one database in your PostgreSQL server, you may want to make a logical backup of all of the databases at the same time.

How to do it...

Our recommendation is that you repeat exactly what you do for one database for each database in your cluster. You can run individual dumps in parallel if you want to speed things up.

Once this is complete, dump the global information using the following command:

```
pg_dumpall -g
```

How it works...

To back up all databases, you may be told that you only need to use the pg_dumpall utility. The following are four good reasons why you shouldn't do that:

- If you use pg_dumpall, the only output produced will be in a script file. Script files can't benefit from all the features of archive files, such as parallel and selective restore of pg_restore. By making your backup in this way, you will immediately deprive yourself of flexibility and versatility at restore time.

- The pg_dumpall utility produces dumps of each database, one after another. This means that pg_dumpall is slower than running multiple pg_dump tasks in parallel, one against each database.

- When using pg_dumpall, dumps of individual databases are not consistent to a single point in time. As we pointed out in the *Hot logical backup of one database* recipe, if you start the dump at 4:00 and it ends at 7:00, then you cannot be sure exactly what time the dump relates to; it could be any time between 4:00 and 7:00.

- Options for pg_dumpall and pg_dump are similar in many ways. pg_dump has more options and, therefore, gives you more flexibility.

See also

If you are making a logical backup of all of your databases for disaster recovery purposes, you should look at the *hot physical backup* recipes in this chapter instead.

Backup of database object definitions

Sometimes, it's useful to get a dump of the object definitions that make up a database. This is useful for comparing what's in the database against the definitions in a data or object-modeling tool. It's also useful to make sure that you can recreate objects in the correct schema, tablespace, and database with the correct ownership and permissions.

How to do it...

There are several important commands to note here:

- The basic command to dump the definitions for every database of your PostgreSQL instance is as follows:

   ```
   pg_dumpall --schema-only > myscriptdump.sql
   ```

   This includes all objects, including roles, tablespaces, databases, schemas, tables, indexes, triggers, constraints, views, functions, ownership, and privileges.

- If you want to dump PostgreSQL role definitions, use the following command:

   ```
   pg_dumpall --roles-only > myroles.sql
   ```

- If you want to dump PostgreSQL tablespace definitions, use the following command:

   ```
   pg_dumpall --tablespaces-only > mytablespaces.sql
   ```

- If you want to dump both roles and tablespaces, use the following command:

   ```
   pg_dumpall --globals-only > myglobals.sql
   ```

The output is a human-readable script file that can be re-executed to recreate each of the databases.

The short form for the --globals-only option is -g, which we have already seen in a previous recipe, *Hot logical backups of all databases*. Similar abbreviations exist for --schema-only (-s), --tablespaces-only (-t), and --roles-only (-r).

There's more...

In PostgreSQL, the word **schema** is also used to organize a set of related objects of a database in a logical container, similar to a directory. It is also known as a **namespace**. Be careful that you don't confuse what is happening here. The --schema-only option makes a backup of the database schema – that is, the definitions of all objects in the database (and in all namespaces). To make a backup of the data and definitions in just one namespace and one database, use pg_dump with the -n option. To make a backup of only the definitions, in just one namespace and one database, use pg_dump with both -n and --schema-only together.

You can also take advantage of a previously generated archive file (see the *Hot logical backups of one database* recipe) and generate a script file using pg_restore, as follows:

```
pg_restore --schema-only mydumpfile > myscriptdump.sql
```

A standalone hot physical backup

Hot physical backup is an important capability for databases.

Physical backup allows us to get a completely consistent view of the changes to all databases at once. Physical backup also allows us to back up even while DDL changes are being executed on the database. Apart from resource constraints, there is no additional overhead or locking with this approach.

Physical backup procedures used to be slightly more complex than logical backup procedures, but since version 10, some defaults have been changed, making them easier; after these changes, making a backup with pg_basebackup has become very easy, even with default settings.

In this recipe, we will first describe the easiest method, which is to use the pg_basebackup utility, and then provide a lower-level equivalent process to explain physical backups in more detail and describe the changes required for additional features, such as differential backup or a parallel file copy.

Getting ready

You just need to decide upon a directory where you want to place backups and make sure it exists – for instance, /var/lib/postgresql/standalone.

How to do it...

Just log in as the postgres user and run the pg_basebackup utility, as follows:

```
pg_basebackup -D /var/lib/postgresql/backup/standalone -c fast -P -R
```

This command asks for an immediate checkpoint, progress reporting, and the addition of connection settings to postgresql.auto.conf. Once it returns, the /var/lib/postgresql/backup/standalone directory will contain a PostgreSQL data directory whose contents are an exact copy of the contents of the PostgreSQL server that you have backed up.

How it works...

By default, pg_basebackup connects to the database using the same defaults as other utilities based on libpq, such as psql. This normally means that you don't need to specify connection information such as the database user, the hostname, and the port. But, if you are running a server that uses non-default values for those settings, then you can specify them with familiar syntax. For example, take the following options:

```
-h myhost  -U myuser -p 5433
```

If you add them to the previous pg_basebackup command, then pg_basebackup will use the myhost host, the myuser user, and port 5433.

Under the hood, pg_basebackup proceeds in a way that is equivalent to the following sequence of steps:

1. Create an empty directory as the target of the standalone backup:

    ```
    rm -rf /var/lib/postgresql/backup/standalone
    mkdir -p /var/lib/postgresql/backup/standalone/pg_wal
    ```

2. Start streaming WAL into /var/lib/postgresql/backup/standalone/pg_wal with the following command:

    ```
    pg_receivewal -D /var/lib/postgresql/backup/standalone/pg_wal
    ```

3. Connect to the server and ask it to start the backup, as follows:

    ```
    select pg_backup_start('standalone');
    ```

    This last step can take a while because PostgreSQL performs a checkpoint before returning to ensure that the data files copied in the next step include all of the latest data changes. See the *Understanding and controlling crash recovery* recipe from earlier in this chapter for more details about checkpoints.

    Depending on system configuration and workload, a checkpoint can take a long time, even several minutes. This time is part of the backup duration, which, in turn, affects the amount of WAL files needed for the backup; it can be a good idea to reduce the duration of this checkpoint by issuing a CHECKPOINT command just before pg_backup_start is issued in *step 3* and then by starting the backup in fast mode, as follows:

    ```
    select pg_backup_start('standalone', fast := true);
    ```

    fast mode means that the checkpoint included in pg_backup_start runs as quickly as possible, irrespective of its impact on the system; this should not be a problem because most of the shared buffers will have been written already by the CHECKPOINT command that was issued previously.

4. Make a base backup – copy the data files (excluding the content of the pg_wal and pg_replslot directories) using the following commands:

    ```
    tar -cvf- \
        --directory=$PGDATA \
        --exclude="pg_wal/*" --exclude="pg_replslot/*" . \
        | tar -xf- --directory=/var/lib/postgresql/backup/standalone
    ```

5. From the same connection as before, stop the backup as follows:
   ```
   select * from pg_backup_stop(wait_for_archive := true), current_
   timestamp;
   ```

6. Stop archiving by hitting *Ctrl + C* in the terminal session where `pg_receivewal` is still running.

pg_backup_stop returns three values. The second field, `labelfile`, must be written to a file called `backup_label` in `/var/lib/postgresql/backup/standalone/`. If the third field, `spcmapfile`, is not empty, it should be written to a file called `tablespace.map`. These files are necessary for the backup to work and their content must be written byte for byte with no changes.

There's more...

After the backup is finished, remember to store the contents of `/var/lib/postgresql/backup/standalone` somewhere safe. A safe place is definitely not on the same server.

This procedure ends with a directory populated with a copy of the database. It is imperative to remember to copy it somewhere safe. It contains everything that you need to recover.

The backup taken with this procedure only allows you to restore to a single point in time. That point is the time of the `pg_backup_stop()` function.

A physical backup takes a copy of all files in the database (*step 4* – the *base backup*). That alone is not sufficient as a backup, and you need the other steps as well. A simple copy of the database produces a time-inconsistent copy of the database files. To make the backup time consistent, we need to add all of the changes that took place from the start to the end of the backup. That's why we have *steps 3* and *5* to bracket our backup step.

PostgreSQL only supports non-exclusive backups, allowing users to run multiple backups concurrently but, more importantly, to perform the `pg_backup_start()` and `pg_backup_stop()` functions on a read-only standby server.

The changes that are made are put in the `standalone/pg_wal` directory as a set of archived transaction logs or WAL files by the `pg_receivewal` command started in *step 2*.

If your PGDATA does not contain configuration files, such as `postgresql.conf` and `pg_hba.conf`, you might have to manually copy them before performing a recovery. Remember that standard Debian and Ubuntu installations keep configuration files outside PGDATA, specifically under `/etc/postgresql`.

The important thing to understand in this recipe is that we need both the base backup and the appropriate archived WAL files to allow us to recover. Without both of these, we have nothing. Most of these steps are designed to ensure that we really will have the appropriate WAL files in all cases.

As an alternative to WAL streaming, it is possible to configure file-based WAL archiving, which works in *push* mode, without requiring inbound access to the database server. However, streaming WAL archiving has the advantage of transferring WAL as soon as it is produced, without waiting for the 16 MB WAL segment to be completed, which usually results in little or no data loss, even in the event of a disaster.

We describe this procedure only for the purpose of illustrating how pg_basebackup works. If you want to copy files more efficiently, as explained in the *Improving the performance of backup/recovery* and *A standalone hot physical backup* recipes, then you should use software that is specialized in backup and recovery, rather than writing your own scripts. In this book, we cover Barman, software written by EDB developers, which has become very popular among PostgreSQL users. It is an open source tool used in most of the remaining recipes in this book.

Hot physical backups with Barman

The main motivation to start a new open source project for disaster recovery of PostgreSQL databases was the lack (back in 2011) of a simple and standard procedure for managing backups and, most importantly, recovery. Disasters and failures in **Information and Communication Technology (ICT)** will happen.

As a database administrator, your duty is to plan for backups and the recovery of PostgreSQL databases and perform regular tests in order to sweep away stress and fear, which typically follow those unexpected events. **Barman**, which stands for **Backup and Recovery Manager**, is definitely a tool that you can use for these purposes.

Barman hides most of the complexity of working with PostgreSQL backups. For more information on the underlying technologies, you can refer to other recipes in this chapter: *Understanding and controlling crash recovery*, *Planning backups*, and *Recovery to a point in time*. It is important to be aware of how Barman works underneath if you need to address issues with installation, configuration, and recovery.

Barman 3.9.0 is currently available only for Linux systems (but will likely run on other UNIX systems too) and is written in Python. It supports all PostgreSQL versions starting from 10.

Among its main features worth citing are remote backup, remote recovery, multiple server management, backup catalogs, incremental backups, retention policies, WAL streaming, compression of WAL files, parallel copy (backup and restore), backing up from a standby server, and geo-redundancy.

For the sake of simplicity, in this recipe, we will assume the following architecture:

- One Linux server named db1, running your PostgreSQL production database server.
- One Linux server named backup1, running Barman for disaster recovery of your PostgreSQL database server.
- WAL streaming is configured from PostgreSQL to Barman.
- Both servers are in the same LAN and, for better business continuity objectives, the only resource they share is the network.

Later on, we will see how easy it is to add more PostgreSQL servers (such as bon) to our disaster recovery solution on backup1 with Barman.

Getting ready

Although Barman can be installed via sources or through pip – Python's main package manager – the easiest way to install Barman is by using the software package manager of your Linux distribution.

Currently, EDB maintains packages for modern RHEL, Rocky, Debian, and Ubuntu systems. If you are using a different distribution or another UNIX system, you can follow the instructions written in the official documentation of Barman, available at https://docs.pgbarman.org/.

In this book, we will cover the installation of Barman 3.9.0 (currently the latest stable release) on Rocky Linux 8 and Ubuntu 20.04 LTS Linux servers.

If you are using RHEL or Rocky 8 on the backup1 server, you need to install the following repositories:

- Fedora's **Extra Packages Enterprise Linux (EPEL)**, available at https://fedoraproject.org/wiki/EPEL/
- The PostgreSQL Global Development Group RPM repository, available at https://yum.postgresql.org/

Then, as root, type in the following:

```
yum install barman
```

If you are using Ubuntu on backup1, you need to install the APT PostgreSQL repository, following the instructions available at https://apt.postgresql.org/. Then, as root, type in the following:

```
apt install barman
```

From now on, we will assume the following:

- PostgreSQL is running on db1 and listening to the default port (5432).
- Barman is installed on backup1.
- You have created a superuser called barman in your PostgreSQL server on db1 that can only connect from the backup1 server (see the *Enabling access for network/remote users* recipe in *Chapter 1, First Steps*, and *The PostgreSQL superuser* recipe in *Chapter 6, Security*).
- The barman system user on backup1 can connect as the barman database user to the PostgreSQL instance on db1 without having to type in a password.

How to do it...

We will start by looking at Barman's main configuration file:

1. Log in as root on backup1 and open the /etc/barman.conf file for editing. This file contains global options for Barman. Once you are familiar with the main configuration options, we recommend at least that you set the default WAL compression method by uncommenting the following line:

    ```
    compression = gzip
    ```

2. Add the configuration file for the db1 server. Create the db1.conf file, containing the following lines, in the /etc/barman.d directory:

    ```
    [db1]
    description = "PostgreSQL database on db1"
    active = off
    backup_method = postgres
    archiver = off
    streaming_archiver = on
    slot_name = "barman_backup1"
    conninfo = "host=db1 dbname=postgres user=barman"
    streaming_conninfo = "host=db1 dbname=postgres user=streaming_barman"
    ```

3. You have just added the db1 server to the list of Postgres servers managed by Barman. The server is temporarily inactive until the configuration is completed. You can verify this by typing barman list-server, as follows:

```
[root@backup1]# barman list-server
db1 - PostgreSQL database on db1 (inactive)
```

In this recipe, you will be executing commands such as barman list-server as root. However, Barman will run its own commands using the barman system user (or, more generally, the user specified in the configuration file by the barman_user option).

4. The next task is to initialize the directory layout for the db1 server through the check command:

```
[root@backup1]# barman check db1
Server db1 (inactive):
        WAL archive: FAILED (please make sure WAL shipping is setup)
        PostgreSQL: OK
        superuser or standard user with backup privileges: OK
        PostgreSQL streaming: OK
        wal_level: OK
        replication slot: FAILED (replication slot 'barman_backup1' doesn't exist. Please execute 'barman receive-wal --create-slot db1')
(...)
[root@backup1]# echo $?
0
```

As you can see, the returned value is 0, meaning that there is no reason to worry; the server is marked as inactive, meaning that we are still configuring it, so if there are failures, then they are expected.

In fact, you are advised to add this command to your monitoring infrastructure as, among other things, it ensures that the required libpq connection to the database server is working properly, as well as continuously archiving.

5. Let's filter out **OK** so we display a to-do list:

   ```
   [root@backup1]# barman check db1 | grep -v OK
   Server db1 (inactive):
           WAL archive: FAILED (please make sure WAL shipping is setup)
           replication slot: FAILED (replication slot 'barman_backup1'
   doesn't exist. Please execute 'barman receive-wal --create-slot
   db1')
   ```

6. Now, we can mark the server as active by changing the following line in /etc/barman.d/db1.conf on backup1:

   ```
   active = on
   ```

7. Then, we can create the replication slot, as previously suggested by Barman itself:

   ```
   [root@backup1 ~]# barman receive-wal --create-slot db1
   Creating physical replication slot 'barman_backup1' on server 'db1'
   Replication slot 'barman_backup1' created
   ```

8. At this point, we can ask PostgreSQL to switch to the next WAL file:

   ```
   [root@backup1 ~]# barman switch-wal db1
   The WAL file 000000010000000000000002 has been closed on server
   'db1'
   ```

9. Now, we run the barman check command again, and we repeat it until all the checks pass.
10. Initially, the check command will raise an error like this:

    ```
    WAL archive: FAILED (please make sure WAL shipping is setup)
    ```

    This simply means, *"I did not archive any WAL file for this server yet, so I am not sure whether you have already configured WAL archiving."*

    This error will cease when the first WAL file has been fully streamed, archived, and compressed, which requires a run of the cron job installed by Barman, which is executed at the start of every minute. So, you should not have to wait more than a couple of minutes, and then this check will return 0:

    ```
    [root@backup1]# barman -q check db1
    [root@backup1]# echo $?
    0
    ```

Everything is good! PostgreSQL on db1 is now regularly streaming WAL files to Barman on backup1.

11. Once you have set up continuous archiving, in order to add disaster recovery capabilities to your database server, you need at least one full base backup. Taking a full base backup in Barman is as easy as typing barman backup db1. Barman initiates the physical backup procedure and waits for the checkpoint to happen, before copying data files from db1 to backup1 using pg_basebackup:

```
[root@backup1 ~]# barman backup db1
Starting backup using postgres method for server db1 in /var/lib/barman/db1/base/20231003T103940
Backup start at LSN: 0/3000060 (000000010000000000000003, 00000060)
Starting backup copy via pg_basebackup for 20231003T103940
(...)
```

12. Note that the command ends with the following warning:

```
WARNING: IMPORTANT: this backup is classified as WAITING_FOR_WALS,
meaning that Barman has not received yet all the required WAL files
for the backup consistency.
This is a common behaviour in concurrent backup scenarios, and
Barman automatically set the backup as DONE once all the required
WAL files have been archived.
Hint: execute the backup command with '--wait'
```

You don't need to worry, because WAL files are expected to arrive soon anyway, but if you want to include the wait in the backup command, then you can add the option suggested in the warning.

13. You can see that the new backup is now listed in the catalog:

```
[root@backup1 ~]# barman list-backup db1
db1 20231003T103940 - Tue Oct  3 10:39:42 2023 - Size: 23.5 MiB - WAL Size: 0 B
```

It is worth noting that, during the backup procedure, your PostgreSQL server is available for both read and write operations. This is because PostgreSQL natively implements hot backup, a feature that other DBMS vendors might make you pay for.

From now on, your db1 PostgreSQL server is continuously backed up on backup1. You can now schedule weekly backups (using the barman user's cron) and manage retention policies so that you can build a catalog of backups, covering you for weeks, months, or years of data, allowing you to perform recovery operations at any point in time between the first available backup and the last successfully archived WAL file.

How it works...

Barman is a Python application that wraps PostgreSQL core technology for continuous backup and PITR. It also adds some practical functionality, focused on helping the database administrator to manage disaster recovery of one or more PostgreSQL servers.

When devising Barman, we decided to keep the design simple and not use any daemon or client/server architecture. Maintenance operations are simply delegated to the barman cron command, which is mainly responsible for archiving WAL files (moving them from the incoming directory to the WAL file and compressing them) and managing retention policies.

If you have installed Barman through RPM or APT packages, you will notice that maintenance is run every minute through cron:

```
[root@backup1 ~]# cat /etc/cron.d/barman
# m h  dom mon dow   user      command
  * *   *   *   *    barman    [ -x /usr/bin/barman ] && /usr/bin/barman -q cron
```

Barman follows the *convention over configuration* paradigm and uses an INI format configuration file with options operating at two different levels:

- **Global options:** These are options specified in the [barman] section, and are used by any Barman command and for every server. Several global options can be overridden at the server level.
- **Server options:** These are options specified in the [SERVER_ID] section, used by server commands. These options can be customized at the server level (including overriding general settings).

  The SERVER_ID placeholder (such as db1) is fundamental, as it identifies the server in the catalog (therefore, it must be unique). Similarly, commands in Barman are of two types:

- **Global commands:** These are general commands, not tied to any server in particular, such as a list of the servers managed by the Barman installation (list-server) and maintenance (cron).

- **Server commands**: These are commands executed on a specific server, such as diagnostics (check and status), backup control (backup, list-backup, delete, and show-backup), and recovery control (recover, which is discussed in the next recipe, *Recovery with Barman*).

The previous sections of this recipe showed you how to add a server (db1) to a Barman installation on the backup1 server. You can easily add a second server (db2) to the Barman server on backup1. All you have to do is create the db2.conf file in the /etc/barman.d directory and repeat the steps outlined in the *How it works...* section, as you have done for db1.

There's more...

Every time you execute the barman backup command for a given server, you make a full base backup (a more generic term for this is a periodical full backup). Once completed, this backup can be used as a base for any recovery operation from the start time of the backup to the last available WAL file for that server (provided there is continuity among all of the WAL segments).

As we mentioned earlier, by scheduling daily or weekly automated backups, you end up having several periodic backups for a server. In Barman's jargon, this is known as the backup catalog, and it is one of our favorite features of this tool.

We already saw how to list all the available backups for a given server through the list-backup command. You might also want to get familiar with show-backup, which gives you detailed information on a specific backup regarding the server, base backup time, the WAL archive, and context within the catalog (for example, the last available backup):

```
[root@backup1 ~]# barman show-backup db1 20231003T103940
```

Rather than the full backup ID (20231003T103940), you can use a few synonyms, such as the following:

- last *or* latest: This refers to the latest available backup (the last in the catalog).
- first *or* oldest: This refers to the oldest available backup (the first in the catalog).

For the show-backup command, however, we will use a real and concrete example, taken directly from one of our customer's installations of Barman on a 16.4 TB Postgres 9.4 database:

```
Backup 20180930T130002:
  Server Name            : skynyrd
  Status                 : DONE
  PostgreSQL Version     : 90409
  PGDATA directory       : /srv/pgdata
```

```
    Base backup information:
      Disk usage           : 16.4 TiB (16.4 TiB with WALs)
      Incremental size     : 5.7 TiB (-65.08%)
      Timeline             : 1
      Begin WAL            : 000000010000035880000063
      End WAL              : 00000001000035A0000000A2
      WAL number           : 6208
      WAL compression ratio: 79.15%
      Begin time           : 2018-09-30  13:00:04.245110+00:00
      End time             : 2018-10-01  13:24:47.322288+00:00
      Begin Offset         : 24272
      End Offset           : 11100576
      Begin XLOG           : 3588/63005ED0
      End XLOG             : 35A0/A2A961A0
    WAL information:
      No of files          : 3240
      Disk usage           : 11.9 GiB
      WAL rate             : 104.33/hour
      Compression ratio    : 76.43%
      Last available       : 00000001000035AD0000004A
    Catalog information:
      Retention Policy     : not enforced
      Previous Backup      : 20180923T130001
      Next Backup          : - (this is the latest base backup)
```

As you can see, Barman is a production-ready tool that can be used in large, business-critical contexts, as well as in basic Postgres installations. It provides good RPO outcomes, allowing you to limit potential data loss to a single WAL file, or even less when WAL streaming is configured.

Finally, note that Barman also supports parallel and incremental backups, through the `rsync` method, which can dramatically reduce disk usage as well as backup and recovery time. For further information, please refer to the *Improving performance of physical backup/restore* recipe later in this chapter.

Manually performing each step of this procedure is a great way to gain a clear understanding of PostgreSQL's backup and restore infrastructure. However, to reduce the chance of human error, it is good practice to use a dedicated tool rather than rely on complex activities that must be performed by a human operator, or on custom-developed scripts that will not likely have the maturity of a tool that is used in production already in many installations.

While our preference goes with Barman, as we are its creators and main developers, there are other third-party tools that are specialized in managing hot physical backups, such as the following:

- **pgBackRest**: http://www.pgbackrest.org/
- **pghoard**: https://github.com/Aiven-Open/pghoard
- **wal-e**: https://github.com/wal-e/wal-e
- **wal-g**: https://github.com/wal-g/wal-g

Barman is distributed under GNU GPL v3 terms and is available for download at https://www.pgbarman.org/.

For further and more detailed information, refer to the following:

- The `man barman` command, which gives the man page for the Barman application
- The `man 5 barman` command, which gives the man page for the configuration file
- The `barman help` command, which gives a list of the available commands
- The official documentation for Barman, which is publicly available at https://docs.pgbarman.org/
- The mailing list for community support at https://www.pgbarman.org/support/

Recovery of all databases

Recovery of a complete database server, including all of its databases, is an important feature. This recipe covers how to execute a recovery in the simplest way possible.

Some complexities are discussed here, though most are covered in later recipes.

Getting ready

Find a suitable server on which to perform the restore.

Before you recover onto a live server, always make another backup. Whatever problem you thought you had can get worse if you aren't prepared.

Physical backups (including Barman ones) are more efficient than logical ones, but they are subject to additional restrictions.

To be precise, a single instance of Barman can manage backups of several servers having different versions of PostgreSQL. However, when it comes to recovery, the same requirements for the PITR technology of PostgreSQL apply – in particular, the following:

- You must recover on a server with the same hardware architecture and PostgreSQL major version.

- You will restore the entire PostgreSQL instance, with all its databases.

Actually, with backups, you don't *get ready* when you need to use them; you must *be ready before* you need them, so preparation is everything. This also means that you will have been aware of those requirements before the failure.

How to do it...

Here, we'll provide four distinct examples, depending on what type of backup was taken.

Logical – from the custom dump taken with pg_dump -F c

The procedure is as follows:

1. Restoring all databases means simply restoring each individual database from each dump you took. Confirm that you have the correct backup before you restore:

   ```
   pg_restore --schema-only -v dumpfile -f - 2>/dev/null |\ head | grep Started
   ```

2. Reload the global objects from the script file, as follows:

   ```
   psql -f myglobals.sql
   ```

3. Reload all databases. Create the databases using parallel tasks to speed things up. This can be executed remotely without the need to transfer dump files between systems. Note that there is a separate `dumpfile` for each database:

   ```
   pg_restore -C -d postgres -j 4 dumpfile
   ```

Logical – from the script dump created by pg_dump -F p

As in the previous method, this can be executed remotely without needing to transfer `dumpfile` between systems:

1. Confirm that you have the correct backup before you restore. If the following command returns nothing, then it means that the file is not timestamped, and you'll have to identify it in a different way:

   ```
   head myscriptdump.sql | grep Started
   ```

2. Reload `globals` from the script file, as follows:

   ```
   psql -f myglobals.sql
   ```

3. Reload all scripts, as follows:

```
psql -f myscriptdump.sql
```

Logical – from the script dump created by pg_dumpall

In order to recover a full backup generated by pg_dumpall, you need to execute the following steps on a PostgreSQL server that has just been initialized:

1. Confirm that you have the correct backup before you restore. If the following command returns nothing, then it means that the file is not timestamped, and you'll have to identify it in a different way:

```
head myscriptdump.sql | grep Started
```

2. Reload the script in full:

```
psql -f myscriptdump.sql
```

Physical – from a standalone backup

If you made the backup following the *A standalone hot physical database backup* recipe, then recovery is very easy:

1. Restore the backup files in the new data directory on the target server.
2. Confirm that you have the correct backup before you restore:

```
$ cat backup_label
START WAL LOCATION: 0/12000020 (file 000000010000000000000012)
CHECKPOINT LOCATION: 0/12000058
BACKUP METHOD: streamed
BACKUP FROM: primary
START TIME: 2023-06-03 19:53:23 BST
LABEL: standalone
START TIMELINE: 1
```

3. Verify that all file permissions and ownerships are correct and that the links are valid. This should already be the case if you are using the postgres user ID everywhere, which is recommended.
4. Start the server.

This procedure is so simple because, in the *A standalone hot physical database backup* recipe, we gift-wrapped everything for you. That also helped you to understand that you need both a base backup and the appropriate WAL files.

Physical – with Barman

If you made your backup according to the *Hot physical backups with Barman* recipe, then you can restore it using the barman recover command.

In this example, we are making the following assumptions:

- We have a new server called db2 on which we want to restore the latest backup of the PostgreSQL instance running on db1.
- The barman user on the backup host can connect via SSH as the postgres user to the db2 host.

These are the steps for recovering a Barman backup of the db1 server to db2:

1. Decide the location of the data directory where you want to restore PostgreSQL – for instance, /var/lib/pgsql/16/data.
2. Check whether the target data directory is empty; if not, ask yourself why (for example, *"Am I accidentally connected to the wrong host?"*). If you don't find a good answer, stop here; otherwise, make sure that it's okay to empty it, and then do it.
3. Connect as the barman user on backup1 and issue the following command:

```
barman recover db1 last /var/lib/pgsql/16/data \
    --remote-ssh-command 'ssh postgres@db2'
```

This command will use the latest available backup for the db1 server and prepare everything you need to restore your server in the PostgreSQL destination directory (/var/lib/pgsql/16/data), as shown in the following output:

```
Starting remote restore for server db1 using backup 20231003T103940
Destination directory: /var/lib/pgsql/16/data
Remote command: ssh postgres@db2
Copying the base backup.
Generating recovery configuration
Identify dangerous settings in destination directory.
```

At this point, Barman might find something in your settings that is considered potentially dangerous. It doesn't mean that it definitely is; it just means that this setting could accidentally be used in the wrong way, so you must use your knowledge of how PostgreSQL works to decide whether there could actually be a problem.

For instance, you could get the following output:

```
WARNING
You are required to review the following options as potentially
dangerous
postgresql.conf line 822: include_dir = 'conf.d'
```

In any case, the restore will finish with a message like the following:

```
Recovery completed (start time: 2023-10-03 14:12:29.314034, elapsed
time: 1 second)
Your PostgreSQL server has been successfully prepared for recovery!
```

4. Before you start the server, it is a good idea to inspect the contents of /var/lib/pgsql/16/data; its contents should look very similar to what was in the db1 server just before you started the recovery.

5. You are also strongly encouraged to review the contents of the postgresql.conf file before starting the server, even though Barman takes care of disabling or removing some potentially dangerous options and detecting others (as in the example warning reported previously).

    The most critical option is archive_command, which is preemptively set to false and is good for a disposable instance or when testing recovery. Barman does this because, in most cases, you don't want the instance you restored to start archiving to the original location. The goal is to make you think before you activate archiving. If you are restoring a backup because you want to create a permanent database server, then you need to consider your archiving strategies – for example, you might want to add the new database server to the Barman server by repeating the steps outlined in the previous recipe.

6. When you are satisfied with your checks, you can start Postgres in /var/lib/pgsql/16/data the usual way. For example, on Rocky Linux 8, you can execute the following command:

```
sudo systemctl start postgresql-16
```

7. Look at the PostgreSQL logs to verify that you do not have any problems, and then also check `ps -axf`.

Now, you have a PostgreSQL instance running on db2, with a copy of all the databases that are hosted on db1 and the same contents that those databases had at the point in time when the last WAL file was closed, which is usually a few minutes in the past.

How it works...

A logical backup is taken by asking PostgreSQL to print a description of each object and its contents, in the form of a sequence of SQL commands. The restore procedure for each database consists simply of issuing those SQL commands on an empty database.

A physical backup is a copy of the files inside the data directory. The format used by PostgreSQL to store data in the filesystem allows those files to be copied even while they are being written, as long as the correct procedure is followed. The restore procedure consists of creating a new PostgreSQL instance whose data directory will initially contain a copy of those files. These procedures are implemented in `pg_basebackup`, as well as in third-party utilities such as Barman.

When executed with the `--remote-ssh-command` option, the `barman recover` command uses that command to connect to the remote server (similar to what the `ssh-command` configuration option does in the backup phase but in reverse – see the *Hot physical backups with Barman* recipe for more information) and perform the restore. Internally, Barman relies on `rsync` for this operation.

There's more...

You can start and stop the server once recovery has started without any problems. It will not interfere with the recovery.

You can connect to the database server while it is recovering and run queries if that is useful. This is known as **hot standby** mode and is discussed in *Chapter 12, Replication and Upgrades*.

Barman allows you to perform two types of recovery:

- **Local recovery**: This involves restoring a PostgreSQL instance on the same server where Barman is running.
- **Remote recovery**: This involves restoring a PostgreSQL instance on a different server, through the network, as we just did.

Note that the terms *local* and *remote* are relative to the host where Barman is installed, which is where you execute the `barman recover` command.

A common reason for performing a local recovery is to test your backup or to extract some data from the backup – for instance, to recover from user error. This is described in the *Recovery to a point in time* recipe later in the chapter.

If you are using tablespaces and you are unable (or unwilling) to use the exact same directory paths when restoring the backup, you can use a feature known as **tablespace mapping** in pg_basebackup or **tablespace relocation** in Barman, where you can indicate the desired path for each tablespace. For more details, please refer to the user manual links at the end of this recipe or barman help recover.

In this example, we only recover those transactions that have been committed inside a WAL file that is already closed. While this is the default mode of operation of barman recover, it can be changed by adding the --get-wal option, which produces two effects:

- Barman recover does not copy WAL files; instead, it configures an appropriate restore_command on the recovered instance, which will then fetch WAL files from Barman on demand.
- The restore_command will also consider a partial WAL file when recovering transactions.

In that case, Barman will also recover those transactions whose commit record is in the WAL file currently being streamed. For further information, look at the *Barman client utilities* section of the Barman documentation at https://docs.pgbarman.org/.

Finally, another important use case for Barman is to regularly create copies of the server to be used for purposes such as development, staging, or business intelligence. These environments do not normally require extremely current data – for instance, a snapshot taken on the previous day could be enough.

This recipe has covered only a few aspects of the recovery process in Barman. For further and more detailed information, refer to the following links:

- User manuals, available here:
    - https://www.postgresql.org/docs/16/app-pgbasebackup.html
    - https://docs.pgbarman.org/
- Mailing lists for community support:
    - for pg_basebackup
    - https://www.pgbarman.org/support/
- The EDB blog at https://www.enterprisedb.com/blog/

Recovery to a point in time

If your database suffers a problem at 3:22 p.m. and your backup was taken at 4:00 a.m., you're probably hoping there is a way to recover the changes made between those two times. What you need is known as **PITR**.

Regrettably, if you've made a backup with the pg_dump utility at 4:00 a.m., then you won't be able to recover to any other time. As a result, the term PITR has become synonymous with the physical backup and restore technique in PostgreSQL.

Getting ready

If you have a backup made with the pg_dump utility, then give up all hope of using that as a starting point for a PITR. It's a frequently asked question, but the answer is still *no*. The reason it gets asked is exactly why we are pleading with you to plan your backups ahead of time.

First, you need to decide the point in time you would like to recover to. If the answer is *as late as possible*, then you don't need to do a PITR at all – just recover until the end of the transaction logs.

How to do it...

How do you decide what point to recover to? The point where we stop recovery is known as the recovery target. The most straightforward way is to indicate a timestamp, as in this example:

```
barman recover db1 last /var/tmp/pitr --get-wal \
    --target-time '2023-06-01 16:59:14.27452+01'
```

You will have noticed that we are recovering the backup into a path that is normally used to hold temporary files, instead of using a standard PostgreSQL data directory path. That's because the files will be restored by the barman user, as we are not using --ssh-command, and the permissions of that user do not allow writing inside a subdirectory of /var/lib/pgsql, which is accessible only to the postgres user.

Restoring this backup as a temporary directory is not necessarily wrong; in fact, it is appropriate because we are running a PITR exercise, meaning that we are creating an instance that will only live for the time required to extract the specific data we need.

For the same permission reasons, it is convenient to change the ownership of the restored backup to the postgres user, using this command:

```
chown -R postgres: /var/tmp/pitr
```

You might have to review the configuration files to change those settings that make sense only to the original database server (for instance, disable SSL in case it uses certificates that are not available on the Barman host).

After that, you can start PostgreSQL and run queries in hot standby mode, for monitoring recovery progress or extracting data, as explained in the *Recovery of a dropped/damaged table* recipe.

Finally, when you no longer need this instance, you should remember to stop PostgreSQL and remove the data directory.

 You need to be careful to specify the time zone of the target so that it matches the time zone of the server that wrote the log. This might differ from the time zone of the current server, so be sure to double-check them.

How it works...

Barman simply uses the PITR feature of PostgreSQL, which allows a user to specify a target time by setting a parameter in the `postgresql.conf` file, as shown in this example:

```
recovery_target_time = '2023-06-01 16:59:14.27452+01'
```

More generally, Barman supports three ways to define the recovery target:

- `--target-time TARGET_TIME`: The target is a timestamp.
- `--target-xid TARGET_XID`: The target is a transaction ID.
- `--target-name TARGET_NAME`: The target is a named restore point, which was created previously with the `pg_create_restore_point(name)` function.

When executed with one of these options, Barman will generate the appropriate PostgreSQL configuration so that the server will stop recovery at the requested point.

Barman also needs to create a `recovery.signal` file inside the data directory; if this file exists, PostgreSQL will start in **targeted recovery mode**, meaning that recovery will end when the database has reached the state it was in at the point of time indicated by the target. PITR works by applying individual WAL records. These correspond to individual block changes, so there are many WAL records for each transaction. The final part of any successful transaction is a commit WAL record, though there are abort records as well. Each transaction completion record has a timestamp that allows us to decide whether or not to stop at that point.

There's more...

You can also define a recovery target using a transaction ID (xid), though finding out which xid to use is somewhat difficult, and you may need to refer to external records, if they exist. Using a **Log Sequence Number (LSN)** is also possible, and equally tricky; in both cases, you can get an idea of what transaction ID, or LSN, to use, by inspecting the contents of a given WAL file with the pg_waldump utility, which is part of PostgreSQL.

Another practical way, which rarely applies after an unexpected disaster, is to define a recovery target with a label, formally known as a named restore point. A restore point is created with the pg_create_restore_point() function and requires superuser privileges. For example, let's say you have to perform a critical update of part of the data in your database. As a precaution, before you start the update, you can execute the following query as a superuser:

```
SELECT pg_create_restore_point('before_critical_update');
```

Then, you can use the before_critical_update label in the recovery_target_name option.

Finally, you can simply stop as soon as the recovery process becomes consistent by specifying recovery_target = 'immediate' in place of any other recovery target parameter.

The recovery target is specified in the server configuration and cannot change while the server is running. If you want to change the recovery target, you can shut down the server, edit the configuration, and then restart the server. However, be careful – if you change the recovery target and recovery is already past the new point, it can lead to errors. If you define a recovery_target_timestamp that has already been passed, then the recovery will stop almost immediately, though this will be after the correct stopping point. If you define recovery_target_xid or recovery_target_name parameters that have already been passed, then the recovery will just continue until the end of the logs. Restarting a recovery from the beginning using a fresh restore of the base backup is always the safest option.

Once a server completes the recovery, it will assign a new timeline. Once a server is fully available, we can write new changes to the database. Those changes might differ from the changes we made in a previous future history of the database. So, we differentiate between alternate futures using different timelines. If we need to go back and run the recovery again, we can create a new server history using the original or subsequent timelines. The best way to think about this is that it is exactly like a sci-fi novel – you can't change the past, but you can return to an earlier time and take a different action instead. However, you'll need to be careful to not get confused.

The timeline is represented by a 32-bit integer that constitutes the first eight characters in the name of a WAL file; therefore, changing the timeline means using a new series of filenames. There are cases where this is important – for instance, if you restore a backup and start that server as a new server while the original server is still running, then it's convenient that both servers archive the WAL they produce without disturbing each other. In other words, if you made a backup, then you want to be able to restore it as many times as you want, and you don't want the restored instances overwriting some files in the original backup.

By default, when recovery reaches the target, then recovery is paused and the server can be accessed with read-only queries, exactly like a hot standby replica. You can change this behavior with the --target-action option, which by default is set to pause.

This corresponds to setting recovery_target_action in the PostgreSQL configuration, as discussed in the *Delaying, pausing, and synchronizing replication* recipe in *Chapter 12, Replication and Upgrades*.

The pg_dump utility cannot be used as a base backup for a PITR. The reason for this is that a log replay contains the physical changes to data blocks, not the logical changes based on primary keys. If you reload the pg_dump utility, the data will likely go back into different data blocks, so the changes won't correctly reference the data.

See also

PostgreSQL can pause, resume, and stop recovery dynamically while the server is up. This allows you to use the hot standby feature to locate the correct stopping point more easily. You can trick hot standby into stopping recovery, which may help. See the *Delaying, pausing, and synchronizing replication* recipe in *Chapter 12, Replication and Upgrades*, on managing hot standby. This procedure is also covered by the Barman command-line utility, as mentioned in the *Hot physical backup with Barman* recipe.

You can use the pg_waldump utility to print the content of WAL files in a human-readable way. This can be very valuable to locate the exact transaction ID or timestamp, or when a certain change was committed – for instance, if we want to stop recovery right before that. pg_waldump is part of PostgreSQL and is described here: https://www.postgresql.org/docs/16/pgwaldump.html.

Recovery of a dropped/damaged table

You may drop or even damage a table in some way. Tables could be damaged for physical reasons, such as disk corruption, or they could also be damaged by running poorly specified UPDATE or DELETE commands, which update too many rows or overwrite critical data.

Recovering from this backup situation is a common request.

How to do it...

The methods of this approach differ, depending on the type of backup you have available. If you have multiple types of backup, you have a choice.

Logical – from the custom dump taken with pg_dump -F c

If you've taken a logical backup using the pg_dump utility in a custom file, then you can simply extract the table you want from the dumpfile, like so:

```
pg_restore -t mydroppedtable dumpfile -f - | psql
```

Alternatively, you can directly connect to the database using -d. If you use this option, then you can allow multiple jobs in parallel with the -j option.

When working with just one table, as in this case, this is useful only if there are things that can be done at the same time – that is, if the table has more than one index and/or constraint. More details about parallel restore are available in the *Improving the performance of backup/recovery* recipe later in this chapter.

Note that PostgreSQL can also use multiple jobs when creating one B-tree index. This is controlled by an entirely different set of parameters; see the *Maintaining indexes* recipe in *Chapter 9*, *Regular Maintenance*, for more details.

The preceding command tries to recreate the table and then load data into it. Note that the pg_restore -t option does not dump any of the indexes on the selected table. This means that we need a slightly more complex procedure than would first appear, and the procedure needs to vary, depending on whether we are repairing a damaged table or putting back a dropped table.

To repair a damaged table, we want to replace the data in the table in a single transaction. There isn't a specific option to do this, so we need to do the following:

1. Dump the data of the table (the -a option) to a script file, as follows:

   ```
   pg_restore -a -t mydamagedtable dumpfile -f - > mydamagedtable.sql
   ```

2. Edit a script named repair_mydamagedtable.sql with the following code:

   ```
   BEGIN;
   TRUNCATE mydamagedtable;
   \i mydamagedtable.sql
   COMMIT;
   ```

3. Then, run it using the following command:

   ```
   psql -f repair_mydamagedtable.sql
   ```

If you've already dropped a table, then you need to perform these steps:

1. Create a new database in which to work and name it `restorework`, as follows:

   ```
   CREATE DATABASE restorework;
   ```

2. Restore the complete schema (using the `-s` option) to the new database, like this:

   ```
   pg_restore -s -d restorework dumpfile
   ```

3. Now, dump only the definitions of the dropped table in a new file. It will contain CREATE TABLE, indexes, and other constraints and grants. Note that this database has no data in it, so specifying `-s` is optional, as follows:

   ```
   pg_dump -t mydroppedtable -s restorework > mydroppedtable.sql
   ```

4. Now, recreate the table on the main database:

   ```
   psql -f mydroppedtable.sql
   ```

5. Now, reload only the data into the `maindb` database:

   ```
   pg_restore -t mydroppedtable -a -d maindb dumpfile
   ```

If you've got a very large table, then the fourth step can be a problem because it builds indexes as well. If you want, you can manually edit the script in two pieces – one before the load (preload) and one after the load (post-load). There are some ideas for that at the end of this recipe.

Logical – from the script dump

Once you have located the PostgreSQL server on which you will prepare and verify the data to restore (the staging server), you can proceed like so:

1. Reload the script in full on the staging server, as follows:

   ```
   psql -f myscriptdump.sql
   ```

2. From the recovered database server, dump the table, its data, and all of the definitions of the dropped table into a new file:

   ```
   pg_dump -t mydroppedtable -F c mydatabase > dumpfile
   ```

3. Now, recreate the table in the original server and database, using parallel tasks to speed things up (here, we will pick two parallel jobs as an example):

```
pg_restore -d mydatabase -j 2 dumpfile
```

 The last step can be executed remotely without having to transfer `dumpfile` between systems. Just add connection parameters to `pg_restore`, as in the following example: `pg_restore -h remotehost -U remoteuser ....`

The only way to extract a single table from a script dump without doing all of the preceding steps is to write a custom script to read and extract only those parts of the file that you want. This can be complicated because you may need certain SET commands at the top of the file, the table, and data in the middle of the file, and the indexes and constraints on the table are near the end of the file. Writing a custom script can be very complex. The safer route is to follow the recipe we just described.

Physical

To recover a single table from a physical backup, you first need to recreate a PostgreSQL server from scratch, usually in a confined environment. Typically, this server is called the **recovery server** if dedicated to recovery drills and procedures, or the **staging server** if used for a broader set of cases, including testing. Then, you need to proceed as follows:

1. Recover the database server in full, as described in the previous recipes on physical recovery, including all databases and all tables. You may wish to stop at a useful point in time, in which case, you can look at the *Recovery to a point in time* recipe earlier in this chapter.

2. From the recovered database server, dump the table, its data, and all the definitions of the dropped table into a new file, as follows:

```
pg_dump -t mydroppedtable -F c mydatabase > dumpfile
```

3. Now, recreate the table in the original server and database using parallel tasks to speed things up. This can be executed remotely without needing to transfer `dumpfile` between systems:

```
pg_restore -d mydatabase -j 2 dumpfile
```

How it works...

Restoring a single table from a logical backup is relatively easy, as each logical object is backed up separately from the others, and its data and metadata can be filtered out.

However, a physical backup is composed of a set of binary data files in a complex storage format that can be interpreted by a PostgreSQL engine.

This means that the only way to extract individual objects from it, at present, is to restore the backup on a new instance and then make a logical dump, as explained in the previous recipe – there's no way to restore a single table from a physical backup in just a single step.

See also

The pg_dump and pg_restore utilities are able to split the dump into three parts: pre-data, data, and post-data. Both commands support a section option that's used to specify which section(s) should be dumped or reloaded.

Recovery of a dropped/damaged database

Recovering a complete database is also required sometimes. It's actually a lot easier than recovering a single table. Many users choose to place all of their tables in a single database; in that case, this recipe isn't relevant as a whole server physical backup restore can be used.

How to do it...

The methods differ, depending on the type of backup you have available. If you have multiple types of backup, you have a choice.

Logical – from the custom dump -F c

Recreate the database in the original server using parallel tasks to speed things along. This can be executed remotely without needing to transfer dumpfile between systems, as shown in the following example, where we use the -j option to specify four parallel processes:

```
pg_restore -h myhost -d postgres --create -j 4 dumpfile
```

Logical – from the script dump created by pg_dump

Recreate the database in the original server. This can be executed remotely without needing to transfer dump files between systems, as shown here, where we must create the empty database first:

```
createdb -h myhost myfreshdb
psql -h myhost -f myscriptdump.sql myfreshdb
```

Logical – from the script dump created by pg_dumpall

There's no easy way to extract the required tables from a script dump. You need to operate on a separate PostgreSQL server for recovery or staging purposes, and then follow these steps:

1. Reload the script in full, as follows:

   ```
   psql -f myscriptdump.sql
   ```

2. Once the restore is complete, you can dump the tables in the database by following the *Hot logical backups of one database* recipe.
3. Now, recreate the database on the original server, as described for logical dumps earlier in this recipe.

Physical

To recover a single database from a physical backup, you need to work on a separate PostgreSQL server (for recovery or staging purposes), and then you must follow these steps:

1. Recover the database server in full, as described in the previous recipes on physical recovery, including all databases and all tables. You may wish to stop at a useful point in time, in which case, you can look at the *Recovery to a point in time* recipe from earlier in this chapter.
2. Once the restore is complete, you can dump the tables in the database by following the *Hot logical backups of one database* recipe.
3. Now, recreate the database on the original server, as described for logical dumps earlier in this recipe.

Extracting a logical backup from a physical one

Once you have a physical backup, you can extract a logical backup from it, applying some of the recipes that we have already seen.

This recipe is quite short because it is essentially a combination of recipes that we have already described. Nevertheless, it is important because it clarifies that you don't need to worry about extracting logical backups if you already have physical ones.

Getting ready

You just need to decide whether you want to extract a logical backup corresponding to a specific point in time or simply to the latest available snapshot.

How to do it...

First, perform a PITR, as indicated in the *Recovery to a point in time* recipe earlier in this chapter. If you want a logical backup corresponding to the latest available snapshot, just omit the `--target-time` clause. Then, follow the *Hot logical backups of one database* recipe to take a logical backup from the temporary instance.

Finally, remember to stop the temporary instance and delete its data files.

There's more...

You can also extract other kinds of logical backups – for example, global metadata only or a logical backup of all databases; you just need to change the second half of this recipe accordingly.

Improving the performance of logical backup/recovery

Performance is often a concern in any medium-sized or large database.

Backup performance is often a delicate issue because resource usage may need to be limited to remain within certain boundaries. There may also be a restriction on the maximum runtime for the backup – for example, a backup that runs every Sunday.

Again, restore performance may be more important than backup performance, even if backup is the more obvious concern.

In this recipe, we will discuss the performance of logical backup and recovery; the physical case is quite different and is examined in the next recipe.

Getting ready

If performance is a concern or is likely to be, then you should read the *Planning backups* recipe first.

How to do it...

You can use the `-j` option to specify the number of parallel processes that `pg_dump` should use to perform the database backup. This requires that you use the `-F d` option, which selects the *directory* format, where every table is backed up into a separate data file.

You can use the `-j` option to specify the number of parallel processes that `pg_restore` should use to restore the backup, similar to what `pg_dump` supports. There is one important difference from `pg_dump` – namely, that this is compatible with both the directory format (as in `-F d`) and the custom format (as in `-F c`).

You'll have to be careful about how you select the degree of parallelism to use. A good starting point is the number of CPUs on the server. Be very careful that you don't overflow the available memory when using parallel restore. Each job will use memory up to the value of `maintenance_work_mem`, so the whole restore can begin swapping when it hits larger indexes later in the restore. Plan the size of `shared_buffers` and `maintenance_work_mem` according to the number of jobs specified.

Whether you use `psql` or `pg_restore`, you can speed up the program by assigning `maintenance_work_mem = 128MB` or more, either in `postgresql.conf` or on the user that will run the restore. If neither of those ways is easily possible, you can specify the option using the `PGOPTIONS` environment variable, as follows:

```
export PGOPTIONS ="-c work_mem = 128000"
```

This will then be used to set that option value for subsequent connections.

If you are running archiving or streaming replication, then transaction log writes can create a significant burden while restoring a logical backup. This can be mitigated by increasing the size of the WAL buffer and making checkpoints less frequent for the duration of the recovery operation.

Set `wal_buffers` between 16 MB and 64 MB, and then set `max_wal_size` to a large value, such as 20 GB, so that it has room to breathe.

If you aren't running archiving or streaming replication, or you've turned it off during the restore, then you'll be able to minimize the amount of transaction log writes. In that case, you may wish to use the single transaction option, as that will also help to improve performance.

Whatever you do, make sure that you run `ANALYZE` afterward on every object that was created. This will happen automatically if `autovacuum` is enabled. It often helps to disable `autovacuum` completely while running a large restore, so double-check that you have it switched on again after the restore. The consequence of skipping this step will be extremely poor performance when you start your application again, which can easily make everybody panic.

How it works...

Logical backup and restore involve moving data out of and into the database. That's typically going to be slower than physical backup and restore. Particularly with a restore, rebuilding indexes and constraints takes time, even when run in parallel. Plan ahead and measure the performance of your backup and restore techniques so that you have a chance when you need your database back in a hurry.

There's more...

Compressing backups is often considered a way to reduce the size of the backup for storage. Even mild compression can use large amounts of CPU. In some cases, this might offset network transfer costs, so there isn't any hard rule as to whether compression is always good.

By default, the custom dump format for logical backups will be compressed. Even when compressed, the objects can be accessed individually if required.

Using `--compress` with script dumps will result in a compressed text file, just as if you had dumped the file and then compressed it. Access to individual tables is not possible.

Improving the performance of physical backup/recovery

Physical backups are quite different from logical ones, and this difference also extends to the options available to make them faster.

In both cases, it is possible to use multiple parallel processes, although for quite different reasons. Physical backups are mostly constrained by network and storage bandwidth, meaning that the benefit of parallelism is limited, although not marginally. Usually, there is little benefit in using more than four parallel processes, and you can expect to reduce backup time to 40–60% of what it is with a single thread. And, in any case, the more threads you use, the more it will impact the current system.

Incremental backup and restore are currently available only for physical backups. Although, in theory, it is possible to implement incremental behavior for logical backup/restore, in practice, this feature does not exist yet. Perhaps this is because physical backups are by nature faster and lighter than logical ones and, therefore, more suitable for addressing higher demands.

Getting ready

Make sure that you understand the limitations of parallel and incremental backup and restore, which are only available for some tools and might require specific operation modes and choices.

For instance, parallel backup and restore are supported by Barman through the `-j` option but not by `pg_basebackup`, so you need to have configured Barman's `rsync` backup method. A similar restriction applies to incremental backup and restore.

How to do it...

In the following example, which intentionally resembles a recipe that we saw earlier, we are taking a parallel backup of the db1 server using four parallel jobs:

```
[root@backup1 ~]# barman backup db1 -j 4
```

We can restore it in parallel on the (remote) db2 server with similar syntax:

```
[root@backup1 ~]# barman recover db1 last \
    /var/lib/pgsql/16/data \
    --remote-ssh-command 'ssh postgres@db2' -j 4
```

If we want to take an incremental backup, then we can add the reuse-backup option, as shown in this example:

```
[root@backup1 ~]# barman backup db1 --reuse-backup=link
```

The process of restoring a backup is automatically incremental, provided that the rsync backup method is used; this is because Barman will copy files using rsync, whose algorithm is able to efficiently reuse existing files and transmit only the differences. If you want to force a non-incremental restore, you just need to empty the target directory before you run barman recover.

How it works...

A physical backup and restore is completely up to you. Copy those files as fast as you like and in any way you like. Put them back in the same way or a different way.

If you set backup_method=postgres in the Barman configuration, then pg_basebackup will be used for taking backups, and Barman will have the same restrictions: all files will be copied in full and all by the same process.

Conversely, if you set backup_method=rsync, then Barman will make a backup that is incremental compared to the latest existing backup for the same server and will deduplicate any file that is unchanged. This implies that the backup will take less disk space and will complete in a shorter time as well.

When restoring a backup, Barman always uses rsync, irrespective of settings. In other words, a restore is always incremental, in the sense that any files existing in the target directory are reused.

Parallel backup and restore in Barman is actually a consequence of the **parallel copy** feature, which is activated by the -j N switch and applies to both barman backup and barman recover. This feature is implemented by splitting the list of files in N sublists and running N rsync processes in parallel, one for each sublist.

There's more...

Remember that your backup speed may be bottlenecked by your disks or your network. Some larger systems have dedicated networks in place, solely for backups.

Compressing backups is a popular technique for reducing the size of the backup in storage. The actual extent of the reduction depends on the kind of data being backed up and is also affected by the algorithm and the options being used; in other words, there isn't a hard rule on what is the best level of compression, and you need to find your own best compromise between disk usage, backup time, and network transfer costs. Compression of WAL files from physical backups is a common practice. In Barman, you can activate it with a configuration setting in /etc/barman.conf globally or in /etc/barman.conf.d/ files, as shown in this example:

```
compression = gzip
```

Note that there are a number of possible choices other than gzip.

Physical backups can be compressed in various ways, depending on the exact backup mechanism used.

Using multiple processes is known as pipeline parallelism. If you're using a physical backup, then you can copy the data in multiple streams, which also allows you to take advantage of parallel compression/decompression.

See also

If making a backup is an expensive operation, then a way around this is to make the backup from a replica instead, which offloads the cost of the backup operation away from the master. Look at the recipes in *Chapter 12, Replication and Upgrades*, to see how to set up a replica.

Validating backups

In this recipe, we will use the data checksum feature to detect data corruption caused by I/O malfunctioning in advance.

It is important to discover such problems as soon as possible. For instance, we want a chance to recover lost data from one of our older backups, or we may want to stop data errors before they spread to the rest of the database when new data depends on existing data.

Getting ready

This feature is disabled by default, since it results in some overhead; it can be enabled when the cluster is initialized by using the --data-checksums option of the initdb utility, or on an existing cluster, with pg_checksums --enable.

Also, before trying this recipe, you should be familiar with how to make backups and how to restore them afterward, which are the subjects of most of this chapter.

How to do it...

First, check whether data checksums are enabled:

```
postgres=# SHOW data_checksums ;
 data_checksums
----------------
 on
(1 row)
```

If not, then you need to stop the cluster and enable checksums. This will require some downtime, so you need to wait for the next maintenance window and then run the following command:

```
$ pg_checksums --enable
```

Once data checksums are enabled, if you are making a backup with pg_basebackup, then checksums are verified while pages are read from data files. Let's look at an example:

```
$ pg_basebackup -D backup2
```

If nothing goes wrong, then the backup finishes with no output – we know already that pg_basebackup operates by default in no-news-is-good-news mode. Conversely, if a checksum fails, then the return code is non-zero, and we get a warning like the following:

```
WARNING:  checksum verification failed in file "./base/16385/16388", block 0: calculated 246D but expected C938
pg_basebackup: checksum error occurred
```

In the (unlikely) case that you have a good reason for skipping this check, you can use the no-verify-checksums option.

When a physical backup is made without pg_basebackup, there is no PostgreSQL utility that can verify checksums while the backup is being made; the check must be carried out afterward by running the pg_checksums utility against the actual files in the data directory.

Unfortunately, this utility requires the data directory to be in a clean shutdown state, which is not the case when hot physical backups are made. Therefore, we need to restore the backup to a temporary directory and then carry out a recovery process, as described in the *Recovery to a point in time* recipe previously – for instance, by using the following settings in postgresql.conf:

```
recovery_target = 'immediate'
recovery_target_action = shutdown
```

The immediate target means that the recovery will stop as soon as the data directory becomes consistent, and then PostgreSQL will shut down, which is the specified target action.

Once we have a clean data directory, we just run pg_checksums against the temporary directory, as follows:

```
$ pg_checksums -D tempdir1
```

Should any checksum fail, you will see output like the following:

```
pg_checksums: checksum verification failed in file "tempdir1/base/16385/16388", block 0: calculated checksum 246D but block contains C938
Checksum operation completed
Files scanned:   1226
Blocks scanned: 3852
Bad checksums: 1
Data checksum version: 1
```

How it works...

When the data checksum feature is enabled, each page header includes a 16-bit checksum of its contents and block number, which is updated when the page is flushed to disk.

If enabled, data checksums are verified every time a block is read from disk to shared buffers, as well as when pg_basebackup is used to perform a backup.

Since the checksum is computed and added to the block when flushing to disk, a failure must be caused by a change inside the block that occurred while the block was not cached in the shared buffers; conversely, a change occurring while the block was cached in the shared buffers will be overwritten at the next flush.

There's more...

In our example, we have shown a case where the checksum fails. The checksum mismatch will also be detected when a query causes PostgreSQL to attempt reading that block into the shared buffers.

In that case, the query will fail with an error, which is good because it protects the user from inadvertently using corrupt data:

```
postgres=# SELECT * FROM t;
WARNING:  page verification failed, calculated checksum 42501 but expected 37058
ERROR:  invalid page in block 0 of relation base/16385/16388
```

If we want to intentionally load corrupt data – for example, to attempt some repair activities – we can temporarily disable the checksum, as in the following example:

```
postgres=# SET ignore_checksum_failure = on;
postgres=# SELECT * FROM t;
WARNING:  page verification failed, calculated checksum 42501 but expected 37058
 x
----
 88
(1 row)
```

We can see that the warning is still displayed, but we can proceed to read the data.

If the data corruption results in an invalid page format, the user will get the same error, irrespective of the value of `ignore_checksum_failure`. This is intentional: this parameter eliminates the risk of undetected failures. In other words, a page with an invalid format does not need checksums to be detected, nor can it be read or amended within SQL queries.

As you would expect, only a superuser can change the `ignore_checksum_failure` parameter.

Learn more on Discord

To join the Discord community for this book – where you can share feedback, ask questions to the author, and learn about new releases – follow the QR code below:

https://discord.gg/pQkghgmgdG

12

Replication and Upgrades

Replication isn't magic, though it can be pretty cool! It's even cooler when it works, and that's what this chapter is all about.

Replication requires understanding, effort, and patience. There are a significant number of points to get right. Our emphasis here is on providing simple approaches to get you started, as well as some clear best practices on operational robustness.

PostgreSQL has included some form of native or in-core replication since version 8.2, though that support has steadily improved over time. External projects and tools have always been a significant part of the PostgreSQL landscape, with most of them being written and supported by very skilled PostgreSQL technical developers. Some people with a negative viewpoint have observed that this weakens PostgreSQL or emphasizes shortcomings. Our view is that PostgreSQL has been lucky enough to be supported by a huge range of replication tools, which together offer a wide set of supported use cases from which to build practical solutions. This view extends throughout this chapter on replication, with many recipes using tools that are not part of the core PostgreSQL project yet.

All tools mentioned in this chapter are actively enhanced by current core PostgreSQL developers. The pace of change in this area is high, as can be observed in each new release that brings improvements and new features for replication, and it is likely that some of the restrictions mentioned here could well be removed by the time you read this book. Double-check the documentation for each tool or project.

Which technique is the best? This is a question that gets asked many times. The answer varies depending on the exact circumstances. In many cases, people use one technique on one server and a different technique on other servers.

Even the developers of particular tools use other tools when appropriate. Use the right tools for the job. All the tools and techniques listed in this chapter have been recommended by us at some point, in relevant circumstances. If something isn't mentioned here by us, that could imply that it is less favorable for various reasons, and there are some tools and techniques that we would personally avoid altogether in their present form or level of maturity.

This chapter was originally written by Simon Riggs for many previous editions of this book. In his words:

> "I must also confess to being the developer or designer of many parts of the basic technology presented here. That gives me some advantages and disadvantages over other authors. It means I understand some things better than others, which hopefully translates into better descriptions and comparisons. It may also hamper me by providing too narrow a focus, though the world is big, and this book is already long enough!"

This book, and especially this chapter, covers technology in depth. As a result, we face the risk of minor errors, such as configuration problems, permissions, or even order in the execution of steps. We've gone to a lot of trouble to test all of our recommendations but, just as with software, we have learned that books can be buggy too. We hope our efforts to present actual commands, rather than just words, will be appreciated by you.

In this chapter, we will cover the following recipes:

- Replication concepts
- Replication best practices
- Setting up streaming replication
- Setting up streaming replication security
- Hot Standby and read scalability
- Managing streaming replication
- Using repmgr
- Using replication slots
- Setting up replication with TPA
- Setting up replication with CloudNativePG
- Monitoring replication

- Performance and synchronous replication (sync rep)
- Delaying, pausing, and synchronizing replication
- Logical replication
- EDB Postgres Distributed
- Archiving transaction log data
- Upgrading minor releases
- Major upgrades in-place
- Major upgrades online

Replication concepts

In this recipe, we do not solve any specific replication problem—or, rather, we try to prevent the generic problem of getting confused when discussing replication. We do that by clarifying in advance the various concepts related to replication.

Indeed, replication technology can be confusing. You might be forgiven for thinking that people have a reason to keep it that way. Our observation is that there are many techniques, each with its own advocates, and their strengths and weaknesses are often hotly debated.

There are some simple, underlying concepts that can help you understand the various options available. The terms used here are designed to avoid favoring any particular technique, and we've used standard industry terms whenever available.

Topics

Database replication is the term we use to describe technology that's used to maintain a copy of a set of data on a remote system.

There are usually two main reasons for you wanting to do this, and those reasons are often combined:

- **High availability (HA)**: Reducing the chances of data unavailability by having multiple systems, each holding a full copy of the data.
- **Data movement**: Allowing data to be used by additional applications or workloads on additional hardware. Examples of this are **Reference Data Management (RDM)**, where a single central server might provide information to many other applications, and systems for **business intelligence (BI)**/reporting.

Of course, both of these topics are complex areas, and there are many architectures and possibilities for implementing each of them.

What we will talk about here is HA, where there is *no transformation* of the data. We simply copy the data from one PostgreSQL database server to another. So, we are specifically avoiding all discussions of popular keywords such as **extract, transform, and load** (**ETL**) tools, **enterprise application integration** (**EAI**) tools, inter-database migration, and data warehousing strategies. Those are valid topics in **information technology** (**IT**) architecture; it's just that we don't cover them in this book.

Basic concepts

Let's look at the basic database cluster architecture. Typically, individual database servers are referred to as nodes. The whole group of database servers involved in replication is known as a cluster. That is the common usage of the term, but be careful—the term **cluster** has two other quite separate meanings elsewhere in PostgreSQL. Firstly, **cluster** is sometimes used to refer to the database instance, though we prefer the term **database server**. Secondly, there is a command named `cluster`, designed to sort data in a specific order within a table.

A database server that allows a user to make changes is known as a **primary** or **master** or may be described as a source of changes.

A database server that only allows read-only access is known as a **standby** or as a read replica. A standby server is an exact copy of its upstream node and, therefore, is *standing by*, meaning that it can be quickly activated and replace the upstream node should it fail (for instance).

A key aspect of replication is that data changes are captured on a master and then transferred to other nodes. In some cases, a node may send the changes it receives to other nodes, which is a process known as **cascading** or **relay**. Thus, the master is a sending node, but a sending node does not need to be a master.

Replication is often categorized by whether more than one master node is allowed, in which case, it will be known as multi-master replication. There is a significant difference between how single-master and multi-master systems work, so we'll discuss that aspect in more detail later. Each has its advantages and disadvantages.

History and scope

PostgreSQL didn't always have in-core replication. For many years, PostgreSQL users needed to use one of many external packages to provide this important feature.

Slony was the first package to provide useful replication features. **Londiste** was a variant system that was somewhat easier to use. Both of those systems provided single-master replication based on triggers. Another variant of this idea was the **bucardo** package, which offered multi-master replication using triggers. Single-master replication is unidirectional, whereas multi-master replication has multiple sources.

Trigger-based replication has now been superseded by transaction log-based replication, which provides considerable performance improvements. There is some discussion regarding exactly how much difference that makes, but log-based replication is approximately twice as fast, though many users have reported much higher gains. Trigger-based systems also have considerably higher replication lag. Lastly, triggers need to be added to each table involved in replication, making these systems more time-consuming to manage and more sensitive to production problems. These factors taken together mean that trigger-based systems are usually avoided for new developments, and we take the decision not to cover them at all in the latest edition of this book.

Outside the world of PostgreSQL, there are many competing concepts, and there is a lot of research being done on them. This is a practical book, so we've mostly avoided comments on research or topics concerning computer science.

The focus of this chapter is replication technologies that are part of the core software of PostgreSQL or will be in the reasonably near future. The first of these is known as **streaming replication (SR)**, introduced in PostgreSQL 9.0, but it is based on earlier file-based mechanisms for physical transaction log replication. In this book, we refer to this as **physical SR (PSR)** because we take the transaction log (often known as the **write-ahead log (WAL)**) and ship that data to the remote node. The WAL contains an exact physical copy of the changes made to a data block, so the remote node is an exact copy of the primary. Therefore, the remote node cannot execute transactions that write to the database because we want to keep applying the WAL from the upstream node; this type of node is known as a standby.

Starting with PostgreSQL 9.4, we introduced an efficient mechanism for reading the transaction log (WAL) and transforming it into a stream of changes; this is a process known as **logical decoding**. This was then the basis for a later, even more useful mechanism, known as **logical SR (LSR)**. This allows a receiver to replicate data without needing to keep an exact copy of the data blocks, as we do with PSR. This has significant advantages, which we will discuss later.

PSR requires us to have only a single master node due to the physical copy of data blocks, though it allows multiple standbys. LSR can be used for all the same purposes as PSR. It just has fewer restrictions and allows a great range of additional use cases. Crucially, LSR can be used as the basis of multi-master clusters.

PSR and LSR are sometimes known as **physical log SR (PLSR)** and **logical log SR (LLSR)**. Those terms are sometimes used when explaining the differences between transaction log-based and trigger-based replication.

Practical aspects

Since we refer to the transfer of replicated data as streaming, it becomes natural to talk about the flow of data between nodes as if it were a river or stream. Cascaded data can flow through a series of nodes to create complex architectures. From the perspective of any node, it may have downstream nodes that receive replicated data from it and/or upstream nodes that send data to it. Practical limits need to be understood to allow us to understand and design replication architectures.

After a transaction commits on the primary, the time taken to transfer data changes to a remote node is usually referred to as the **latency** or **replication delay**. Once the remote node has received the data, changes must then be applied to the remote node, which takes an amount of time known as the **apply delay**. The total time a record takes from the primary to a downstream node is the replication delay plus the apply delay. Be careful to note that some authors describe those terms differently and sometimes confuse the two, which is easy to do. Also note that these delays will be different for any two nodes.

Replication delay is best expressed as an interval (in seconds), but that is much harder to measure than it first appears. Since PostgreSQL 14, the delays of particular phases of replication are given with the lag columns on `pg_stat_replication`. These are derived from sampling the message stream and interpolating the current delay from recent samples.

All forms of replication are initialized in roughly the same way. First, you enable change capture and then make a full replica of the dataset on the remote node, which we refer to as the **base backup** or the **initial copy**. After that, you begin applying the changes, starting from the point immediately before the base backup started and continuing with any changes that occurred while the base backup was taking place. As a result, the replication delay immediately following the initial copy task will be equal to the duration of the initial copy task. The remote node will then begin to catch up with the primary, and the replication delay will begin to reduce. The time taken to get the lowest replication delay possible is known as the **catch-up interval**. If the primary is busy generating new changes, which can increase the time it takes for the new node to catch up, you should try to generate new nodes during quieter periods, if any exist. Note that, in some cases, the catch-up period will be too long to be acceptable. Be sure to include this understanding in your planning and monitoring. The faster and more efficient your replication system is, the easier it will be to operate in the real world. *Performance matters!*

Either replication will copy all tables or, in some cases, you can copy a subset of tables, in which case we call it **selective replication**. If you choose selective replication, you should note that the management overhead increases roughly as the number of objects managed increases. Replicated objects are often manipulated in groups known as **replication sets** to help minimize the administrative overhead.

Data loss

By default, PostgreSQL provides **asynchronous replication** (**async rep**), where data is streamed out whenever convenient for the server. If replicated data is acknowledged back to the user prior to committing, we refer to that as **sync rep**.

With sync rep, the replication delay *directly* affects the elapsed time of transactions on the primary. With async rep, the primary may continue at full speed, though this opens up a possible risk that the standby may not be able to keep pace with the primary. All replications must be monitored to ensure that a significant lag does not develop, which is why we must be careful to monitor the replication delay.

Sync rep guarantees that data is written to at least two nodes before the user or application is told that a transaction has committed. You can specify the number of nodes and other details that you wish to use in your configuration.

Single-master replication

In single-master replication, if the primary dies, one of the standbys must take its place. Otherwise, we will not be able to accept new write transactions. Thus, the designations of primary and standby are just roles that any node can take at some point. To move the primary role to another node, we perform a procedure named **switchover**. If the primary dies and does not recover, then the more severe role change is known as a **failover**. In many ways, these can be similar, but it helps to use different terms for each event.

We use the term **clusterware** for software that manages the cluster. Clusterware may provide features such as automatic failover and, in some cases, load balancing.

The complexity of failover makes single-master replication harder to configure correctly than many people would like it to be. The good news is that from an application perspective, it is safe and easy to retrofit this style of replication to an existing system. Or, put another way, since application developers don't really worry about HA and replication until the very end of a project, single-master replication is frequently the best solution.

Multinode architectures

Multinode architectures allow users to write data to multiple nodes concurrently. There are two main categories—tightly coupled and loosely coupled:

- **Tightly coupled database clusters**: These allow a single image of the database, so there is less perception that you're even connected to a cluster at all. This consistency comes at a price—the nodes of the cluster cannot be separated geographically, which means if you need to protect against site disasters, then you'll need additional technology to allow **disaster recovery** (**DR**). Clustering requires replication as well.

- **Loosely coupled database clusters**: These have greater independence for each node, allowing us to spread nodes out across wide areas, such as across multiple continents. You can connect to each node individually. There are two benefits of this. The first is that all data access can be performed quickly against local copies of the data. The second benefit is that we don't need to work out how to route read-only transactions to one or more standby nodes and read/write transactions to the primary node.

Multi-master replication

An example of a loosely coupled system would be **EDB Postgres Distributed** (**PGD**). PGD does not utilize a global transaction manager, and each node contains data that is eventually consistent between nodes. This is a performance optimization since tests have shown that trying to use tightly coupled approaches catastrophically limits performance when servers are geographically separated.

In its simplest multi-master configuration, each node has a copy of similar data. You can update data on any node, and the changes will flow to other nodes. This makes it ideal for databases that have users in many different locations, which is probably the case with most websites. Each location can have its own copy of the application code and database, giving fast response times for all your users, wherever they are located.

It is *possible* to make changes to the same data at the same time on different nodes, causing write conflicts. While these could become a problem, the reality is that it is also *easily possible* to design applications that do not generate conflicts in normal running, especially if each user is modifying their own data (for example, in social media or retail).

We need to understand where conflicts might arise so that we can resolve them. On a single node, any application that allows concurrent updates to the same data will experience poor performance because of contention.

The negative effect of contention will get much worse on multi-master clusters. In addition, the ability to write on multiple nodes forces us to implement conflict resolution in any case, to resolve data differences between nodes. Therefore, with some thought and planning, we can use multi-master technologies very effectively in the real world.

In fact, the word *conflict* has a negative connotation that does not match an objective cost/benefit analysis, at least in some cases. If the conflict resolution logic is compatible with the application model, then a conflict is nothing more than a small amount of unnecessary work that does no harm, and the application will be faster by accepting sporadic conflicts rather than trying to prevent them.

Visit https://en.wikipedia.org/wiki/Replication_(computing) for more information on this.

Other approaches to replication

This book covers in-database replication only. Replication is also possible in the application layer (that is, above the database) or in the **operating system (OS)** layers (that is, below the database):

- **Application-level replication**: For example, **HA-JDBC** and rubyrep
- **OS-level replication**: For example, **Distributed Replicated Block Device (DRBD)**

None of these approaches is very satisfying, since core database features cannot easily integrate with them in ways that truly work. From a **system administrator's (sysadmin's)** perspective, they work, but not very well from the perspective of a database architect.

Replication best practices

Some general best practices for running replication systems are described in this recipe.

Getting ready

Reading a list of best practices should be the very first thing you do when designing your database architecture. So, the best way to get ready for it is to avoid doing anything and start straight away with the next section, *How to do it...*.

How to do it...

Here are some best practices for replication:

- Use the latest release of PostgreSQL. Replication features are changing fast, with each new release improving on the previous in major ways based on our real-world experience. The idea that earlier releases are somehow more stable, and thus more easily usable, is definitely not the case for replication.

- Use similar hardware and OSs on all systems. Replication allows nodes to switch roles. If we switch over or fail over to different hardware, we may get performance issues, and it will be hard to maintain a smoothly running application.

- Configure all systems identically as far as possible. Use the same mount points, directory names, and users; keep everything the same where possible. Don't be tempted to make one system more important than others in some way. It's just a **single point of failure** (**SPOF**) and gets confusing.

- Give systems/servers good names to reduce confusion. Never, ever call one of your systems primary and the other standby. When you do a switchover, you will get very confused! Try to pick system names that have nothing to do whatsoever with their role. Replication roles will inevitably change; system names should not. If one system fails and you add a new system, never reuse the name of the old system; pick another name, or it will be too confusing. Don't pick names that relate to something in the business. Colors are also a bad choice because if you have two servers named Yellow and Red, you then end up saying things such as *There is a red alert on server Yellow*, which can easily be confusing. Don't pick place names, either. Otherwise, you'll be confused trying to remember that London is in Edinburgh and Paris is in Rome. Make sure that you use names, rather than **Internet Protocol** (**IP**) addresses.

- Set the application_name parameter to be the server name in the replication connection string. Set the cluster_name parameter to be the server name in the postgresql.conf file.

- Make sure that all tables are marked as LOGGED (the default). UNLOGGED and TEMPORARY tables will not be replicated by either **PSR** or **LSR**.

- Keep the system clocks synchronized. This helps you keep sane when looking at log files that are produced by multiple servers. You should automate this rather than do it manually, but however you do it—for instance, by relaying on **Network Time Protocol** (**NTP**) servers—make sure it works.

- Use a single, unambiguous time zone. Use **Coordinated Universal Time** (**UTC**) or something similar. Don't pick a time zone that has **Daylight Savings Time** (**DST**), especially in regions that have complex DST rules. This just leads to (human) confusion with replication, as servers are often in different countries, and time zone differences vary throughout the year. Do this even if you start with all your servers in one country because, over the lifetime of the application, you may need to add new servers in different locations. Think ahead.

- Monitor each of the database servers. If you want HA, then you'll need to check regularly that your servers are operational. We speak to many people who would like to regard replication as a one-shot deal. Think of it more as a marriage and plan for it to be a happy one!

- Monitor the replication delay between servers. All forms of replication are only useful if the data is flowing correctly between the servers. Monitoring the time it takes for the data to go from one server to another is essential for understanding whether replication is working for you. Replication can be bursty, so you'll need to watch to make sure it stays within sensible limits. You may be able to set tuning parameters to keep things low, or you may need to look at other factors.

The important point is that your replication delay is directly related to the amount of data you're likely to lose when running async rep. Be careful here because it is the replication delay, not the apply delay, that affects data loss. A long apply delay may be more acceptable as a result.

As described previously, your initial replication delay will be high, and it should reduce to a lower and more stable value over a period of time. For large databases, this could take days, so be careful to monitor it during the catch-up period.

There's more...

The preceding list doesn't actually say this explicitly, but you should use the same major version of PostgreSQL for all systems. With PSR, you are required to do that, so it doesn't even need to be said.

We've heard people argue that it's OK to have dissimilar systems and even that it's a good idea because, if you get a bug, it only affects one node. I'd say that the massive increase in complexity is much more likely to cause problems.

Setting up streaming replication

Physical replication is a technique used by many database management systems. The primary database node records changes in a transaction log (WAL), and then the log data is sent from the primary to the standby, where the log is replayed.

In PostgreSQL, PSR transfers WAL data directly from the primary to the standby, giving us integrated security and a shorter replication delay.

There are two main ways to set up streaming replication: with or without an additional archive. We present how to set it up without an external archive, as this is simpler and generally more efficient. However, there is one downside related to the retention of WAL at the primary, suggesting that the simpler approach may not be appropriate for larger databases, which is explained later in this recipe.

Getting ready

If you haven't read the *Replication concepts* and *Replication best practices* recipes at the start of this chapter, go and read them now. Note that streaming replication refers to the master node as the primary node, and the two terms can be used interchangeably.

How to do it...

You can use the following procedure for base backups:

1. Identify your primary and standby nodes and ensure that they have been configured according to the *Replication best practices* recipe. In this recipe, we assume that `host1` and `host2` are the primary and the standby, respectively.

2. Configure replication security. Create or confirm the existence of a replication user on the primary node:

   ```
   CREATE USER repuser
     REPLICATION
     LOGIN
     CONNECTION LIMIT 2
     ENCRYPTED PASSWORD 'changeme';
   ```

3. Allow the replication user on the standby node to authenticate on the primary node. The following example allows access from the standby node using password authentication encrypted with **SCRAM-SHA-256**; you may wish to consider other options. First, add the following line to `pg_hba.conf` on the primary node:

   ```
   Host replication repuser host2 scram-sha-256
   ```

4. Then, ensure that the client password file for the `postgres` user on the standby node contains the following line, as explained in the *Avoiding hardcoding your password* recipe in *Chapter 1, First Steps*:

   ```
   host1:5432:replication:repuser:changeme
   ```

5. Set the logging options in `postgresql.conf` on both the primary and the standby so that any replication connection attempts and associated failures are logged (this is not needed, but we recommend it, especially the first time when configuring replication):

   ```
   log_connections = on
   ```

6. Take a base backup of the primary node from the standby node:

   ```
   pg_basebackup -d 'host=host1 user=repuser' -D /path/to_data_dir -R -P
   ```

7. Start the standby server on host2:

   ```
   pg_ctl start -D /path/to_data_dir
   ```

8. Carefully monitor the replication delay until the catch-up period is over. During the initial catch-up period, the replication delay will be much higher than we would normally expect it to be.

How it works...

pg_basebackup will perform a base backup and populate the directory indicated with -D, and then configure the files in the newly created data directory as a standby of the upstream specified with the -d option, which is what we requested with the -R option. The -P option will enable a progress display, which can be quite useful if the base backup takes a long time.

Multiple standby nodes can connect to a single primary; max_wal_senders must be set to the number of standby nodes, plus at least 1. The default value of 10 is enough unless you are planning a large number of standbys. You may wish to set up an individual user for each standby node, though it may be sufficient just to set the application_name parameter in primary_conninfo if you only want to know which connection is used by which standby node. The architecture for streaming replication is this: on the primary, one WALSender process is created for each standby that connects to the streaming replication. On the standby node, a WALReceiver process is created to work cooperatively with the primary. Data transfer has been designed and measured to be very efficient, and data is typically sent in 8,192-byte chunks, without additional buffering at the network layer.

Both WALSender and WALReceiver will work continuously on any outstanding data and will be replicated until the queue is empty. If there is a quiet period, then WALReceiver will sleep for a while.

The standby connects to the primary using native PostgreSQL libpq connections. This means that all forms of authentication and security work for replication, just as they do for normal connections; just specify replication as the database name, which PostgreSQL will interpret as follows: it will not connect to a database called replication, but it will apply these settings to establish a PSR connection, which replicates all databases at once. Note that, for replication sessions, the standby is the client and the primary is the server if any parameters need to be configured.

Using standard PostgreSQL libpq connections also means that normal network port numbers are used, so no additional firewall rules are required. You should also note that if the connections use **Secure Sockets Layer** (**SSL**), then encryption costs will slightly increase the replication delay and the **central processing unit** (**CPU**) resources required.

There's more...

If the connection between the primary and standby drops, it will take some time for that to be noticed across an indirect network. To ensure that a dropped connection is noticed as soon as possible, you may wish to adjust the timeout settings.

The standby will notice that the connection to the primary has dropped after wal_receiver_timeout milliseconds. Once the connection is dropped, the standby will retry the connection to the sending server every wal_retrieve_retry_interval milliseconds. Set these parameters in the postgresql.conf file on the standby.

A sending server will notice that the connection has dropped after wal_sender_timeout milliseconds, set in the postgresql.conf file on the sender. Once the connection is dropped, the standby is responsible for re-establishing the connection.

Data transfer may stop if the connection drops or the standby server or the standby system is shut down. If replication data transfer stops for any reason, it will attempt to restart from the point of the last transfer. Will that data still be available? It depends on how long the standby was disconnected. If the requested WAL file has been deleted in the meantime, then the standby will no longer be able to replicate data from the primary, and you will need to rebuild the standby from scratch.

In order to avoid this scenario, there are a few options; the easiest one is now to use **replication slots**, which reserve WAL files for use by disconnected nodes. When using replication slots, it is important to watch that WAL files don't build up, causing *out-of-disk-space* errors—for instance, if one standby is disconnected for a long time and its slot prevents the deletion of old WAL files while new WAL is being produced. The amount of space taken by the WAL should be monitored, and the slot should be dropped if space reaches a critical limit. As in many cases, simple monitoring of basic measures such as available disk space can be very effective in preventing a wide range of problems with timely alerts.

There are --create-slot and --slot options in pg_basebackup, respectively, for creating a replication slot and for using it to set up the standby.

When using replication slots, we recommend setting `max_slot_wal_keep_size` to a positive value, which will define the maximum lag allowed for replication slots. Any slots that fall beyond that limit will be marked as invalid, meaning that they will no longer be considered for WAL retention. The default is -1, meaning that there is no limit.

For example, if you set `max_slot_wal_keep_size = '1GB'` and a standby is lagging more than 1 **gigabyte (GB)**, then its replication connection might break when the next checkpoint removes old WAL files in which case, that standby must be rebuilt from scratch, but this is normally preferable to breaking the primary (and all its standbys) because its `pg_wal` directory fills. If this parameter is so good, why is it not enabled by default? Because a reasonable value should be the maximum available disk space minus some allowance to let the checkpoint clear old WAL files. This depends on the workload and disk layout, and hence it is best estimated by the user.

In some cases, using a replication slot is not the best choice because it effectively means that `pg_wal` on the upstream server is used as a long-term storage solution for a large number of old WAL files for the convenience of standby nodes. A better practice for that scenario is to configure `restore_command` on the standby so that it can fetch files from the backup server (for example, Barman). The standby will no longer need a replication slot to retain WAL on the primary and will be able to retrieve WAL files from Barman instead. Barman itself will still use a replication slot, and the primary server will then be vulnerable to a prolonged failure of Barman's connection, but this will be appropriate because a production system should not be considered healthy if its backup function is failing for a long time.

The `--max-rate` option can be used to throttle the base backup taken by `pg_basebackup`, which could be desirable – for instance, if the overall network bandwidth is limited and is shared with other important services.

Setting up streaming replication security

Streaming replication is at least as secure as normal user connections to PostgreSQL.

Replication uses standard `libpq` connections, so we have all the normal mechanisms for authentication and SSL support, and all the firewall rules are similar.

Replication must be specifically enabled on both the sender and standby sides. Cascading replication does not require any additional security.

When performing a base backup, the `pg_basebackup`, `pg_receivewal`, and `pg_recvlogical` utilities will use the same type of `libpq` connections as a running, streaming standby. You can use other forms of base backup, such as `rsync`, though you'll need to set up the security configuration manually.

> **NOTE**
>
> Standbys are identical copies of the primary, so all users exist on all nodes with identical passwords. All of the data is identical (eventually), and all the permissions are the same too. If you wish to control access more closely, then you'll need different pg_hba.conf rules on each server to control this. Obviously, if your config files differ between nodes, then failover will be slightly more dramatic unless you've given that some prior thought.

Getting ready

Identify or create a user/role to be used solely for replication. Decide what form of authentication will be used. If you are going across data centers or the wider internet, take this very seriously.

How to do it...

On the primary, perform these steps:

1. Enable replication by setting a specific host access rule in pg_hba.conf.
2. Give the selected replication user/role the REPLICATION and LOGIN attributes:

   ```
   ALTER ROLE replogin REPLICATION;
   ```

   Alternatively, you can create it using this command:

   ```
   CREATE ROLE replogin REPLICATION LOGIN;
   ```

On the standby, perform these steps:

1. Request replication by setting primary_conninfo in postgresql.conf.
2. If you are using SSL connections, use sslmode=verify-full.
3. Enable per-server rules, if any, for this server in pg_hba.conf.

How it works...

Streaming replication connects to a virtual database called replication. We do this because the WAL data contains changes to objects in all databases, so in a way, we aren't just connecting to one database—we are connecting to all of them.

Streaming replication connects similarly to a normal user, except that instead of a normal user process, we are given a WALSender process.

You can set a connection limit on the number of replication connections in two ways:

- At the role level, you can do it by issuing the following command:

  ```
  ALTER ROLE replogin CONNECTION LIMIT 2;
  ```

- By limiting the overall number of WALSender processes using the `max_wal_senders` parameter

Always allow one more connection than you think is required to allow for disconnections and reconnections.

There's more...

You may notice that the WALSender process may hit 100% CPU if you use SSL with compression enabled and write lots of data or generate a large WAL volume from things such as **data definition language** (DDL) or vacuuming. You can disable compression on fast networks when you aren't paying per-bandwidth charges by using `sslcompression=0` in the connection string specified for `primary_conninfo`. Note that security can be compromised if you use compression since the data stream is easier to attack.

Hot Standby and read scalability

Hot Standby is the name for the PostgreSQL feature that allows us to connect to a standby node and execute read-only queries. Most importantly, Hot Standby allows us to run queries while the standby is being continuously updated through either file-based or streaming replication.

Hot Standby allows you to offload large or long-running queries or parts of your read-only workload to standby nodes. Should you need to switch over or fail over to a standby node, your queries will keep executing during the promotion process to avoid any interruption of service.

You can add additional Hot Standby nodes to scale the read-only workload. There is no hard limit on the number of standby nodes, as long as you ensure that enough server resources are available and parameters are set correctly—10, 20, or more nodes are easily possible.

There are two main capabilities provided by a Hot Standby node. The first is that the standby node provides a secondary node in case the primary node fails. The second capability is that we can run queries on that node. In some cases, these two aspects can come into conflict with each other and can result in queries being canceled. We need to decide the importance we attach to each capability ahead of time so that we can prioritize between them.

In most cases, the role of standby will take priority: queries are good, but it's OK to cancel them to ensure that we have a viable standby. If we have more than one Hot Standby node, it may be possible to nominate one node as standby and dedicate the others to serving queries, without any regard for their capability to act as standbys.

Standby nodes are started and stopped using the same server commands as primary servers, which were covered in earlier chapters.

Getting ready

Hot Standby can be used with physical replication as well as with **point-in-time recovery** (**PITR**).

The parameters required by Hot Standby are enabled by default on all recent PostgreSQL versions, so there is nothing you need to do in advance unless you have changed them explicitly (in which case, if you have disabled this feature, you will know that already).

How to do it...

On the standby node, changes from the primary are read from the transaction log and applied to the standby database. Hot Standby works by emulating running transactions from the primary so that queries on the standby have the visibility information they need to respect **multi-version concurrency control** (**MVCC**). This makes the Hot Standby mode particularly suitable for serving a large workload of short or fast `SELECT` queries. If the workload is consistently short, then few conflicts will delay the standby and the server will run smoothly.

Queries that run on the standby node see a version of the database that is slightly behind the primary node. We describe this behavior as the cluster being **eventually consistent**. *How long is "eventually"?* That time is exactly the replication delay plus the apply delay, as discussed in the *Replication concepts* section. You may also request that standby servers delay the application of the changes they receive from their upstreams; see the *Delaying, pausing, and synchronizing replication* recipe later on in this chapter for more information.

Resource contention (CPU, I/O, and so on) may increase the apply delay. If the server is busy applying changes from the primary, then you will have fewer resources to use for queries. This also implies that if there are no changes arriving, then you'll get more query throughput. If there are predictable changes in the write workload on the primary, then you may need to throttle back your query workload on the standby when they occur.

Replication apply may also generate conflicts with running queries. Conflicts may cause the replay to pause, and eventually, queries on the standby may be canceled or disconnected. Conflicts that can occur between the primary and queries on the standby can be classified based on their causes:

- Locks, such as access-exclusive locks
- Cleanup records
- Other special cases, such as dropping tablespaces

If cancellations do occur, they will throw either error or fatal-level errors. These will be marked with code—SQLSTATE 40001 SERIALIZATION FAILURE. The application can be programmed to detect this error code and then resubmit the same **SQL** code, given the nature of the error.

There are two sources of information for monitoring the number of conflicts. The total number of conflicts in each database can be seen using this query:

```
SELECT datname, conflicts FROM pg_stat_database;
```

You can drill down further to look at the types of conflict using the following query:

```
SELECT datname, confl_tablespace, confl_lock, confl_snapshot,
confl_bufferpin, confl_deadlock
FROM pg_stat_database_conflicts;
```

Tablespace conflicts are the easiest to understand: if you try to drop a tablespace that someone is still using, then you're going to get a conflict. Don't do that!

Lock conflicts are also easy to understand. If you wish to run certain commands on the primary—such as ALTER TABLE ... DROP COLUMN, for instance—then you must lock the table first to prevent all types of access because of the way that command is implemented. While it will leave the database in a consistent state when it completes, it is not designed to preserve that consistency at all times while it is running, meaning that another session reading that table while that command runs could get inconsistent results. For that reason, the lock request is sent to the standby server as well, and the standby will then prevent those reads, meaning that it will cancel standby queries that are currently accessing that table after a configurable delay.

On HA systems, making DDL changes to tables that cause long periods of locking on the primary can be unacceptable. You may want the tables on the standby to stay available for reads during the period in which changes are being made on the primary, even if that means that the standby might delay the application of changes when it runs a conflicting query. To do that, temporarily set these parameters on the standby: max_standby_streaming_delay= -1 and max_standby_archive_delay= -1. Then, reload the server. As soon as the first lock record is seen on the standby, all further changes will be held. Once the locks on the primary are released, you can reset the original parameter values on the standby, which will then allow changes to be made there.

Note that max_standby_streaming_delay is used when the standby is streaming WAL, which is usually the case while replication is running normally, while max_standby_archive_delay is used when WAL files are fetched using restore_command, which is the case when the standby has fallen behind considerably and is fetching older WAL from the archive (for example, Barman). There are two separate settings because the extent of what is an acceptable lag can differ between those scenarios.

Setting the max_standby_streaming_delay and max_standby_archive_delay parameters to -1 is very timid and may not be useful for normal running if the standby is intended to provide HA. No user query will ever be canceled if it conflicts with applying changes, which will cause the apply process to wait indefinitely. As a result, the apply delay can increase significantly over time, depending on the frequency and duration of queries and the frequency of conflicts. To work out an appropriate setting for these parameters, you need to understand more about the other types of conflicts, though there is also a simple way to avoid this problem entirely.

Snapshot conflicts require some understanding of the internal workings of MVCC, which many people find confusing. To avoid snapshot conflicts, you can set hot_standby_feedback = on in the standby's postgresql.conf file.

In some cases, this could cause table bloat on the primary, so it is not set by default. If you don't wish to set hot_standby_feedback = on, then you have further options to consider; you can set an upper limit with max_standby_streaming_delay and max_standby_archive_delay, as explained previously. Other conflict types (buffer pin, deadlocks, and so on) are possible, but they are rare.

Finally, if you want a completely static standby database with no further changes applied, then you can do this by modifying the configuration so that neither restore_command nor primary_conninfo is set but standby_mode is on, and then restarting the server. You can come back out of this mode, but only if the archive contains the required WAL files to catch up; otherwise, you will need to reconfigure the standby from a base backup again.

If you attempt to run a non-read-only query, then you will receive an error marked with SQLSTATE 25006 READ ONLY TRANSACTION. This could be used by the application (if aware) to redirect SQL to the primary, where it can execute successfully.

How it works...

Changes made by a transaction on the primary will not be visible until the commit is applied to the standby. So, for example, we have a primary and a standby with a replication delay of 4 seconds between them. A long-running transaction may write changes to the primary for 1 hour.

How long does it take before those changes are visible on the standby? With Hot Standby, the answer is 4 seconds after the commit on the primary. This is because changes made during the transaction on the primary are streamed while the transaction is still in progress, and in most cases, they are already applied on the standby when the commit record arrives.

You may also wish to use the remote_apply mode; see the *Delaying, pausing, and synchronizing replication* recipe later in this chapter.

Hot Standby can also be used when running a PITR, so the WAL records that are applied to the database need not arrive immediately from a live database server. We can just use file-based recovery in that case, not streaming replication.

Finally, query performance has been dramatically improved in Hot Standby over time, so it's a good idea to upgrade for that reason alone.

Managing streaming replication

Replication is great, provided that it works. Replication works well if it's understood, and it works even better if it's tested.

Getting ready

You need to have a plan for the objectives for each individual server in the cluster. *Which standby server will be the failover target?*

How to do it...

Switchover is a controlled switch from the primary to the standby. If performed correctly, there will be no data loss. To be safe, simply shut down the primary node cleanly, using either the smart or fast shutdown modes. Do not use the immediate shutdown mode because you will almost certainly lose data that way.

Failover is a forced switch from the primary node to a standby because of the loss of the primary. So, in that case, there is no action to perform on the primary; we presume it is not there anymore.

Next, we need to promote one of the standby nodes to be the new primary. A standby node can be triggered into becoming a primary node with the pg_ctl promote command.

The standby will become the primary only once it has fully caught up. If you haven't been monitoring replication, this could take some time.

Once the ex-standby becomes a primary, it will begin to operate all normal functions, including archiving files, if configured. Be careful and verify that you have all the correct settings for when this node begins to operate as a primary.

It is likely that the settings will be different from those on the original primary from which they were copied.

Note that we refer to this new server as **a primary**, not **the primary**. It is up to you to ensure that the previous primary doesn't continue to operate—a situation known as **split-brain**. You must be careful to ensure that the previous primary stays down.

Management of complex failover situations is not provided with PostgreSQL, nor is automated failover. Situations can be quite complex with multiple nodes, and appropriate clusterware is recommended and used in many cases to manage this.

There's more...

When following a switchover from one node to another, it is common to think of performing a switchover back to the old primary server, which is sometimes called failback or switchback.

Once a standby has become a primary, it cannot go back to being a standby again. So, with log replication, there is no explicit switchback operation. This is a surprising situation for many people and there is a repeated question, but it is quick to work around. Once you have performed a switchover, all you need to do is the following:

- Reconfigure the old primary node again, repeating the same process as before to set up a standby node
- Switch over from the current to the old primary node

The important part here is that if we perform the first step without deleting the files on the old primary, it allows `rsync` to go much faster. When no files are present on the destination, `rsync` just performs a copy. When similarly named files are present on the destination, then `rsync` will compare the files and send only the changes. So, the `rsync` we perform on a switchback operation performs much less data transfer than in the original copy. It is likely that this will be enhanced in later releases of PostgreSQL. There are also ways to avoid this, as shown in the `repmgr` utility, which will be discussed later.

The `pg_rewind` utility has been developed as a way to perform an automated switchback operation. It performs a much faster switchback when there is a large database with few changes to apply. To allow correct operation, this program can only run on a server that was previously configured with the `wal_log_hints = on` parameter or initialized with checksums enabled.

Using that parameter can cause more I/O on large databases, so while it improves performance for switchback, it has a considerable overhead for normal running. If you think you would like to run pg_rewind, then make sure you work out how it behaves ahead of time. Trying to run it for the first time in a stressful situation when the server is down is a bad idea.

If all goes wrong, then please remember that pg_resetwal is not your friend. It is specifically designed to remove WAL files, destroying your data changes in the process. Always back up WAL files before using it.

PostgreSQL provides a recovery_end_command utility that was used to clean up after switchover or failover with older versions when replication was based on copying WAL files to a third location (archive) that needed to be maintained; this is largely unnecessary nowadays.

See also

Clusterware may provide additional features, such as automated failover, monitoring, or ease of management of replication:

- The repmgr utility is designed to manage PostgreSQL replication and failover and is discussed in more detail in the *Using repmgr* recipe.
- The pgbouncer utility is designed to allow session pooling and routing of requests to multiple backend nodes.

Using repmgr

As we stated previously, replication is great, provided that it works. It works well if it's understood, and it works even better if it's tested. This is a great reason to use the repmgr utility.

repmgr is an open source tool that was designed specifically for PostgreSQL replication. To get additional information about repmgr, visit https://www.repmgr.org/.

The repmgr utility provides a **command-line interface (CLI)** and a management process (daemon) that's used to monitor and manage PostgreSQL servers involved in replication. The repmgr utility easily supports more than two nodes with automatic failover detection.

Getting ready

Install the repmgr utility from binary packages on each PostgreSQL node.

Set up replication security and network access between nodes according to the *Setting up streaming replication security* recipe.

How to do it...

The `repmgr` utility provides a set of single command-line actions that perform all the required activities on one node:

1. To register a new cluster with `repmgr` with the current node as its primary, use the following command:

    ```
    repmgr primary register
    ```

2. To add an existing standby to the cluster with `repmgr`, use the following command:

    ```
    repmgr standby register
    ```

3. Use the following command to request `repmgr` to create a new standby for you by copying `node1`. This will fail if you specify an existing non-empty data directory:

    ```
    repmgr standby clone node1 -D /path/of_new_data_directory
    ```

4. To switch from one primary to another one, run this command on the standby that you want to make a primary:

    ```
    repmgr standby switchover
    ```

5. To reuse an old primary as a standby, use the `rejoin` command:

    ```
    repmgr node rejoin -d 'host=node2 user=repmgr'
    ```

6. To promote a standby to be the new primary, use the following command:

    ```
    repmgr standby promote
    ```

7. To request a standby to follow a new primary, use the following command:

    ```
    repmgr standby follow
    ```

8. Check the status of each registered node in the cluster, like this:

    ```
    repmgr cluster show
    ```

9. Request a cleanup of monitoring data, as follows. This is relevant only if `--monitoring-history` is used:

    ```
    repmgr cluster cleanup
    ```

10. Create a `witness` server for use with auto-failover voting, like this:

    ```
    repmgr witness create
    ```

The preceding commands are presented in a simplified form. Each command also takes one of these options:

- `--verbose`: This is useful when exploring new features
- `-f`: This specifies the path to the `repmgr.conf` file

For each node, create a `repmgr.conf` file containing at least the following parameters. Note that the node_id and node_name parameters need to be different on each node:

```
node_id=2
node_name=beta
conninfo='host=beta user=repmgr'
data_directory=/var/lib/pgsql/16/data
```

Once all the nodes are registered, you can start the repmgr daemon on each node, like this:

```
repmgrd -d -f /var/lib/pgsql/repmgr/repmgr.conf &
```

If you would like the daemon to generate monitoring information for that node, you should set `monitoring_history=yes` in the `repmgr.conf` file.

Monitoring data can be accessed using this:

```
repmgr=# select * from repmgr.replication_status;
-[ RECORD 1 ]-------------+----------------------------
primary_node_id           | 1
standby_node_id           | 2
standby_name              | node2
node_type                 | standby
active                    | t
last_monitor_time         | 2023-10-24 16:28:41.260478+09
last_wal_primary_location | 0/6D57A00
last_wal_standby_location | 0/5000000
replication_lag           | 29 MB
replication_time_lag      | 00:00:11.736163
apply_lag                 | 15 MB
communication_time_lag    | 00:00:01.365643
```

How it works...

repmgr works with all supported PostgreSQL versions. It supports the latest features of PostgreSQL, such as cascading, sync rep, and replication slots. It can use pg_basebackup, allowing you to clone from a standby. The use of pg_basebackup also removes the need for rsync and key exchange between servers. Also, cascaded standby nodes no longer need to follow.

There's more...

The default behavior for the repmgr utility is manual failover.

The repmgr utility also supports automatic failover capabilities. It can automatically detect failures of other nodes and then decide which server should become the new primary by voting among all of the still-available standby nodes. The repmgr utility supports a witness server to ensure that there is an odd number of voters in order to get a clear winner in any decision.

Using replication slots

Replication slots allow you to define your replication architecture explicitly. They also allow you to track the details of nodes even when they are disconnected. Replication slots work with both PSR and LSR, though they operate slightly differently.

Replication slots ensure that data required by a downstream node persists until the node receives it. They are crash-safe, so if a connection is lost, the slot still continues to exist. By tracking data on downstream nodes, we avoid these problems:

- When a standby disconnects, the feedback data provided by hot_standby_feedback is lost. When the standby reconnects, it may be sent cleanup records that result in query conflicts. Replication slots remember the standby's xmin value even when disconnected, ensuring that cleanup conflicts can be avoided.
- When a standby disconnects, knowledge of which WAL files were required is lost. When the standby reconnects, we may have discarded the required WAL files, requiring us to regenerate the downstream node completely (assuming that this is possible). Replication slots ensure that nodes retain the WAL files needed by all downstream nodes.

Replication slots are required by LSR and for any other use of logical decoding. Replication slots are optional with PSR.

Getting ready

This recipe assumes that you have already set up replication according to the earlier recipes, either via manual configuration or by using repmgr.

A replication slot represents one link between two nodes. At any time, each slot can support one connection. If you draw a diagram of your replication architecture, then each connecting line is one slot. Each slot must have a unique name. The slot name must contain only lowercase letters, numbers, and underscores.

As we discussed previously, each node should have a unique name, so a suggestion would be to construct the slot name from the two node names that it links. For various reasons, there may be a need for multiple slots between two nodes, so additional information is also required for uniqueness. For two servers called alpha and beta, an example of a slot name would be alpha_beta_1.

For LSR, each slot refers to a single database rather than the whole server. In that case, slot names could also include database names.

How to do it...

If you set up replication with repmgr, then you just need to set the following in the repmgr.conf file:

```
use_replication_slots = yes
```

For manual setup, you need to follow these steps:

1. Ensure that max_replication_slots > 0 on each sending PostgreSQL node; the default of 10 is usually enough.

2. For PSR slots, you first have to create a slot on the sending node with SQL like this, which will then display its LSN after creation:

    ```
    SELECT (pg_create_physical_replication_slot
    ('alpha_beta_1', true)).lsn;
    wal_position
    ----------------
    0/5000060
    ```

3. Monitor the slot in use with the following query:

    ```
    SELECT * FROM pg_replication_slots;
    ```

4. Set the primary_slot_name parameter on the standby using the unique name that you assigned earlier:

    ```
    primary_slot_name = 'alpha_beta_1'
    ```

Note that slots can be removed using the following query when you don't need them anymore:

```
SELECT pg_drop_replication_slot('alpha_beta_1');
```

There's more...

Replication slots can be used to support applications where downstream nodes are disconnected for extended periods of time. Replication slots prevent the removal of WAL files, which are needed by disconnected nodes. Therefore, it is important to be careful that WAL files don't build up and cause *out-of-disk-space* errors due to leftover physical replication slots with no currently connected standby.

See also

See the *Logical replication* recipe for more details on using slots with LSR.

Setting up replication with TPA

The careful reader will note that the *PostgreSQL with TPA* recipe in *Chapter 1, First Steps*, is a good prerequisite for this recipe. In fact, TPA does not have a simple option to skip setting up replication, given that most users need it; so we had to manually delete the replication part from the initial `config.yml` file.

Hence, this recipe is almost identical to the *PostgreSQL with TPA* recipe in *Chapter 1, First Steps*, except that we don't do any manual deletion.

In order to avoid wasting space in a large book, we assume that you have read that recipe already, so we don't need to provide a detailed description of the directory structure of a TPA cluster again or an example of how a successful deploy completes.

Getting ready

This is exactly the same as the *PostgreSQL with TPA* recipe in *Chapter 1, First Steps*. Those are one-off activities, so if you did them already, then you are good to go.

How to do it...

First, create the cluster configuration using the `tpaexec configure` command as follows:

```
tpaexec configure mysecondcluster --architecture M1 \
--platform docker --enable-repmgr --postgresql 16
```

This command creates a directory named mysecondcluster, with all the initial TPA configuration.

Note that, this time, you will get different hostnames because TPA selects them at random from a list of suitable words. If you don't want TPA to do that, you can use the --hostnames-from option to specify a file with the hostnames that will be used. This time, the first PostgreSQL instance is called uproar (instead of kennel).

You can then deploy the cluster with the usual command:

```
tpaexec deploy mysecondcluster
```

After a few minutes, the deploy ends and you can connect to the cluster as usual, again:

```
$ cd mysecondcluster
$ ssh -F ssh_config uproar
[root@uproar ~]# su - postgres
```

At this point, we can do the same things that we did in the *Chapter 1* recipe, like connecting to PostgreSQL with psql, or connecting to the Barman node and printing the list of backups. But we can also do new things: for instance, we can inspect replication, which we didn't have in *Chapter 1, First Steps*.

We are already logged in one of the instances as the postgres user, so we can just issue repmgr cluster show to display an overview of the cluster:

```
postgres@uproar:~ $ repmgr cluster show --compact
 ID | Name    | Role    | Status    | Upstream | Location | Prio. | TLI
----+---------+---------+-----------+----------+----------+-------+-----
  1 | uproar  | primary | * running |          | default  | 100   | 1
  2 | quaff   | standby |   running | uproar   | default  | 100   | 1
  4 | knavery | standby |   running | quaff    | default  | 100   | 1
```

How it works...

TPA processes the config.yml file to compile an Ansible inventory of the roles and the specific settings that have been assigned to each target instance. Then it runs a number of tasks, which encode the best practices learned by us and our colleagues in assisting a very large number of PostgreSQL users.

The actual tasks performed on each instance depend on which roles apply to that particular instance. This means that config.yml is usually short; if that is not the case, it is usually because the cluster has a large number of nodes, or the user chooses to apply a significant number of non-default settings.

In this case, we have four nodes:

- Node 1 (uproar) with one role: `primary`
- Node 2 (quaff) with one role: `replica`
- Node 3 (unfold) with three roles: `barman`, `log-server`, `monitoring-server`
- Node 4 (knavery) with one role: `replica`

There's more...

Many other things are possible with TPA, but here we want to provide an introduction that is short enough to reach lots of people.

If the TPA recipes in this book lit the sparkle of your curiosity, please follow the official TPA documentation:

`https://www.enterprisedb.com/docs/tpa/latest/`

Setting up replication with CloudNativePG

This will probably be one of the simplest recipes in this book, and it is because the automation achieved through the cloud-native operator is such that setting up replication is reduced to increase the value of the integer that describes how many instances you want to deploy in your cluster.

Getting ready

We assume for this recipe that you already have a Kubernetes environment with the CloudNativePG operator, as described in the *PostgreSQL in Kubernetes* recipe in *Chapter 1, First Steps*. In that recipe, we wrote our first manifest to deploy a single instance. In this recipe, we will increase the number of instances of the same manifest.

How to do it...

Adding more instances is as simple as editing the number of instances in the `spec` section of the manifest file, `cluster-example.yaml`. In this case, we will have 3 instances, meaning 1 primary node and 2 replicas:

```
apiVersion: postgresql.cnpg.io/v1
kind: Cluster
metadata:
  name: cluster-example
spec:
```

```
    instances: 3
    primaryUpdateStrategy: unsupervised

    # Require 1Gi of space
    storage:
      size: 1Gi
```

How it works...

The **CloudNativePG** operator will apply the new manifest that specifies three PostgreSQL instances. It will observe that only one instance exists in the current cluster, and therefore, the cluster does not match what is defined in the manifest; thus, it will create two new pods with the corresponding PVCs to store the database files and WAL. The new pods will be clones of the current primary node connecting to the primary through PSR using a replication slot.

If at any point in time a pod fails, the operator will restart the pod to recover the amount of desired instances. This is not done by using external replication management tools such as repmgr. The CloudNativePG operator does it natively, hence its name, by communicating with Kubernetes.

There's more...

Note that we have defined an unsupervised primaryUpdateStrategy. This means that the operator will decide how to switch over and update the pods to minor version upgrades. Alternatively, one can specify it as supervised, in which case a human DBA will perform the switchover operations to upgrade the primary and standby pods.

Regarding the future of PostgreSQL clusters, recent work by the CloudNativePG team has managed to create new standby instances of 4.5 TB in 2 minutes using Kubernetes volume snapshots. This kind of progress makes us believe that more and more projects will adopt the CloudNativePG operator for their developments.

Monitoring replication

Monitoring the status and progress of your replication is essential. We'll start by looking at the server status and then query the progress of replication.

Getting ready

You'll need to start by checking the state of your server(s).

Check whether a server is up by using pg_isready or another program that uses the PQping() **application programming interface** (**API**) call. You'll get one of the following responses:

- PQPING_OK (return code 0): The server is running and appears to be accepting connections.
- PQPING_REJECT (return code 1): The server is running but is in a state that disallows connections (startup, shutdown, or crash recovery) or a standby that is not enabled with Hot Standby.
- PQPING_NO_RESPONSE (return code 2): The server could not be contacted. This might indicate that the server is not running, there is something wrong with the given connection parameters (for example, wrong port number), or there is a network connectivity problem (for example, a firewall blocking the connection request).
- PQPING_NO_ATTEMPT (return code 3): No attempt was made to contact the server—for example, invalid parameters.

NOTE

At present, pg_isready does not differentiate between a primary and a standby, though this may change in later releases, nor does it specify whether a server is accepting write transactions or only read-only transactions (a standby or a primary connection in read-only mode).

You can find out whether a server is a primary or a standby by connecting and executing this query:

```
SELECT pg_is_in_recovery();
```

A true response means *this server is in recovery*, meaning it is running in Hot Standby mode.

There is another state that may be important to monitor in replication: while the server is paused. The paused state doesn't affect user queries, but replication will not progress at all when paused. We will see more about this in the *Delaying, pausing, and synchronizing replication* recipe later in this chapter.

You can also check whether replay is paused by executing this query:

```
SELECT pg_is_wal_replay_paused();
```

How to do it...

The rest of this recipe assumes that Hot Standby is enabled. Actually, this is not an absolute requirement, but it makes things much, much easier.

Both repmgr and pgpool provide replication monitoring facilities. Munin plugins are available for graphing replication and apply delay.

Replication works by processing the WAL transaction log on servers other than the one where it was created. You can think of WAL as a single, serialized stream of messages. Each message in the WAL is identified by an 8-byte integer known as an LSN. For historical reasons (and for readability), we show this as two separate 4-byte **hexadecimal** (**hex**) numbers; for example, the LSN value 00000XXX0YYYYYYY is shown as XXX/YYYYY.

You can compare any two LSNs using pg_wal_lsn_diff(). In some column and function names, prior to PostgreSQL 10, an LSN was referred to as a **location**, a term that's no longer in use. Similarly, the WAL was referred to as an **xlog** or **transaction log**.

To understand how to monitor progress, you need to understand a little more about replication as a transport mechanism. The stream of messages flows through the system like water through a pipe, and at certain points of the pipe, you have a meter that displays the total amount of bytes (LSNs) that have flown via that point at that time. You can work out how much progress has been made by measuring the LSN at two different points in the pipe; the difference will be equal to the number of bytes that are in transit between those two points. You can also check for blockages in the pipe, as they will cause all downstream LSNs to stop.

Our pipe begins on the primary, where new WAL records are inserted into WAL files. The current insert LSN can be found using this query:

```
SELECT pg_current_wal_insert_lsn();
```

However, WAL records are not replicated until they have been written and synced to the WAL files on the primary. The LSN of the most recent WAL write is given by this query on the primary:

```
SELECT pg_current_wal_lsn();
```

Once written, WAL records are then sent to the standby. The recent status can be found by running this query on the standby (this and the later functions return NULL on a primary):

```
SELECT pg_last_wal_receive_lsn();
```

Once WAL records have been received, they are written to WAL files on the standby. When the standby has written those records, they can then be applied to it. The LSN of the most recent apply is found using this standby query:

```
SELECT pg_last_wal_replay_lsn();
```

Remember that there will always be timing differences if you run status queries on multiple nodes. What we really need is to see all of the information on one node. A view called `pg_stat_replication` provides the information that we need:

```
SELECT pid, application_name /* or other unique key */
,pg_current_wal_insert_lsn() /* WAL Insert lsn */
,sent_lsn /* WALSender lsn */
,write_lsn /* WALReceiver write lsn */
,flush_lsn /* WALReceiver flush lsn */
,replay_lsn /* Standby apply lsn */
,backend_start /* Backend start */
FROM pg_stat_replication;
-[ RECORD 1 ]----------+------------------------
pid                    | 16496
application_name       | pg_basebackup
pg_current_wal_insert_lsn | 0/80000D0
sent_lsn               |
write_lsn              |
flush_lsn              |
replay_lsn             |
backend_start          | 2023-10-27 15:25:42.988149+00

-[ RECORD 2 ]----------+--------------------
pid                    | 16497
application_name       | pg_basebackup
pg_current_wal_insert_lsn | 0/80000D0
sent_lsn               | 0/80000D0
write_lsn              | 0/8000000
flush_lsn              | 0/8000000
replay_lsn             |
backend_start          | 2023-10-27  15:25:43.18958+00
```

Each row in this view represents one replication connection. The preceding snippet shows the output from a `pg_basebackup` that is using the option

--wal-method=stream. The first connection that's shown is the base backup, while the second session is streaming WAL changes. Note that the replay_lsn value is NULL, indicating that this is not a standby.

This view is possible because standby nodes send regular status messages to their upstream to let it know how far they have progressed. If you run this query on the primary, you'll be able to see all the directly connected standbys. If you run this query on a standby, you'll see values representing any cascaded standbys, but nothing about the primary or any of the other standbys connected to the primary. Note that because the data has been sent from a remote node, the values displayed are not exactly in sync; they will each refer to a specific instant in the (recent) past. It is very likely that processing will have progressed beyond the point being reported, but we don't know that for certain. That's just physics. Welcome to the world of distributed systems!

Starting from PostgreSQL 14, replication delay times are provided directly using sampled message timings to provide the most accurate viewpoint of current delay times. Use this query:

```
SELECT pid, application_name /* or other unique key */
    ,write_lag, flush_lag, replay_lag
FROM pg_stat_replication;
```

Finally, there is another view, called pg_stat_wal_receiver, that provides information about the current standby node; this view returns zero rows on the primary and one row on a standby. pg_stat_wal_receiver contains connection information to allow you to connect to the primary server and detailed state information on the WALReceiver process.

There's more...

The pg_stat_replication view shows only the currently connected nodes. If a node is supposed to be connected but it isn't, then there is no record of it at all, anywhere. If you don't have a list of the nodes that are supposed to be connected, then you'll just miss it.

Replication slots give you a way to define which connections are supposed to be present. If you have defined a slot and it is currently connected, then you will get one row in pg_stat_replication for the connection and one row in pg_replication_slots for the corresponding slot; they can be matched via the **process identifier** (**PID**) of the receiving process, which is the same. To find out which slots don't have current connections, you can run this query:

```
SELECT slot_name, database, age(xmin), age(catalog_xmin)
FROM pg_replication_slots
WHERE NOT active;
```

To find details of currently connected slots, run something like this:

```
SELECT slot_name
FROM pg_replication_slots
JOIN pg_stat_replication ON pid = active_pid;
```

Performance and synchronous replication (sync rep)

Sync rep allows us to offer a confirmation to the user that a transaction has been committed and fully replicated on at least one standby server. To do that, we must wait for the transaction changes to be sent to at least one standby, and then have that feedback returned to the primary.

The additional time taken for the message's round trip will add elapsed time for the commit of write transactions, which increases in proportion to the distance between servers. PostgreSQL offers a choice to the user as to what balance they would like between durability and response time.

Getting ready

The user application must be connected to a primary to issue transactions that write data. The default level of durability is defined by the synchronous_commit parameter. That parameter is user-settable, so it can be set for different applications, sessions, or even individual transactions. For now, ensure that the user application is using this level:

```
SET synchronous_commit = on;
```

We must decide which standbys should take over from the primary in the event of a failover. We do this by setting a parameter called synchronous_standby_names.

NOTE

You will need to configure at least three nodes to use sync rep correctly. This is the short story, which you probably know already. For completeness, let's explain the full story, which is slightly more nuanced.

When enabling sync rep, as in the preceding example, you are requesting that a transaction is considered committed only if it is stored at least on two different nodes, so you have the guarantee that each transaction is safe even if one node suddenly fails.

Based on your request, if you only have two nodes, A and B, and (say) node B is down, then you cannot commit that transaction. This is not a limitation of the software, but simply the logical consequence of your request: you only have node A left, so there is no way to place a transaction on two different nodes.

So, either (1) you wait until you have two nodes or (2) you accept the (tiny) risk of losing the transaction after commit, should node A fail. Most people prefer (2) over (1), and if they do not like (1) or (2), then they choose (3): to spend a bit more money and add a third node, C.

How to do it...

Make sure that you have set the `application_name` parameter on each standby node. Decide the order of servers to be listed in the `synchronous_standby_names` parameter. Note that the standbys named must be directly attached standby nodes, or else their names will be ignored. Sync rep is not possible for cascaded nodes, though cascaded standbys may be connected downstream. An example of a simple four-node configuration of nodeA (primary), nodeB, nodeC, and nodeD (standbys) would be set on nodeA, as follows:

```
synchronous_standby_names = 'nodeB, nodeC, nodeD'
```

If you want to receive replies from the first two nodes in a list, then we would specify this using the following special syntax:

```
synchronous_standby_names = '2 (nodeB, nodeC, nodeD)'
```

If you want to receive replies from any two nodes, known as **quorum commit**, then use the following syntax:

```
synchronous_standby_names = 'any 2 (nodeB, nodeC, nodeD)'
```

Set synchronous_standby_names on all of the nodes, not just the primary.

You can see the sync_state value of connected standbys by using this query on the primary:

```
SELECT
application_name
,state                   /* startup, backup, catchup or streaming */
,sync_priority           /* 0, 1 or more */
,sync_state              /* async, sync or potential */
FROM pg_stat_replication
ORDER BY sync_priority;
```

There are a few columns here with similar names, so be careful not to confuse them.

The sync_state column is just a human-readable form of sync_priority. When sync_state is async, the sync_priority value will be zero (0). Standby nodes that are mentioned in the synchronous_standby_names parameter will have a non-zero priority that corresponds to the order in which they are listed. The standby node with a priority of one (1) will be listed as having a sync_state value of sync. We refer to this node as the **sync standby**. Other standby nodes configured to provide feedback are shown with a sync_state value of potential and a sync_priority value of more than 1.

If a server is listed in the synchronous_standby_names parameter but is not currently connected, then it will not be shown at all by the preceding query, so it is possible that the node is shown with a lower actual priority value than the stated order in the parameter. Setting wal_receiver_status_interval to 0 on the standby will disable status messages completely, and the node will show as an async node, even if it is named in the synchronous_standby_names parameter. You may wish to do this when you are completely certain that a standby will never need to be a failover target, such as a test server.

The state for each server is shown as one of startup, catchup, or streaming. When another node connects, it will first show as startup, though only briefly before it moves to catchup. Once the node has caught up with the primary, it will move to streaming, and only then will sync_priority be set to a non-zero value.

Catch-up typically occurs quickly after a disconnection or reconnection, such as when a standby node is restarted. When performing an initial base backup, the server will show as backup. After this, it will stay for an extended period at catchup. The delay at this point will vary according to the size of the database, so it could be a long period. Bear this in mind when configuring the sync rep.

When a new standby node moves to the streaming mode, you'll see a message like this in the primary node log:

```
LOG standby $APPLICATION_NAME is now the synchronous
standby with priority N
```

How it works...

Standby servers send feedback messages that describe the LSN of the latest transaction they have processed. Transactions committed on the primary will wait until they receive feedback saying that their transaction has been processed. If there are no standbys available for sending feedback, then the transactions on the primary will wait for standbys, possibly for a very long time. That is why we say that you must have at least three servers to sensibly use sync rep. It has probably occurred to you that you could run with just two servers. You can, but such a configuration does not offer any transaction guarantees; it just appears to. Many people are confused on that point, but please don't listen to them!

Sync rep increases the elapsed time of write transactions (on the primary). This can reduce the performance of applications from a user perspective. The server itself will spend more time waiting than before, which may increase the required number of concurrently active sessions.

Remember that when using sync rep, the overall system is still eventually consistent. Transactions committing on the primary are visible first on the standby, and a brief moment later, those changes will be visible on the primary (yes—standby, and then primary). This means that an application that issues a write transaction on the primary followed by a read transaction on the sync standby will be guaranteed to see its own changes.

You can increase performance somewhat by setting the `synchronous_commit` parameter to `remote_write`, though you will lose data if both the primary and standby crash. You can also set the `synchronous_commit` parameter to `remote_apply` when you want to ensure that all changes are committed to the synchronous standbys and the primary before we confirm back to the user. However, this is not the same thing as synchronous visibility—the changes become visible on the different standbys at different times.

There's more...

There is a small window of uncertainty for any transaction that is in progress just at the point at which the primary goes down. This can be handled within the application by checking the return code following a commit operation, rather than just assuming that it has completed successfully, as developers often do.

If the commit fails, it is possible that the server committed the transaction successfully but was unable to communicate that to the client; however, we don't know for certain. PGD resolves this problem, but unfortunately, PostgreSQL does not yet do that. A workaround to resolve that uncertainty is to recheck a unique aspect of the transaction, such as reconfirming the existence of a user ID that was inserted.

If no such object ID exists, we can create a table for this purpose:

```
CREATE TABLE TransactionCheck
(TxnId     SERIAL PRIMARY KEY);
```

During the transaction, we insert a row into that table using this query:

```
INSERT INTO TransactionCheck DEFAULT VALUES RETURNING TxnId;
```

Then, if the commit appears to fail, we can later reread this value to confirm the transaction state as committed or aborted.

Sync rep works irrespective of whether you set up replication with or without `repmgr`, as long as you have the right number of standby nodes. It is enabled by setting the appropriate parameters in PostgreSQL, so nothing needs to be done in the configuration file of `repmgr`.

Delaying, pausing, and synchronizing replication

Some advanced features and thoughts for replication are covered here.

Getting ready

If you have multiple standby servers, you may want to have one or more servers operating in a delayed apply state—for example, one hour behind the primary. This can be useful to help recover from user errors such as mistaken transactions or dropped tables without having to perform a PITR.

How to do it...

Normally, a standby will apply changes as soon as possible. When you set the recovery_min_apply_delay parameter in postgresql.conf, the application of commit records will be delayed by the specified duration. Note that only commit records are delayed, so you may receive Hot Standby cancelations when using this feature. You can prevent that in the usual way by setting hot_standby_feedback to on, but use this with caution since it can cause significant bloat on a busy primary if recovery_min_apply_delay is large.

If something bad happens, then you can hit the **Pause** button, meaning that Hot Standby provides controls for pausing and resuming the replay of changes. Precisely, do the following:

1. To pause replay, issue this query:

    ```
    SELECT pg_wal_replay_pause();
    ```

    Once replay is paused, all queries will receive the same snapshot, which facilitates lengthy repeated analysis of the database, or retrieval of a dropped table.

2. To resume (unpause) processing, use this query:

    ```
    SELECT pg_wal_replay_resume();
    ```

Be careful not to promote a delayed standby. If you have to, because your delayed standby is the last server available, then you should reset recovery_min_apply_delay, restart the server, and let it catch up before you issue a promote action.

There's more...

A standby is an exact copy of the primary. But how do you synchronize things so that the query results you get from a standby are guaranteed to be the same as those you'd get from the primary? Well, that in itself is not possible. It's just the physics of an eventually consistent system. On the one hand, we need our system to be eventually consistent because, otherwise, the synchronization would become a performance bottleneck. And even if we ignored that concern, total consistency would still be impossible because the application cannot guarantee that two different servers are queried at exactly the same time.

What we can reasonably do is synchronize two requests on different servers, meaning that we enforce their ordering—for example, we can issue a write on the primary and then issue a read from a standby in a way that is guaranteed to happen after the write. Such a case can be automatically handled by sync rep, but if we aren't using this feature, then we can achieve a similar behavior by waiting for the standby to catch up with a specific action on the primary (the write). To perform the wait, you need to do the following:

1. On the primary, perform a transaction that writes WAL—for example, create a table or insert a row in an existing table. Make sure you do that with any setting other than `synchronous_commit = off`.

2. On the primary, find the current write LSN using this query:

   ```
   SELECT pg_current_wal_insert_lsn();
   ```

3. On the standby, execute the following query repeatedly, until the LSN value returned is equal to or higher than the LSN value you read from the primary in the previous step:

   ```
   SELECT pg_last_wal_replay_lsn();
   ```

At this point, you know that your transaction has been fully replayed, so you can query the standby and see the effects of the transaction that you performed on the primary.

The following function implements the activity of waiting until we pass a given LSN:

```
CREATE OR REPLACE FUNCTION wait_for_lsn(lsn pg_lsn)
RETURNS VOID
LANGUAGE plpgsql
AS $$
BEGIN

    LOOP
        IF pg_last_wal_replay_lsn() IS NULL OR
           pg_last_wal_replay_lsn() >= lsn THEN
             RETURN;
        END IF;
        PERFORM pg_sleep(0.1);   /* 100ms */
    END LOOP;
END $$;
```

Note that this function isn't ideal since it could be interrupted while waiting due to a Hot Standby conflict. Later releases may contain better solutions.

See also

It is also possible to pause and resume logical replication, except that we use the slightly different terms of *disable* and *enable*, as shown in the following example:

```
ALTER SUBSCRIPTION mysub DISABLE;
ALTER SUBSCRIPTION mysub ENABLE;
```

Logical replication

Logical replication allows us to stream logical data changes between two nodes. By logical, we mean streaming changes to data without referring to specific physical attributes such as a block number or row ID.

These are the main benefits of logical replication:

- Performance is roughly two times better than that of the best trigger-based mechanisms.
- Selective replication is supported, so we don't need to replicate the entire database.
- Replication can occur between different major releases, which can allow a zero-downtime upgrade.

PostgreSQL provides a feature called **logical decoding**, which can be used to stream a set of changes out of a primary server. This allows a primary and, since PostgreSQL 16, a standby to become a sending node in logical replication. The receiving node uses a logical replication process to receive and apply those changes, thereby implementing replication between those two nodes.

So far, we have referred to physical replication simply as **streaming replication**. Now that we have introduced another kind of streaming replication, we have to extend our descriptions and refer either to PSR (physical) or to LSR (logical) when discussing streaming replication. In terms of security, network data transfer, and general management, the two modes are very similar. Many concepts that are used to monitor PSR can also be used to monitor LSR.

When using logical replication, the target systems are fully writable primary nodes in their own right, meaning that we can use the full power of PostgreSQL without restrictions. We can use temporary tables, triggers, different user accounts, and GRANT permissions differently. We can also define indexes differently, collect statistics differently, and run VACUUM on different schedules.

LSR works on a **publish/subscribe (pub/sub)** model, meaning that the sending node publishes changes, and the receiving node receives the changes that it has subscribed to. Because of this, we use the terms *publisher* and *subscriber* to denote, respectively, the sending and receiving nodes.

LSR works on a per-database level, not a whole-server level like PSR, because logical decoding uses the catalog to decode transactions, and the catalog is mostly implemented at the database level. One publishing node can feed multiple subscriber nodes without incurring additional disk write overhead.

Getting ready

Logical replication was introduced in PostgreSQL 10, so it is available on all currently supported PostgreSQL versions.

The procedure goes like this:

1. Identify all nodes that will work together as parts of your replication architecture; for instance, suppose that we want to replicate from node1 to node2.
2. Each LSR link can replicate changes from a single database, so you need to decide which database(s) you want to replicate. Note that you will need one LSR link for each database that you want to replicate.
3. Each LSR link will use one connection and one slot: ensure that the max_replication_slots and max_connections parameters match those requirements.
4. Likewise, each LSR link requires one WAL sender on the publisher: ensure that max_wal_senders matches this requirement.
5. Also, each LSR link requires one apply process on the subscriber: ensure that max_worker_processes matches this requirement.

How to do it...

The following steps have to be repeated once for each replicated database. In these queries, we have used mypgdb as the database name, but you obviously need to replace it with the real name of that database:

1. Dump the database schema from the published database and reload it in the subscriber database:

    ```
    pg_dump --schema-only -o schema.sql -h node1 mypgdb
    psql -1 -f schema.sql -h node2 mypgdb
    ```

2. Publish the changes from all tables with the following statement:

    ```
    CREATE PUBLICATION pub_node1_mypgdb_all
        FOR ALL TABLES;
    ```

3. Subscribe to the changes from all tables with the following statement:

```
CREATE SUBSCRIPTION sub_node1_mypgdb_all
    CONNECTION 'host=node1 dbname=mypgdb'
    PUBLICATION pub_node1_mypgdb_all;
```

Logical replication supports selective replication, which means that you don't need to specify all the tables in the database. You just need to identify the tables to be replicated, and then define publications that correspond to groups of tables that should be replicated together.

The tables that will be replicated may need some preparatory steps as well. To enable logical replication to apply UPDATE and DELETE commands correctly on the target node, we need to define how PostgreSQL can identify rows. This is known as **replica identity**. A **primary key (PK)** is a valid replica identity, so you need not take any action if you have already defined PKs on all your replicated tables. If you want to replicate tables that do not have a PK, it is worth pausing and reviewing them. With this, we mean that you should consider whether those tables have a PK or should be given one. For example, if a table has a column called customer_id that is unique and not null, and that will be updated rarely or never, then it is a valid PK, even if it is not marked as such, so you can make it an official PK.

If you have carried out that review and you still have some tables without a PK that you want to replicate, then you may need to define a replica identity explicitly by using a command like this:

```
ALTER TABLE mytable REPLICA IDENTITY USING INDEX myuniquecol_idx;
```

This means that PostgreSQL will use that index (and the columns it covers) to uniquely identify rows to be deleted or updated.

Tables in a subscriber node must have the same name as in the publisher node and be in the same schema. Tables on the subscriber must also have columns with the same name as the publisher and with compatible data types, to be able to apply incoming changes. Tables must have the same PRIMARY KEY constraint on both nodes. CHECK, NOT NULL, and UNIQUE constraints must be the same or weaker (more permissive) on the subscriber.

Logical replication also supports filtering replication, which means that only certain actions are replicated on the target node; for example, we can specify that INSERT commands are replicated while DELETE commands are filtered away.

How it works...

Logical decoding is very efficient because it reuses the transaction log data (WAL) that was already being written for crash safety. Triggers are not used at all for this form of replication. Physical WAL records are translated into logical changes that are then sent to the receiving node. Only real data changes are sent; no records are generated from changes to indexes, cleanup records from VACUUM, and so on. So, bandwidth requirements are usually reduced, depending on the exact application workload and database setup.

Changes are discarded if the top-level transaction aborts (savepoints and other subtransactions are supported normally). Changes are applied in the order of the transactions that have been committed, meaning that logical replication never breaks because of an inconsistent sequence of activities, which could instead occur with other cruder replication techniques such as statement-based replication.

On the receiving side, changes are applied using direct database calls, leading to a very efficient mechanism. SQL is not re-executed, so volatile functions in the original SQL don't produce any surprises. For example, let's say you make an update like this:

```
UPDATE table
SET
col1 = col1 + random()
,col2 = col2 + random()
WHERE key = value
```

Logical replication will send the final calculated values of col1 and col2, instead of repeating the execution of the functions (and getting different values) when we apply the changes.

PostgreSQL has a mechanism to specify whether triggers should fire or not depending on whether the changes are coming from a client session or via replication, with the default being to fire triggers only for client sessions.

This means that you can define BEFORE ROW triggers that block or filter rows as you wish, with a suitable configuration. For more information, check the documentation for the following:

- The session_replication_role parameter
- The ALTER TABLE ... ENABLE REPLICA TRIGGER syntax

Logical replication will work even if you update one or more columns of the key (or any other replica identity) since it will detect that situation and send the old values of the columns with the changed row values.

A statement that writes many rows results in a stream of single-row changes.

Locks taken at table level (LOCK) or row level (SELECT ... FOR...) are not replicated, nor are SET or NOTIFY commands.

Logical replication doesn't suffer from cancellations of queries on the apply node in the way Hot Standby does. There isn't any need for a feature such as hot_standby_feedback.

Both the publishing and subscribing nodes are primary nodes, so technically, it would be possible for writes (INSERT, UPDATE, and DELETE) and/or row-level locks (SELECT ... FOR...) to be issued on the subscriber database. As a result, it is possible that local changes could lock out, slow down, or interfere with the application of changes from the source node. It is up to the user to enforce restrictions to ensure that this does not occur. You can do this by having a user role defined specifically for replication and then using REVOKE on all access apart from the SELECT privilege to replicated tables, rather than the user role applying the changes.

Data can be read on the apply side while changes are being made. That is just normal, and it's the beautiful power of PostgreSQL's MVCC feature.

The use of replication slots means that if the network drops for some time or if one of the nodes is temporarily offline, replication will automatically pick up again from the precise point at which it stopped. This is because the replication slot keeps track of the last LSN confirmed by the replica.

There's more...

LSR can work alongside PSR in the sense that the same node can have PSR standbys and LSR subscribers at the same time. There are no conflicting parameters; just ensure that all requirements are met for both PSR and LSR.

Logical replication provides cascaded replication.

With LSR and pglogical, neither DDL nor sequences are replicated; only data changes (**data manipulation language**, or **DML**) are sent, including TRUNCATE commands.

Logical replication is one-way only, so if you want multi-master replication, see PGD, which is described in the *EDB Postgres Distributed* recipe. Also, this is currently the only logical replication software that can replicate DDL.

Subscriptions use normal user access security, so there is no need to enable *replication* via pg_hba.conf.

It is also possible to override the synchronous_commit parameter and demand that the server provides sync rep.

Since PostgreSQL 14, the streaming option for CREATE SUBSCRIPTION enables the streaming of in-progress transactions (as opposed to full decoding, then sending), and in PostgreSQL 16, it also supports the application of large transactions by parallel workers, controlled by the max_parallel_apply_workers_per_subscription parameter.

EDB Postgres Distributed

PGD is a project aiming to provide multi-master replication with PostgreSQL. There is a range of possible architectures. The first use case we support is all-nodes-to-all-nodes. PGD will eventually support a range of complex architectures, which will be discussed later.

With PGD, the nodes in a cluster can be distributed physically, allowing worldwide access to data, as well as DR. Each PGD primary node runs individual transactions; there is no globally distributed transaction manager. PGD includes replication of data changes such as DML, as well as DDL changes. New tables are added automatically to replication, ensuring that managing PGD is a low-maintenance overhead for applications.

PGD also provides global sequences, if you wish to have a sequence that works across a distributed system where each node can generate new IDs. The usual *local* sequences are not replicated.

One key advantage of PGD is that you can segregate your write workload across multiple nodes by application, user group, or geographical proximity. Each node can be configured differently, yet all work together to provide access to the same data. Some examples of use cases for this are shown here:

- Social media applications, where users need fast access to their local server, yet the whole database needs a single database view to cater for links and interconnections
- Distributed businesses, where orders are taken by phone in one location and by websites in another location, and then fulfilled in several other locations
- Multinational companies that need fast access to data from many locations, yet wish to see and enforce a single, common view of their data

PGD builds upon the basic technology of logical replication, enhancing it in various ways. We refer heavily to the previous recipe, *Logical replication*.

Getting ready

Currently, PGD can be deployed in several **Always On** architectures, and it was tested in the past on clusters of up to 99 primary nodes. Each of those nodes is a normal, fully functioning PostgreSQL server that can perform both reads and writes.

PGD establishes direct connections between each pair of nodes, forming a mesh of connections. Changes flow directly to other nodes in constant time, no matter how many nodes are in use. This is quite different from circular replication, which is a technique used by other **database management systems** (**DBMSs**) to reduce the number of connections at the expense of latency and (somewhat) simplicity.

All PGD nodes should have pg_hba.conf definitions to allow connection paths between each node pair. It would be easier to have these settings the same on all nodes, but that is not strictly required.

Each node requires an LSR link to all other nodes for each replicated database. So, an eight-node PGD cluster will require seven LSR links per node. Ensure that the settings are configured to allow for this and any possible future expansion. The parameters should be the same on all nodes to avoid confusion. Remember that these changes require a restart.

PGD nodes also require configuring the mechanism for conflict detection:

```
track_commit_timestamp = on
```

The current version of PGD is 5.3, which was released in 2023 and supports PostgreSQL 12 and later. Earlier versions of PostgreSQL were supported by previous versions of BDR, such as 3.6 and 3.7.

For more information on release compatibility, please visit the compatibility matrix in the documentation at https://www.enterprisedb.com/docs/pgd/latest/.

PGD is proprietary software owned and licensed by **EnterpriseDB** (**EDB**).

How to do it...

PGD must be deployed using an open source tool called **TPA**, based on Ansible. It is convenient to use TPA because, internally, it runs the appropriate commands in the right order so that the user doesn't actually need to run those commands directly. TPA is described in other recipes from this book, starting in *Chapter 1, First Steps*.

TPA is free software, released under the GPL v3 license. Therefore, you can download it from public repositories, as explained in the installation instructions:

https://www.enterprisedb.com/docs/tpa/latest/INSTALL/

In this section, we go through some examples of those commands, for the purpose of illustrating how the PGD technology works.

New nodes are created in one of the following ways:

- Using a command-line utility called bdr_init_physical that can convert a physical replica into a PGD node. This utility operates in three modes:
    - Using an existing physical replica
    - Creating a physical replica from a physical backup
    - Creating a physical replica from scratch (bdr_init_physical will take a base backup)
- By running the bdr.join_node_group() function

The time-consuming part is the initial data copy, which, in the first case, is carried out while the node is still a physical replica, possibly using standard methods such as pg_basebackup, a restore from a Barman backup, or a simple file copy, while in the second case, it is included in the function run.

The preceding options for joining a node can be compared in terms of which features they provide—for example, whether the following applied:

- The process can resume without having to restart from scratch if interrupted.
- The data for a single node can be copied using multiple connections in parallel.
- The data for multiple nodes can be copied concurrently.

In the following table, we compare these options:

Join method	Resume after interruption	Parallel data copy	Concurrent data copy
bdr.join_node_group()	No	Yes	No
bdr_init_physical from replica	Depends on the technique used to build the replica		Yes
bdr_init_physical from backup	Yes	Yes	Yes
bdr_init_physical from scratch	No	No	No

Table 12.1: Options available for joining a node

How it works...

PGD optimistically assumes that changes on one node do not conflict with changes on other nodes. Any conflicts are detected and then resolved automatically using a predictable last-update-wins strategy, though custom conflict handlers are supported to allow more precise definitions for particular applications.

Applications that regularly cause conflicts won't run very well on PGD; while conflicts will be resolved automatically as expected, conflicting transactions are generally more expensive than non-conflicting ones because of the extra effort required to resolve the conflict, and also because two conflicting transactions will result in a single transaction being eventually applied. Such applications would also suffer from lock waits and resource contention on a normal database; the effects will be slightly amplified by the distributed nature of PGD, but it is only the existing problems that are amplified. Applications that are properly designed to be scalable and contention-free will work well on PGD.

PGD replicates changes at the row level, though there is an optional feature available to resolve conflicts at the column level, described later. The default mechanism used by BDR has some implications for applications, as shown here:

- Suppose we perform two simultaneous updates on different nodes, like this:

```
UPDATE foo SET col1 = col1 + 1 WHERE key = value;
```

- Then, in the event of a conflict, we will keep only one of the changes (the last change). What we might like in this case is to make the changes additive; PGD provides this alternate behavior using dedicated data types called **conflict-free replicated data types** (**CRDT**).
- Two updates that change different columns on different nodes will still cause replication conflicts. PGD provides an optional feature called column-level conflict resolution, which avoids conflicts altogether in this case.

PGD also supports Eager Replication, meaning that any issues are resolved before commit.

PGD provides tools to diagnose and correct contention problems. Conflicts are logged to a conflict history table with all the necessary details so that they can be identified *ex-post* and removed at the application level. This also enables regular auditing of the conflict resolution logic, to allow a declarative verification.

There's more...

If a primary node fails, you can fail over to either logical or physical standby nodes. Other primary nodes continue processing normally—there is no wait for failover, nor is there the need for complex voting algorithms to identify the best new primary. Failed primary nodes that resume operations later will rejoin the cluster without needing any user action.

Archiving transaction log data

PSR can send transaction log data to a remote node, even if the node is not a full PostgreSQL server, so that it can be archived. This can be useful for various purposes, such as the following:

- Restoring a hot physical backup
- Investigating the contents of previous transactions

Getting ready

Normally, backups should be taken regularly on a production system; if you have configured Barman already, as described in the *Hot physical backups with Barman* recipe of *Chapter 11, Backup and Recovery*, then you are already archiving transaction logs because they are needed to restore a physical backup, so no further action is needed, and you can skip to the *How to do it...* section of the current recipe.

PostgreSQL includes two client tools to stream transaction data from the server to the client. The tools are designed using a pull model; that is, you run the tools on the node you wish the data to be saved on:

- `pg_receivewal` transmits physical transaction log data (WAL files), producing an exact copy of the original WAL files. Replication slots are not required when using this tool but could be useful.
- `pg_recvlogical` transmits the results of the logical decoding of transaction log data, producing a copy of the transformed data rather than reconstructing physical WAL files. A logical replication slot is required for this tool, created with an appropriate logical decoding plugin. Note that, in this case, you must set `wal_level` to `logical`.

You can also configure `archive_command` on the PostgreSQL server; this uses a push model to send complete WAL files to a remote location of your choice.

How to do it...

If you are backing up your PostgreSQL server using Barman, then the WAL is already archived and can be retrieved as follows:

- First, you have to establish the name of the WAL file that you wish to fetch. For instance, you can extract the name of the last WAL file from the metadata for the chosen server:

    ```
    $ barman show-server db1 | grep current_xlog
    current_xlog: 000000010000000000000001B
    ```

- Then, you can download that file in the current directory, as follows:

    ```
    $ barman get-wal -o . -P db1 000000010000000000000001B
    Sending WAL '000000010000000000000001B.partial' for server 'db1' into
    './000000010000000000000001B' file (SSH host: 172.17.0.1)
    ```

Note the message from Barman: the current WAL file has a partial suffix, as it is still being added data from new writes. Barman will download a copy that corresponds to the snapshot it has received so far.

- At this point, you can inspect the content of this WAL file:

    ```
    $ pg_waldump 000000010000000000000001B | tail
    pg_waldump: error: error in WAL record at 0/1B3221F0: invalid record
    length at 0/1B322228: expected at least 24, got 0
    rmgr: Heap        len (rec/tot):     98/    98, tx:      75392, lsn:
    0/1B321F80, prev 0/1B321F40, desc: HOT_UPDATE off 104 xmax 75392
    flags 0x10 ; new off 105 xmax 0, blkref #0: rel 1663/17055/18358 blk
    0
    (...)
    ```

Note the following:

- The error message is not something we should worry about; it just means that the last record in the WAL file does not point to a valid WAL record, which is normal considering that this is a copy of the WAL file that is currently being written.
- The pg_waldump executable might not be in the path of the shell you are using; in that case, you will need to write the full path before running it.

Internally, Barman uses either pg_receivewal or archive_command to receive the WAL from the PostgreSQL server, depending on how it was configured. In either case, you can retrieve and inspect WAL files as in the example we just described.

You can run a standalone `pg_receivewal` process on the archive node, as in this example:

```
pg_receivewal -D /pgarchive/alpha -d "$MYCONNECTIONSTRING" &
```

Note, however, that several users choose Barman, which is also capable of restarting `pg_receivewal` if it crashes and of compressing WAL files when they are stored while returning them uncompressed, as in the preceding example.

Also, you can add the `--slot=slotname` parameter if you want `pg_receivewal` to use a replication slot that you had previously created.

There's more...

The `pg_recvlogical` utility is somewhat different because it prints the contents of the transaction data it receives, rather than just making a copy of the remote WAL file. This utility requires a logical replication slot, and it is able to create one. In this example, we create a new logical slot attached to the `mydb` database:

```
$ pg_recvlogical -d mydb --slot=test1 --create-slot
```

Once a slot exists, we can use it—for example, to display the decoded WAL to stdout:

```
$ pg_recvlogical -d mydb --slot=test1 --start -f -
BEGIN 75811
table bdr.global_consensus_journal: INSERT: log_
index[bigint]:17206 term[bigint]:0 origin[oid]:3643123840
req_id[bigint]:-296718623095749962 req_payload[bytea]:'\
x00000067d925a880000278c7f519d7e700000000114ee5700003bd9500000000114ee570000
40000000d0000000d0000001800000019000000024552b912000000001b36f938000278c7f38
d18fb000000006a9290f200000000114e65f0000278c7f50feb3d00000000' trace_
context[bytea]:'\x736e146a38578a207dbf6e2e01'
COMMIT 75811
(...)
```

While playing with this feature for the first time, try the `--verbose` option, which is supported by all the previous tools.

For more details on logical decoding plugins, refer to the *Logical replication* recipe earlier in this chapter.

Replication monitoring will show `pg_receivewal` and `pg_recvlogical` in exactly the same way as it shows other connected nodes, so there is no additional monitoring required. The default `application_name` parameter is the same as the name of the tool, so you may want to set that parameter to something more meaningful to you.

You can archive WAL files using sync rep by specifying `pg_receivewal --synchronous`. This causes a disk flush (`fsync`) on the client so that WAL data is robustly saved to disk. It then passes status information back to the server to acknowledge that the data is safe (regardless of the setting of the `-s` parameter). There is also a third option, which is faster (and more dangerous)—namely, `pg_receivewal --no-sync`.

See also

- The `pg_waldump` program is an additional server-side utility documented here: https://www.postgresql.org/docs/16/pgwaldump.html
- The `pg_receivewal` program is documented here: https://www.postgresql.org/docs/16/app-pgreceivewal.html
- The `pg_recvlogical` program is documented here: https://www.postgresql.org/docs/16/app-pgrecvlogical.html

Upgrading minor releases

Minor release upgrades are released regularly by all software developers, and PostgreSQL has had its share of corrections. When a minor release occurs, we bump the last number, usually by one. So, the first release of a major release such as 16 is 16.0. The first set of bug fixes is 16.1, then 16.2, and so on.

The PostgreSQL community releases new bug fixes quarterly. If you want bug fixes more frequently than that, you will need to subscribe to a PostgreSQL support company. This recipe is about moving from a minor release to a minor release.

Getting ready

First, get hold of the new release, by downloading either the source or fresh binaries.

How to do it…

In most cases, PostgreSQL aims for minor releases to be simple upgrades. We put in great effort to keep the on-disk format the same for both data/index files and transaction log (WAL) files, but this isn't always the case; some files can change.

The upgrade process goes like this:

1. Read the release notes to see whether any special actions need to be taken for this particular release. Make sure that you consider the steps that are required by all extensions that you have installed.
2. If you have professional support, talk to your support vendor to see whether additional safety checks over and above the upgrade instructions are required or recommended. Also, verify that the target release is fully supported by your vendor on your hardware, OS, and OS release level; it may not be, yet.
3. Apply any special actions or checks; for example, if the WAL format has changed, then you may need to reconfigure log-based replication following the upgrade. You may need to scan tables, rebuild indexes, or perform some other actions. Not every release has such actions, and we try to keep compatibility for minor releases so that they exist only in case they are needed by a bug fix; in any case, watch closely for them because, if they exist, then they are important.
4. If you are using replication, test the upgrade by disconnecting one of your standby servers from the primary.
5. Follow the instructions for your OS distribution and binary packager to complete the upgrade. These can vary considerably.
6. Start up the database server being used for this test, apply any post-upgrade special actions, and check that things are working for you.
7. Repeat *Steps 4 to 6* for other standby servers.
8. Repeat *Steps 4 to 6* for the primary server.

How it works...

Minor upgrades mostly affect the binary executable files, so it should be a simple matter of replacing those files and restarting, but please check.

There's more...

When you restart the database server, the contents of the buffer cache will be lost. The `pg_prewarm` module provides a convenient way to load relation data into the PostgreSQL buffer cache.

You can install the `pg_prewarm` extension that's provided by default, as follows:

```
postgres=# CREATE EXTENSION pg_prewarm;
CREATE EXTENSION
```

You can perform pre-warming for any relation:

```
postgres=# select pg_prewarm('job_status');
 pg_prewarm
------------
          1
```

The return value is the number of blocks that have been pre-warmed.

Major upgrades in-place

PostgreSQL provides an additional supplied program, called pg_upgrade, which allows you to migrate between major releases, such as from 9.6 to 11, or from 11 to 15; alternatively, you can upgrade straight to the latest server version. These upgrades are performed in-place, meaning that we upgrade our database without moving to a new system. That does sound good, but pg_upgrade has a few things that you may wish to consider as potential negatives, which are outlined here:

- The database server must be shut down while the upgrade takes place.
- Your system must be large enough to hold two copies of the database server: old and new copies. If it's not, then you have to use the link option of pg_upgrade, or use the *Major upgrades online* recipe, coming next in this chapter. If you use the link option on pg_upgrade, then there is no pg_downgrade utility. The only option in that case is a restore from backup, and that means extended unavailability while you restore.
- If you copy the database, then the upgrade time will be proportional to the size of the database.
- The pg_upgrade utility does not validate all your additional add-in modules, so you will need to set up a test server and confirm that these work, ahead of performing the main upgrade.

The latest pg_upgrade utility supports versions from PostgreSQL 9.2 onward and allows you to go straight from your current release to the latest release in one hop.

Getting ready

Find out the size of your database (using the *How much disk space does a database use?* recipe in *Chapter 2, Exploring the Database*). If the database is large or you have an important requirement for availability, you should consider making the major upgrade using replication tools as well. Then, check out the next recipe.

How to do it...

Once you are ready to perform major upgrade, follow these steps:

1. Read the release notes for the new server version to which you are migrating, including all of the intervening releases. Pay attention to the **Incompatibilities** section carefully; PostgreSQL changes from release to release. Assume this will take some hours.

2. Set up a test server with the old software release on it. Restore one of your backups on it. Upgrade that system to the new release to verify that there are no conflicts from software dependencies. Test your application. Make sure that you identify and test each add-in PostgreSQL module you were using to confirm that it still works at the new release level.

3. Back up your production server. Prepare for the worst but hope for the best!

4. Most importantly, work out who you will call if things go badly, and exactly how to restore from that backup you just took.

5. Install new versions of all the required software on the production server and create a new database server.

6. Don't disable security during the upgrade. Your security team will do backflips if they hear about this. Keep your job!

7. Now, go and do that backup. Don't skip this step; it isn't optional. Check whether the backup is actually readable, accessible, and complete.

8. Shut down the database servers.

9. Run `pg_upgrade -v` and then run any required post-upgrade scripts. Make sure that you check whether any were required.

10. Start up the new database server and immediately run a server-wide `ANALYZE` operation using `vacuumdb --analyze-in-stages`.

11. Run through your tests to check whether they worked or whether you need to start performing a contingency plan.

12. If all is OK, re-enable wide access to the database server. Restart the applications.

How it works...

The `pg_upgrade` utility works by creating a new set of database catalog tables and then recreating the old objects in the new tables using the same IDs as before.

The `pg_upgrade` utility works easily because the data block format hasn't changed between some releases. Since we can't (always) see the future, make sure you read the release notes.

There's more...

In-place upgrades are also possible with PGD. In fact, this is one of the most important features because it allows true **zero downtime upgrades**, with a simple procedure that does not require additional servers and completes in a short time even with a large database.

In other words, the existing Rolling Upgrades feature can now happen in place, and the other nodes of the cluster keep operating normally while one node in turn is being upgraded.

However, instead of pg_upgrade, you need to use a tool called bdr_pg_upgrade, which will ensure that the additional metadata needed by PGD is preserved during the upgrade.

Major upgrades online

Upgrading between major releases is hard and should be deferred until you have some good reasons and sufficient time to get it right.

You can use replication tools to minimize the downtime required for an upgrade, so we refer to this recipe as an **online upgrade**.

How to do it...

The following general steps should be followed, allowing at least a month for the complete process to ensure that everything is tested and everybody understands the implications:

1. Set up a new release of the software on a new test system.
2. Take a standalone backup from the main system and copy it to the test system.
3. Test the applications extensively against the new release on the test system.

When everything works and performs correctly, then proceed to the next step.

1. Set up a connection pooler to the main database (it may be there already).
2. Set up logical replication for all tables from the old to new database servers, as described in the *Logical replication* recipe earlier in this chapter.
3. Make sure that you wait until all the initial copy tasks have completed for all tables.

At this point, you have a copy of the data that you can use for testing with the next steps.

1. Stop the replication.
2. Retest the application extensively against the new release on live data.

You might have to repeat *Steps 5* to *8* more than once in case you require a new copy of the data—for example, if you want to repeat a test multiple times that affects the contents, or you simply want to test against more recent data.

Then, when you are ready for the final cutover, we can proceed to the next steps.

1. Perform *Steps 5* and *6* again, in order to create a new replica of the production data.
2. Pause the connection pool.
3. Switch the configuration of the pool over to the new system and then reload.
4. Resume the connection pool (so that it now accesses a new server).

The actual downtime for the application is the length of time to execute these last three steps.

How it works...

This recipe allows online upgrades with zero data loss because of the use of the clean switchover process. There's no need for lengthy downtime during the upgrade, and there's a much-reduced risk in comparison with an in-place upgrade, thanks to the ability to carry out extensive testing with less time pressure. It works best with new hardware and is a good way to upgrade the hardware or change the disk layout at the same time.

This procedure is also very useful for those cases where binary compatibility is not possible, such as changing server encoding or migrating the database to a different OS or architecture, where the on-disk format will change as a result of low-level differences, such as endianness and alignment.

Learn more on Discord

To join the Discord community for this book – where you can share feedback, ask questions to the author, and learn about new releases – follow the QR code below:

https://discord.gg/pQkghgmgdG

packt.com

Subscribe to our online digital library for full access to over 7,000 books and videos, as well as industry leading tools to help you plan your personal development and advance your career. For more information, please visit our website.

Why subscribe?

- Spend less time learning and more time coding with practical eBooks and Videos from over 4,000 industry professionals
- Improve your learning with Skill Plans built especially for you
- Get a free eBook or video every month
- Fully searchable for easy access to vital information
- Copy and paste, print, and bookmark content

At www.packt.com, you can also read a collection of free technical articles, sign up for a range of free newsletters, and receive exclusive discounts and offers on Packt books and eBooks.

Other Books You May Enjoy

If you enjoyed this book, you may be interested in these other books by Packt:

Learn PostgreSQL

Luca Ferrari

Enrico Pirozzi

ISBN: 9781837635641

- Gain a deeper understanding of PostgreSQL internals like transactions, MVCC, security and replication
- Enhance data management with PostgreSQL's latest partitioning features
- Choose the right replication strategy for your database

- See concrete examples of how to migrate data from another database, perform backups and restores, monitor your PostgreSQL installation and more
- Ensure security and compliance with schemas and user privileges
- Create customized database functions and extensions
- Get to grips with server-side programming, window functions, and triggers

Mastering PostgreSQL 15, Fifth Edition

Hans-Jürgen Schönig

ISBN: 9781803248349

- Gain a deeper understanding of PostgreSQL internals like transactions, MVCC, security and replication
- Enhance data management with PostgreSQL's latest partitioning features
- Choose the right replication strategy for your database
- See concrete examples of how to migrate data from another database, perform backups and restores, monitor your PostgreSQL installation and more
- Ensure security and compliance with schemas and user privileges
- Create customized database functions and extensions
- Get to grips with server-side programming, window functions, and triggers

Packt is searching for authors like you

If you're interested in becoming an author for Packt, please visit authors.packtpub.com and apply today. We have worked with thousands of developers and tech professionals, just like you, to help them share their insight with the global tech community. You can make a general application, apply for a specific hot topic that we are recruiting an author for, or submit your own idea.

Share your thoughts

Now you've finished PostgreSQL 16 Administration Cookbook, we'd love to hear your thoughts! Scan the QR code below to go straight to the Amazon review page for this book and share your feedback or leave a review on the site that you purchased it from.

https://packt.link/r/1835460585

Your review is important to us and the tech community and will help us make sure we're delivering excellent quality content.

Index

A

Access Control List (ACL) 227, 306
ACID properties 257
actions
  performing, on tables 279-284
Advanced Analytics for Extremely Large European Databases (AXLE) 229
Advanced Encryption Standard (AES) 252
all databases
  logical backups 468, 469
anonymous code block 282
Application Programming Interface (API) 40
apply delay 514
Asymmetric Cryptography 248
asynchronous replication (async rep) 515
atomicity 257
audit log
  managing 230, 231
auto-freezing
  avoiding 387, 388
automatic database maintenance
  controlling 380-387
autonomous transactions 311
average tuple density 81

B

backend 134
Backup and Recovery Manager (Barman) 474
  recovery, types 488
  reference link 475
  using, for physical backups 474-483
backups
  planning 463-465
  validating 503-506
base backup 514
Berkeley Software Distribution (BSD) license 8
biggest tables 76, 77
bloated tables and indexes
  identifying and fixing 392-398
block range index (BRIN index) 452
bulk data changes, with server-side procedures
  creating, with transactions 187-191
business intelligence (BI) 511

C

cascading process 512
catch-up interval 514
Central-Processing Unit (CPU) 237, 522
Certificate Authority (CA) 238
checkpoints 461
client certificate
  using, select database user 242
Cloud Native Computing Foundation (CNCF) 41
CloudNativePG (CNPG) 41, 44
  reference link 44

replication, setting up with 538, 539
cloud security
  setting up, with predefined roles 249-253
cluster 64, 512
clusterware 515
column-level encryption
  using 244-248
  using, for sensitive data 248, 249
Command-Line Interface (CLI) 40, 252
Common Table Expressions (CTEs) 431
complete database server recovery 483-489
complex SQL queries
  simplifying 426-430
conditional psql script
  writing 266, 267
conditional statements 267
configuration parameters
  setting, for database server 93-99
  setting, in programs 99-102
configuration settings, for session
  searching 102-104
conflict-free replicated
    data types (CRDT) 559
connection pool 11, 144
  setting up 144-146
connection service file
  using 32
constraint
  adding concurrently 297-300
contrib modules 110
coud-native monitoring 339, 340
covering indexes 439
crash recovery
  controlling 460-463
cross-tab query 427

customer relationship management (CRM)
    system 414
custom format 466

D

data
  consolidating, with MERGE
      command 195-197
  loading, from flat files 183-187
  loading, from spreadsheet 180-182
  sampling, randomly 176-179
database
  planning 91-93
database access
  auditing 227, 228
  audit log, managing 230, 231
  data changes, auditing 231, 232
  SQL statements, auditing 228, 229
  table access, auditing 230
Database as a Service (DBaaS) 35
database cluster 14, 120
database disk space
  using 71, 72
database extensions
  listing 82, 83
Database Management System
    (DBMS) 78, 557
database memory
  using 72-74
database object definitions
  backups 469, 470
database objects
  names, selecting 154-156
database replication 511
  data movement 511
  high availability (HA) 511

Index 577

databases
  listing, on database server 64-66
Databases as a Service (DBaaS) 249
database server 120, 512
  configuration parameters, setting 93-99
  databases, listing 64-66
  master 512
  primary 512
  standby 512
  starting, manually 121-124
database server files
  locating 57-59
database server's, message log
  locating 60, 61
  working 62
database's system identifier
  locating 63
  working 64
database tables 68-71
database user
  client certificate, using to select 242
  creating 218-220
data changes
  auditing 231, 232
data compression
  types 333
  using 333-335
Data Definition Language (DDL) 227, 257, 422, 525
data directory path 121
Data Manipulation Language (DML) 321
data querying
  update, considering 370
data tables
  columns, adding/removing 284-293

DBA tasks
  pgAdmin, using 271-276
default transaction isolation level 456
Denial-of-Service (DoS) attack 221
design
  deciding on, for multitenancy 136, 137
disaster recovery (DR) 516
disk space, by temporary data
  usage 365-367
Distributed Replicated Block Device (DRBD) 517
Domain Name System (DNS) 239
dropped/damaged database
  physical backups 498
  recovery 497
  recovery, from custom dump -F c 497
  recovery, from script dump created by pg_dump 497
  recovery, from script dump created by pg_dumpall 498
dropped/damaged table
  physical backup 496
  recovery 493
  recovery, from custom dump taken with pg_dump -F c 494, 495
  recovery, from script dump 495, 496
duplicate rows
  IP address range allocation 168, 169
  multiple indexes, creating accidentally 167
  preventing 164-167
  range of time 169
  uniqueness, without indexes 167, 168
duplicates
  identifying and removing 159-164
duplicate SSL connection attempts
  avoiding 241

E

EAI tools 512
EDB Audit Extension (edbaudit) 252
EDB Postgres Advanced Server (EPAS) 248
EDB Postgres Distributed (PGD) 455, 556-558
  working 559
encryption (SSL / GSSAPI)
  used, for connecting 236-238
EnterpriseDB (EDB) 7, 446
entity-relationship model (ERM) 426
enum data type
  definition, modifying 294-297
Equal Probability of Selection (EPS) 179
ETL tools 512
EXPLAIN options 422
EXPLAIN SQL command
  reference link 423
Explain tool (pgAdmin) 272
expression evaluation 449
extensibility 109
Extensible Markup Language (XML) 90, 422
extensions 82, 113
external module, to PostgreSQL
  adding 109-112
  installing, from PGXN 111
  installing, from source code 112
  installing, with software installer 110, 111
external usernames, to database roles
  mapping 243, 244
Extra Packages Enterprise Linux (EPEL) 475
  INI format configuration file, options 480
extrapolation 80

F

failed connection
  troubleshooting 33-35
failover 515
Filesystem Hierarchy Standard (FHS) 58
flat files
  data, loading from 183-187
foreign databases
  objects, accessing 316-319
Foreign Data Wrapper 312
Foreign Key (FK) 117, 434
forks 77
Free Space Map (FSM) 77
functions 86
function side-effects 311

G

General Data Protection Regulation (GDPR) 250
GENERATED data columns
  using 331, 332
generic monitoring tools
  information, obtaining 343
Genetic Query Optimization (GEQO) 431
Geographical Information System (GIS) 5
Graphical User Interface (GUI) 97
grep 347

H

Heap-Only Tuples (HOT) 393, 441
High-Availability (HA) 92
Host-Based Authentication (HBA) file 131

Hot Standby
 feature 6, 525
 mode 488
 read scalability 525
 using 526-528
 using, with physical replication 526
 working 528

HyperText Markup Language (HTML) report 424

I

Icinga 339
identifier (ID) 424
idle in-transaction sessions
 terminating 358
indexes 438-440
 maintaining 403-406
indexes, in PostgreSQL
 standard names 155, 156
index-only scans 439
index-only scans feature 78
index-only scan technique 78
Information and Communication Technology (ICT) 474
information schema 69
initdb utility 58
initial copy 514
initialization fork 77
installed extensions
 managing 114-117
installed module/extension
 using 113, 114
Internet Assigned Numbers Authority (IANA) 13, 120
Internet Protocol (IP) address 231, 518

I/O statistics
 monitoring 362-364

J

Java Database Connectivity (JDBC) 33, 100
JavaScript Object Notation (JSON) 422
jobs
 scheduling, for regular background execution 276-279
Join Pushdown 319
JSON data types
 examples 199-203
 usage, deciding 197, 198
Just-In-Time (JIT) compilation 449
 expression evaluation 449
 tuple deforming 449
 using 449, 452

K

Key Management System (KMS) 252
Kubernetes
 PostgreSQL, working with 41-43

L

large tables, with table partitioning
 dealing with 191-194
latency 514
Lightweight Directory Access Protocol (LDAP) 32
 client, setting up to use 236
 integrating with 235
 User Name Map feature, replacing 236
locking problems 422
log_destination parameter 63

logical backups 484
  extracting, from physical backup 498
  of all databases 468, 469
  of one database 465-468

logical backup/recovery
  performance, improving 499-501

logical decoding 513, 551

logical log SR (LLSR) 514

logical replication 551-553
  working 554, 555

logical SR (LSR) 513

Log Sequence Number (LSN) 492

long-running queries
  monitoring 351

M

maintenance
  planning 409-411

major upgrades
  performing 565, 566
  performing online 567, 568

master database server 512

materialized views
  creating 329
  using 329, 330, 433

MERGE command
  data, consolidating with 195-197

Minikube 41

minor releases
  upgrading 563, 564

modules 82

monitoring tools
  PostgreSQL information, providing to 341

multi-master replication 516

multinode architectures, replication 516

multiple client certificates
  using 242

multiple PgBouncer
  running on same port, to use multiple cores 150-152

multiple schemas
  using 137-140

multiple servers
  accessing, with same host and port 148, 149
  running, on one system 142, 143

multitenancy 136

Multiversion Concurrency Control (MVCC) 3, 78, 125, 371, 387, 392 526
  design 422

Munin 339

N

Nagios 339

namespace 470

Network Interface Cards (NICs) 16

network/remote users
  access, enabling for 15-18

Network Time Protocol (NTP) servers 518

new connections
  preventing 131, 132

NOLOGIN users
  forcing, to disconnect 222

non-default settings
  used, for searching parameters 104, 105

non-materialized view 330

number of rows returned
  reducing 423, 424

O

object dependencies 84-87
object-relational database management system (ORDBMS) 52
object-relational mappers (ORMs) 424
objects
　accessing, in other foreign databases 316-319
　accessing, in other PostgreSQL databases 311-316
　moving, between schemas 302, 303
　moving, between tablespaces 308-311
objects, with quoted names
　handling 156-158
obscure table
　usage, monitoring 359, 360
offending database sessions
　terminating 357, 358
one database
　logical backups 465-468
ON_ERROR_STOP variable 262
Online Transaction Processing (OLTP) 350, 453
online upgrades
　performing 567, 568
Optimal Flexible Architecture (OFA) 59
optimistic locking
　using, to avoid long lock waits 455, 456
order of magnitude (OOM) 441
Out-Of-Memory (OOM) 107

P

parallel copy feature 502
parallel query
　using 446-449

parameters
　searching, with non-default settings 104, 105
　setting, for particular groups of users 106, 107
partial indexes 439, 444
partitioned tables 455
partitioning
　reference link 453
　time-series tables, creating with 452-454
partition keys
　candidates, searching for 194, 195
password
　changing securely 29
　keeping, in secure password file 30, 31
peer authentication 12
performance mailing list
　reference link 458
performance problem report
　reference link 458
performance problems
　reporting 457
performance-related information
　reference link 458
permission group 213
Personally Identifiable Information (PII) 230
pgAdmin 127
　used, for real-time viewing 343, 344
　using, for DBA tasks 271-276
　version 4 GUI tool, using 18-23
PgBouncer 120, 144, 150
　configuring 144-146
　SHOW commands 147, 148
pg_buffercache view 74
pg_database 67, 68

pg_dump utilities 497

pgfincore 130

pg_hint_plan 446

pg_relation_size() 81

pg_restore utilities 497

pg_size_pretty() function 74, 76

pg_stat_reset() 359

pg_statviz extension
  used, for tracking important metrics over time 374-377

pg_upgrade utility 566

physical backup 471-474, 485-488
  logical backup, extracting from 498
  with Barman 474-483

physical backup/recovery
  performance, improving 501-503

physical log SR (PLSR) 514

physical replication 519

physical SR (PSR) 513

PL/Proxy 315, 316

Point-in-Time Recovery (PITR) 4, 252, 463, 490-493

PostGIS extension 5

Postgres 52

Postgres Advanced Server (EPAS) 446

Postgres Distributed (PGD) 516

Postgres Enterprise Manager (PEM) 360

PostgreSQL 1, 2, 52, 120
  commercial support 7
  concurrency 6
  downloading 8-10
  extensibility 5
  features 2
  in cloud, as DBaaS 35-40
  message log, monitoring 345-347

  objectives 2, 3
  performance 6
  popularity 7
  reference link 52
  research and development funding 8
  robustness 4
  scalability 6
  security 4
  SQL and NoSQL data models 6
  superuser 207, 208
  superuser-like attributes 208
  Trusted Postgres Architect (TPA), using 45-49
  user friendly 5
  users 3
  working, with Kubernetes 41-43

PostgreSQL Audit Extension (pgaudit) 252

PostgreSQL databases
  objects, accessing 311-316

PostgreSQL data directory 59, 60

PostgreSQL Extension Network (PGXN) 110, 316, 374
  modules, installing from 111, 112

PostgreSQL Foreign Data Wrapper 312

PostgreSQL Global Development Group (PGDG) 110

PostgreSQL information
  providing, to monitoring tools 341

PostgreSQL release support policy
  reference link 55

PostgreSQL security
  overview 206
  typical user roles 207

PostgreSQL server
  connecting to 11-14

postmaster 120

Index

predefined roles
    used, for setting up cloud security 249-253
Pretty Good Privacy (PGP) 245
primary database server 512
primary key (PK) 455, 553
Principle Of Least Privilege (POLP) 206
private databases
    providing, to users 140, 141
Privileged Access Management (PAM) 243
Procedural Language/PostgreSQL (PL/
        pgSQL) 424
process identifier (PID) 543
programs
    configuration parameters, setting in 99-102
progress, of commands
    monitoring 352-354
Prometheus 339
psql error
    investigating 267-269
psql prompt
    setting, with useful information 269-271
psql query and scripting tool 24
    using 24-28
psql script, that exits on first error
    writing 261, 262
psql variables
    query output, placing into 264, 265
    using 263, 264
Public-Key Cryptography 248
publish/subscribe (pub/sub) model 551
Python Languages (PLs) 228
Python Package Index (PyPI) 375

Q

queries
    active or blocked status, checking 354-356
    catching, that only run for
        few milliseconds 350
    real-time performance, analyzing 372-374
    run time, considerations 368, 369
    run time, considering when
        they run alone 370
    run time, monitoring when running for the
        second time 371
    speeding up, without
        rewriting them 434, 435
queries, from ps
    monitoring 352
query
    blocking information, monitoring 356, 357
    execution, considerations 349, 350
    forcing, to use index 443-446
    reasons for not using index,
        discovering 441, 443
query, in psql
    executing, repeatedly 348, 349
query output
    placing, into psql variables 264, 265

R

Random-Access Memory (RAM) 91, 421
read the fine manual (RTFM) 90, 91
Recovery Point Objective (RPO) 465
recursive_worktable_factor parameter
    setting 435
Reference Data Management (RDM) 511
referential integrity 84
Regular Expression (regex) 243

relay process 512
replica identity 553
replication
  application-level replication 517
  best practices 517-519
  concepts 511, 512
  database replication 511, 512
  data loss 515
  history 512, 513
  monitoring 539-543
  multi-master replication 516
  multinode architectures 516
  OS-level replication 517
  other approaches 517
  practical aspects 514
  reference link 517
  scope 513
  setting up, with CloudNativePG (CNPG) 538, 539
  setting up, with TPA 536, 537
  single-master replication 515
replication delay 514
replication sets 515
replication slots 522, 534
  using 534-536
repmgr utility
  using 531-533
  working 534
Row-Level Security (RLS) 4, 206

S

schema
  rewriting 441
schema-level privileges
  using 302

schemas 137
  adding/removing 300, 301
  objects, moving between 302, 303
SCRAM 4
SCRAM-SHA-256 206, 520
script, that either succeeds entirely or fails entirely
  writing 257-261
Secure Sockets Layer (SSL) 522
Security Support Provider Interface (SSPI) 243
selective replication 515
sensitive data
  column-level encryption, using 248, 249
sequential scan 78
serializable transaction isolation level 456
serialization failure 268
server
  restarting, quickly 129, 130
  stopping, in emergency 125, 126
  stopping, safely and quickly 124, 125
server authenticity
  checking 239, 240
server configuration checklist 107, 108
server configuration files
  reloading 126-128
server log 61
server-side procedures
  used, for creating bulk data changes with transactions 187-191
server type 52, 53
server uptime 56
server version 53, 54
service unit 121

Index

set, of data
  unique key, searching 169-172
set-returning functions
  using, for some parts of queries 434
shared_buffers configuration parameter
  reference link 422
single-master replication 515
single point of failure (SPOF) 518
Single Sign-On (SSO) 243
slow SQL statements
  finding 414-418
  reasons 418-421
snapshot export feature 467
socket 15
software installer
  used, for installing modules 110, 111
solid-state drives (SSDs) 445
source code
  modules, installing from 112
split-brain situation 530
spreadsheet
  data, loading from 180-182
SQL statements
  auditing 228, 229
SQL table
  access, auditing 230
SSL
  client, setting up to use 238, 239
SSL certificates
  using, to authenticate 240, 241
SSL key and certificate
  obtaining 238
standby database server 512
statement_timeout
  using, to clean up queries that take too long to run 358
static scripting 279
streaming replication (SR) 513, 551
  logical log SR 514
  logical SR 513
  managing 529, 530
  physical log SR 514
  physical SR 513
  security, setting up 523, 524
  setting up 519, 520
  working 521-523
switchover 515
symlink 46
sync rep 515, 544
  delaying 549, 550
  pausing 549, 550
  performance 544-547
  working 547, 548
sync standby 546

T

table
  usage, monitoring 360-362
table and index bloat 371
table constraint 297
table disk space
  using 74, 75
table rows 77, 79
  estimating 79-81
TABLESAMPLE clause 425
tablespaces 303
  adding/removing 303-307
  mapping 489
  objects, moving between 308-311

pg_wal directory, putting on separate device 307, 308
tablespace-level tuning 308
targeted recovery mode 491
template databases 66
temporary file
usage, logging 368
usage, monitoring 368
temporary tables
heavy users, actions 390-392
test data
generating 172-175
generating, tools 175
The Oversized Attribute Storage Technique (TOAST) 384
The PostgreSQL License (TPL) 8
Third Normal Form (3NF) 6
time-series partitioning 440
time-series tables
creating, with partitioning 452-454
Transaction Identifier (TID) 229
transaction isolation modes 268
transaction log data
archiving 560-562
transaction log (xlog) 541
Transport Layer Security (TLS) 252
troubleshooting
failed connection 33-35
Trusted Postgres Architect (TPA) 57, 93
replication, setting up with 536, 537
using 45-49
tuple deforming 449

U

Uniform Resource Identifier (URI) format 12
United States Department of Defense (US DoD) 229
unused indexes
removing, carefully 407-409
searching 406, 407
updatable views
creating 319-328
user access, to specific columns
granting 214-216
user access, to specific rows
granting 216-218
user access, to tables
database creation scripts 210
default search path 211
granting 212-214
revoking 208-210
views, securing 211, 212
user attributes
not inheriting 235
user, computer connection
checking 348
user connection
checking 347, 348
limiting 220, 221
preventing, temporarily 220
user, from database
removing, without dropping 222, 223
user groups
parameters, setting 106, 107
User Name Map feature
replacing 236
user, PostgreSQL session
logged in, considerations 233-235

users
  backup privileges, assigning to 226
  providing, with private databases 140, 141
  removing, from database server 134, 135
  restricting, to only one
      session each 132, 133
  superuser powers, limiting 224-226

user's database access
  revoking 221

users, with secure password
  checking 223, 224

V

VACUUM command 398-402

VACUUM maintenance process 81

view 86
  materialized views, using 329, 330
  non-materialized view 330
  updatable views, creating 319-328
  using, that contains TABLESAMPLE 440

Virtual Private Network (VPN) 223

Visibility Map (VM) 77

W

WITH queries
  reference link 431

work_mem
  increasing 434

Write-Ahead Logging (WAL) 76, 108, 160, 513

Z

zero downtime upgrades 567

Download a free PDF copy of this book

Thanks for purchasing this book!

Do you like to read on the go but are unable to carry your print books everywhere?

Is your eBook purchase not compatible with the device of your choice?

Don't worry, now with every Packt book you get a DRM-free PDF version of that book at no cost.

Read anywhere, any place, on any device. Search, copy, and paste code from your favorite technical books directly into your application.

The perks don't stop there, you can get exclusive access to discounts, newsletters, and great free content in your inbox daily

Follow these simple steps to get the benefits:

1. Scan the QR code or visit the link below

https://packt.link/free-ebook/9781835460580

2. Submit your proof of purchase
3. That's it! We'll send your free PDF and other benefits to your email directly

www.ingramcontent.com/pod-product-compliance
Lightning Source LLC
LaVergne TN
LVHW080304260326
834688LV00039B/1126